HAZARDOUS WASTE CHEMISTRY, TOXICOLOGY AND TREATMENT

Library of Congress Cataloging-in-Publication Data

Manahan, Stanley E.
  Hazardous waste chemistry, toxicology, and treatment / by Stanley
E. Manahan.

     p. cm.
  Includes bibliographical references.
  1. Hazardous wastes. I. Title.
TD1030.M37   1990
ISBN 0-87371-209-9
628.4'2--dc20   90-33889

LEWIS PUBLISHERS, INC.
121 South Main Street, Chelsea, Michigan 48118

PRINTED IN THE UNITED STATES OF AMERICA

HAZARDOUS WASTE CHEMISTRY, TOXICOLOGY AND TREATMENT

STANLEY E. MANAHAN

Stanley E. Manahan is Professor of Chemistry at the University of Missouri – Columbia, where he has been on the faculty since 1965. He received his A.B. in chemistry from Emporia State University in 1960 and his Ph.D. in analytical chemistry from the University of Kansas in 1965. Since 1968 his primary research and professional activities have been in environmental chemistry and have included development of methods for the chemical analysis of pollutant species, environmental aspects of coal conversion processes, development of coal products useful for pollutant control, hazardous waste treatment, and toxicological chemistry. He teaches courses on environmental chemistry, hazardous wastes, and toxicological chemistry and has lectured on these topics throughout the U.S. as an American Chemical Society Local Section tour speaker. He is also President of ChemChar Research, Inc., a firm working with the development of non-incinerative thermochemical and electrothermo-chemical treatment of mixed hazardous substances containing refractory organic compounds and heavy metals.

Professor Manahan has written books on environmental chemistry (*Environmental Chemistry*, 4th ed., 1984, Lewis Publishers Inc.), toxicological chemistry (*Toxicological Chemistry*, 1989, Lewis Publishers Inc.), applied chemistry, and quantitative chemical analysis. He has been the author or co-author of approximately 70 research articles.

ACKNOWLEDGMENTS

The author would like to acknowledge the assistance of Craig A. Shelley, President of Softshell International, Grand Junction, Colorado, who freely and patiently gave information and backup for the Chemintosh® structure drawing program with which the structures and some illustrations in this book were drawn. Vivian Collier, Senior Coordinating Editor of Lewis Publishers and the rest of the staff of Lewis Publishers have been outstanding to work with in preparing this book. The author sincerely appreciates the opportunity he has had to collaborate with Steven Luff and Paula McClure of Missouri Ingenuity, Inc., through ChemChar Research, Inc., in the acquisition of information and expertise pertaining to hazardous wastes treatment included in this book. The assistance of Anne F. Manahan in producing this work is gratefully acknowledged.

PREFACE

Hazardous Waste Chemistry, Toxicology, and Treatment is designed to provide the reader with the fundamentals of hazardous substances and wastes in relation to chemistry, environmental chemical processes, and toxicology. It is written at a level designed to be useful to a broad spectrum of professionals involved with various aspects of hazardous substances and wastes, including regulation, treatment, remediation, biological effects, chemical phenomena, transport, source reduction, and research.

The first five chapters of the book consist of an introduction and definition of hazardous waste problems. Chapter 2 deals with the fundamentals of chemistry and organic chemistry as these disciplines relate to hazardous materials. Chapter 3 does the same for environmental chemistry and environmental chemical processes pertaining to hazardous wastes in water, air, and soil. Chapter 4 is an introduction to biochemistry, toxicology and toxicological chemistry. Chapter 5 relates specific kinds and classes of hazardous substances to their chemical nature. Frequent reference is made to these chapters in later parts of the book.

Chapters 6-10 discuss the fundamentals of hazardous wastes as inorganic chemicals, organic chemicals, and biohazards. The discussions of inorganic and organic hazardous wastes are each divided into sections on chemistry and toxicology. These chapters are designed to relate hazardous inorganic chemical wastes, hazardous organic chemical wastes, and biohazards to the disciplines of chemistry, biochemistry, and toxicology.

The last five chapters of the book discuss ways of eliminating, treating, and detoxifying hazardous substances and wastes. These chapters are placed in the most desirable order of dealing with hazardous hazardous materials, which is (1) do not produce them; (2) if they are produced, use them; (3) if they cannot be used, treat them; and (4) after treatment, dispose of them. Chapter 11 discusses reduction, recycling, and resource recovery, widely recognized as the most desirable means of eliminating problems with hazardous substances by preventing their production as wastes. Chapter 12 is a discussion of physical treatment processes. Chapter 13 deals with chemical and biochemical treatment of hazardous substances and wastes. Chapter 14 discusses incineration and other means of treating hazardous wastes thermally and Chapter 15 covers immobilization, fixation, and disposal.

The latter chapters of the book are cross-referenced with Chapters 2, 3, and 4 to illustrate how the fundamental principles of chemistry, biochemistry, environmental chemistry, and toxicology are related to hazardous waste problems and their solutions.

Hazardous substances and wastes are discussed with an emphasis upon U.S. regulations, particularly RCRA legislation and the Superfund laws. However, these materials present problems throughout the world so that attention is given in the book to problems, regulations, and treatment measures pertaining to other countries as well.

CONTENTS

Hazardous Substances and Hazardous Wastes

1.1. INTRODUCTION

In general terms a **hazardous substance** is a material with a substantial potential to pose a danger to living organisms, materials, structures, or the environment. Such dangers may consist of explosion or fire hazards, corrosion, toxicity to organisms, or other detrimental effects. A **hazardous waste** is a hazardous substance that has been discarded or otherwise designated as a waste material, or one that may become hazardous by interaction with other substances. Such a waste is a material that has been discarded, abandoned, neglected, or released so that it may pose a danger to living beings, structures, or the environment. In a simple sense a hazardous waste is a material that has been left where it should not be and that may cause harm to you if you encounter it!

Although more exacting legal and regulatory definitions of hazardous substances and hazardous wastes are given in later chapters, hazardous substances and hazardous wastes are discussed in this book in a broad sense that may not fit the legalistic definitions of various laws and regulations of particular governments or agencies thereof.

History of Hazardous Substances

Human exposure to hazardous substances occurred when prehistoric humans inhaled noxious volcanic gases or succumbed to carbon monoxide from inadequately vented fires in cave dwellings sealed too well against Ice-Age cold. There were reports in Ancient Greece of slaves who developed lung disease from weaving mineral asbestos fibers with cloth to make it more degradation-resistant. Some authorities believe that wine containers made of lead were a leading cause of lead poisoning in the Roman Empire. During the 1700s runoff from mine spoils piles began to create serious contamination problems in Europe. As the production of dyes and other organic chemicals developed from the coal tar industry in Germany during the 1800s, pollution and poisoning from coal tar byproducts was observed. By around 1900 the quantity and variety of chemical wastes increased, including wastes such as spent steel and iron pickling liquor, lead battery wastes, chromic wastes, petroleum refinery wastes, radium wastes, and fluoride wastes from aluminum ore refining. As the century progressed, the wastes and hazardous byproducts of manufacturing increased markedly from sources such as chlorinated solvents manufacture, pesticides synthesis, polymers manufacture, plastics, paints, and wood preservatives.

By the 1960s it was obvious that air and water pollution in industrialized countries was a major problem. Rachel Carson's book, "The Silent Spring," published in 1962, warned of environmental damage caused by the indiscriminate use of pesticides and of human health threats, such as cancer. The publication of this book marked the beginning of the modern era of environmental awareness.

Hazardous wastes gained public attention and became a major political issue in the U.S. with problems that surfaced at the Love Canal waste disposal site in Niagara Falls, New York, in the 1970s. Starting around 1940 this site had received about 20,000 metric tons of chemical wastes including significant quantities of at least 80 different chemicals. By 1989 state and Federal governments had spent $140 million to clean up the site and relocate residents.[1]

Other areas containing hazardous wastes that received attention included an industrial site in Woburn, Massachusetts, that had been contaminated by wastes from tanneries, glue-making factories, and chemical companies dating back to about 1850; the Stringfellow Acid Pits near Riverside, California; the Valley of the Drums in Kentucky; and Times Beach, Missouri, an entire town that was abandoned because of contamination by TCDD (dioxin).

Legislation

In response to public concern over environmental and hazardous waste problems, governments in a number of nations have passed legislation to deal with these concerns. For example, the U.S. Congress passed several important laws during the 1970s to control air and water pollution. These included legislation to control air pollution; The Federal Insecticide, Fungicide, and Rodenticide Act of 1972; the Federal Water Pollution Control Acts; the Toxic Substances Control Act of 1976; and the Clean Water Act of 1977. To deal with hazardous substances, the U.S. Congress passed the Resource, Recovery and Conservation Act (RCRA) of 1976, and amended and strengthened it by passage of the Hazardous and Solid Wastes Amendments (HSWA) of 1984. This legislation charged the U.S. Environmental Protection Agency (EPA) with protecting human health and the environment from improper management and disposal of hazardous wastes by issuing and enforcing regulations pertaining to such wastes. Hazardous wastes and their characteristics are required to be listed and they are to be controlled from the time of their origin until their proper disposal or destruction. Firms generating and transporting hazardous wastes are regulated. They must keep detailed records, including reports on their activities and manifests to ensure proper tracking of hazardous wastes through transportation systems. Approved containers and labels must be used, and wastes can only be delivered to facilities approved for treatment, storage, and disposal. There are about 290 million tons of wastes regulated by RCRA. In the U.S. about 3,000 facilities are involved in the treatment, storage, or disposal of RCRA wastes.

Although RCRA deals with hazardous substances from their point of origin to disposal, measures were needed to take care of actual or potential releases of hazardous materials, at uncontrolled or abandoned hazardous waste sites. In the U.S. legislation was passed for that purpose in the form of the Comprehensive Environmental Response, Compensation, and Liability Act (CERCLA) of 1980, commonly known as Superfund. This law was designed to deal with problems created by sites in which chemical wastes had been discarded and that had the potential to endanger people or the surrounding environment.[2] The act requires responsible parties or the government to clean up waste sites. Among CERCLA's major purposes were the following:

- Site identification
- Evaluation of danger from waste sites
- Evaluation of damages to natural resources
- Monitoring of release of hazardous substances from sites
- Removal or cleanup of wastes by responsible parties or government

CERCLA was extended for 5 years by the passage of the Superfund Amendments and Reauthorization Act (SARA) of 1986, legislation with greatly increased scope and $8.5 billion in funding. The Title III section of SARA (its "right-to-know" provisions) is intended to provide information to the general public about what chemicals industrial concerns use in local operations and what substances are released into the environment.[3,4]

SARA greatly increased the scope of its predecessor legislation and is actually longer than CERCLA. Among the important objectives and provisions of SARA are the following:

- Five-fold increase in funding to $8.5 billion for five years
- Alternatives to land disposal that favor permanent solutions reducing volume, mobility, and toxicity of wastes
- Increased emphasis upon public health, research, training, and state and citizen involvement
- Codification of regulations that had been policy under CERCLA
- Mandatory schedules and goals over the lifetime of the legislation
- New procedures and authorities for enforcement
- A new program for leaking underground (petroleum) storage tanks

Improper disposal of wastes continues to be a subject of public and governmental concern. One of the more recent problems to be addressed by legislative action in the U.S. is ocean dumping of sewage sludge. For many years sludge from New York and New Jersey has been disposed in the Atlantic Ocean's 106-Mile Deepwater Municipal Sewage Sludge Disposal Site.[5] In response to this kind of activity, the U.S. Congress passed the Ocean Dumping Ban Act of 1988. Among the act's requirements are an immediate prohibition against new ocean dumping of sewage sludge and stiff penalties for ocean disposal by communities already engaged in the practice starting January 1, 1992.

1.2   CLASSIFICATION OF HAZARDOUS SUBSTANCES AND WASTES

Many specific chemicals in widespread use are hazardous because of their chemical reactivities, fire hazards, toxicities, and other characteristics.[6,7] There are numerous kinds of hazardous substances, usually consisting of mixtures of specific chemicals. A quick overview of these may be had by considering some of the major classes of hazardous materials according to the criteria of the United States Department of Transportation DOT. One of the most obvious of these consists of **explosives**. Examples include Class A explosives, such as dynamite or black powder, that are sensitive to heat and shock; DOT Class B explosives, such as rocket propellant powders, in which contaminants may cause explosion; and Class C explosives, such as ammunition, which are subject to thermal or mechanical detonation. **Compressed gases** and **special forms** of gases may be hazardous. Examples of the former are hydrogen and sulfur dioxide, whereas acetylene gas in acetone solution is an example of the latter. DOT recognizes a wide range of **flammable liquids**, such as gasoline and aluminum alkyls, as hazardous materials. **Flammable solids** are those that burn readily, are water-reactive, or spontaneously combustible; they include substances such as magnesium metal, sodium hydride, and calcium carbide. **Oxidizing materials** include oxidizers (for example, lithium peroxide) that supply oxygen for the combustion of normally nonflammable materials. Examples of **corrosive materials**, which may cause disintegration of metal containers or flesh, are oleum, sulfuric acid and caustic soda. **Poisonous materials** include Class A poisons, such as hydrocyanic acid, which are toxic by inhalation, ingestion, or absorption through the skin; Class B poisons, such as aniline; and etiologic agents, including causative agents of anthrax, botulism, or tetanus. **Radioactive materials** include plutonium, cobalt-60, and uranium hexafluoride.

Characteristics and Listed Wastes

Under the authority of the Resource Conservation and Recovery Act (RCRA) the United States Environmental Protection Agency (EPA) defines hazardous substances in terms of the **characteristics** of **ignitability, corrosivity, reactivity,** and **toxicity.**[8] (Another characteristic that could be used for classifying hazardous substances is etiology, having to do with the potential to cause disease; see Chapter 10, "Biohazards.") Among the substances classified as ignitable are liquids whose vapors are likely to ignite in the presence of ignition sources, nonliquids that may catch fire from friction or contact with water and which burn vigorously or persistently, ignitable compressed gases, and oxidizers. Corrosive substances may exhibit extremes of acidity or basicity or a tendency to corrode steel. Reactive substances are those with a tendency to undergo violent chemical change; an explosive substance is an obvious example. The characteristic of toxicity is defined in terms of a standard extraction procedure followed by chemical analysis for specific substances.

In addition to classification by characteristics, EPA designates more than 450 **listed wastes** which are specific substances or classes of substances known to be hazardous. Each such substance is assigned an EPA hazardous waste number in the format of a letter followed by 3 numerals, where a different letter (F, K, P, or U) is assigned to substances from each of the four following lists:

- Wastes from non-specific sources (F-type wastes)
- Wastes from specific sources (K-type wastes)
- Acute hazardous wastes (P-type wastes)
- Generally hazardous wastes (U-type wastes)

An example of an "F" waste is quenching wastewater treatment sludges from metal heat treating operations where cyanides are used in the process (F012). A typical "K" waste consists of heavy ends from the distillation of ethylene dichloride in ethylene dichloride production (K019). Acute hazardous wastes are mostly specific chemical species such as fluorine (P056) or 3-chloropropanenitrile (P027). Generally hazardous wastes are also predominantly specific compounds such as calcium chromate (U032) or phthalic anhydride (U190).

Additional insight into specific compounds that can be hazardous can be obtained from a U.S. EPA "List of Extremely Hazardous Substances," which lists more than 300 substances.[9] From this list EPA selected 21 chemicals, chosen either for their extreme acute toxicity or high production volume, from which to make a "review of emergency systems for monitoring, detecting, and preventing releases of extremely hazardous substances at representative domestic facilities." The chemicals selected were acrylonitrile, ammonia, benzenearsonic acid, benzotrichloride, chlorine, chloroacetic acid, furan, hydrazine, hydrogen cyanide, hydrogen fluoride, hydrogen sulfide, mechlorethamine, methiocarb, methyl bromide, methyl isocyanate, phosgene, sodium azide, sulfur dioxide, sulfur trioxide, tetraethyl tin, and trichloracetyl chloride.[10]

Some materials that might normally be regarded as hazardous wastes are excluded from RCRA regulation.[11] Some of these materials are specifically covered by other legislation, including nuclear material (Atomic Energy Act) and polychlorinated biphenyls (Toxic Substances Control Act). Other excluded materials are waste dust from cement kilns, byproducts (such as ash) of fossil fuel combustion and stack gas cleanup, mining overburden returned to the mine site, domestic sewage, and return flow from irrigation.

Compared to RCRA, CERCLA gives a rather broad definition of hazardous substances that includes the following:

- Any element, compound, mixture, solution, or substance, release of which may substantially endanger public health, public welfare, or the environment.
- Any element, compound, mixture, solution, or substance in reportable quantities designated by CERCLA Section 102
- Certain substances or toxic pollutants designated by the Federal Water Pollution Control Act
- Any hazardous air pollutant listed under Section 112 of the Clean Air Act
- Any imminently hazardous chemical substance or mixture that has been the subject of government action under Section 7 of the Toxic Substances Control Act (TSCA)
- With the exception of those suspended by Congress under the Solid Waste Disposal Act, any hazardous waste listed or having characteristics identified by RCRA § 3001

Hazardous Wastes

What is a hazardous waste? It has been stated that,[12] "The discussion on this question is as long as it is fruitless." Three basic approaches to classifying hazardous wastes are (1) a qualitative description by origin, type, and constituents, (2) classification by characteristics largely based upon testing procedures; and (3) by means of concentrations of specific hazardous substances. Wastes may be classified by general type such as "spent halogenated solvents" or by industrial sources such as "pickling liquor from steel manufacturing."

Various countries have different definitions of hazardous waste.[13] For example, The Federal Republic of Germany Federal Act on Disposal of Waste (1972, as amended, 1976) mentions special wastes that are, ".....especially hazardous to human health, air, or water, or which are explosive, flammable, or may cause diseases." The United Kingdom's Deposit of Poisonous Waste Act (1972) refers to waste, "....of a kind which is poisonous, noxious, or polluting and whose presence on the land is liable to give rise to an environmental hazard." The Ontario Waste Management Corporation, a provincial crown agency created by the Ontario, Canada, Legislature, defines **special waste** as liquid industrial and hazardous wastes that is unsuitable for treatment or disposal in municipal wastewater treatment systems, incinerators or landfills and that therefore requires special treatment.[14]

Hazardous wastes may be individual substances or mixtures. They are dangerous or potentially dangerous to humans or other living organisms. Direct dangers posed by hazardous wastes include fire, explosion, and toxicity. Some hazardous wastes cause cumulative detrimental effects, such as cancer resulting from repeated or prolonged exposure, some may undergo biomagnification in exposed organisms (through food chains), and many hazardous wastes persist in the environment because they do not degrade.

Radioactive wastes in the U.S. are regulated under the Nuclear Regulatory Commission (NRC) and Department of Energy (DOE). Special problems are posed by **mixed waste** containing both radioactive and chemical wastes.[15]

Hazardous Wastes and Air and Water Pollution Control

Somewhat paradoxically, measures taken to reduce air and water pollution have had a tendency to increase production of hazardous wastes (Figure 1.1). Most water treatment processes yield sludges or concentrated liquors that require stabilization and disposal. Air scrubbing processes likewise produce sludges. Baghouses and precipitators used to control air pollution all yield quantities of solids, some of which are hazardous.

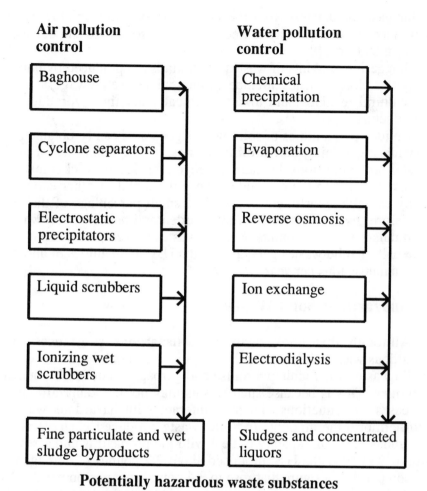

Figure 1.1. Potential contributions of air and water pollution control measures to hazardous wastes production.

Hazardous and Non-Hazardous Wastes

In a nonregulatory sense there is no sharp demarcation between hazardous and nonhazardous wastes. Some wastes such as soluble toxic heavy metal salt wastes are obviously hazardous. By comparison, discarded leaves and tree trimmings, would be regarded as posing no danger. But, if properly treated and immobilized, the heavy metal wastes are of little danger, whereas discarded tree limbs pose a fire hazard under certain circumstances. Materials that by themselves are nonhazardous may interact with hazardous substances to increase the dangers from the latter. For example, soluble humic substances from the decay of tree leaves may solubilize and transport heavy metal ions.[16]

Staggering amounts of wastes of all kinds are produced by human activities. Such wastes include municipal refuse, sewage sludge, agricultural residues, and toxic, chemically reactive byproducts of manufacturing processes.

An idea of quantities of solid wastes generated can be obtained by considering mining and milling wastes.[17] The quantities of such wastes are enormous because large quantities of rock must be removed to get to the ore and because the metal or other economically valuable constituent is usually a small percentage of the ore. Therefore, byproducts such as overburden and beneficiation wastes accumulate in vast amounts. Mining wastes make up somewhat less than half of solid wastes generated in the U.S. in quantities of about 2 billion metric tons per year.

Nonhazardous Solid Wastes

Although this book deals with hazardous substances and hazardous wastes, it is appropriate to consider "nonhazardous" waste (solid waste), the municipal refuse and garbage produced by human activities. This is because such wastes may not be nonhazardous in all cases and situations and may interact with hazardous wastes. Furthermore, the amounts of solid waste produced each year are enormous, and our capacity to deal with the problem is under severe strain.[18] In 1960 the U.S. produced about 2.7 pounds of garbage per person per day, a figure that rose to 3.4 pounds per person per day by 1988. Some of this material can be regarded as hazardous (household hazardous wastes, HHW).[19] Disposal of about 92 percent of municipal refuse in the U.S. is in landfills. However, as total quantities of solid waste have increased, the landfill capacity to handle waste has decreased. When the original RCRA act was passed in 1976 about 30,000 landfills were operating (although many of these were little more than "dumping grounds"). As of 1988 the number of municipal landfills in operation had dropped to about 6,500. As a result, some cities have had to haul municipal refuse as far as 70 miles or more and disposal costs in some areas have reached $75/ton.

The potential of incineration for handling municipal refuse is very high because it can reduce waste mass by 75 percent and volume by 90 percent. However, environmental concern about organic pollutants (particularly dioxins) in stack emissions and heavy metals in incinerator ash have virtually stopped municipal incinerator development in the U.S.

Recycling can certainly reduce quantities of solid waste, perhaps as much as 50 percent, but it is not the panacea claimed by its most avid advocates. The overall solution to the solid waste problem must involve several kinds of measures, particularly (1) reduction of wastes at the source, (2) recycling as much waste as is practical, (3) reducing the volume of remaining wastes by measures such as incineration, (4) treating residual material as much as possible to render it nonleachable and innocuous, and (5) placing the residual material in landfills, properly protected from leaching or release by other pathways.

1.3. PHYSICAL FORMS AND SEGREGATION OF WASTES

Three major categories of wastes based upon their physical forms are **organic materials**, **aqueous wastes**, and **sludges**. These forms largely determine the course of action taken in treating and disposing of the wastes. The **level of segregation**, a concept illustrated in Figure 1.2, is very important in treating, storing, and disposing of different kinds of wastes. It is relatively easy to deal with wastes that are not mixed with other kinds of wastes, that is, those that are highly segregated. For example, spent hydrocarbon solvents can be used as fuel in boilers. However, if these solvents are mixed with spent organochloride solvents, the production of contaminant hydrogen chloride during combustion may prevent fuel use and require disposal in special hazardous waste incinerators. Further mixing with inorganic sludges adds mineral matter and water. These impurities complicate the treatment processes required by producing mineral ash in incineration or lowering the heating value of the material incinerated because of the presence of water. Among the most difficult types of wastes to handle and treat are those with the least segregation, such as "dilute sludge consisting of mixed organic and inorganic wastes," as shown in Figure 1.2.

Concentration of wastes is an important factor in their management. A waste that has been concentrated or preferably never diluted is generally much easier and more economical to handle than one that is dispersed in a large quantity of water or soil. Dealing with hazardous wastes is greatly facilitated when the original quantities of wastes are minimized and the wastes remain separated and concentrated insofar as possible.

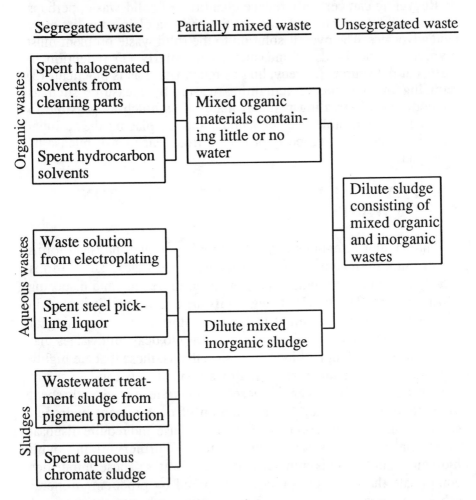

Figure 1.2. Illustration of waste segregation.

1.4. HAZARDOUS WASTE MANAGEMENT: GENERATION, TREATMENT, AND DISPOSAL

Hazardous waste **management** refers to a carefully organized system in which wastes go through appropriate pathways to their ultimate elimination or disposal in ways that protect human health and the environment. Three main aspects of hazardous waste management involve **generation, treatment**, and **disposal**[20] as illustrated in Figure 1.3. This figure is a simplification of the total picture. Under generation, for example, it omits municipal solid wastes and municipal wastewater, which, although not classified as hazardous, certainly have a relationship to the total scheme of hazardous wastes. Also not shown specifically are non-point waste sources, hazardous air emissions, and pesticide applications. Some of the quantities of wastes may be very large; for example, in 1985 RCRA-regulated wastes amounted to 275 million metric tons in the U.S., virtually all from large quantity generators producing more than 1,000 kilograms per month.

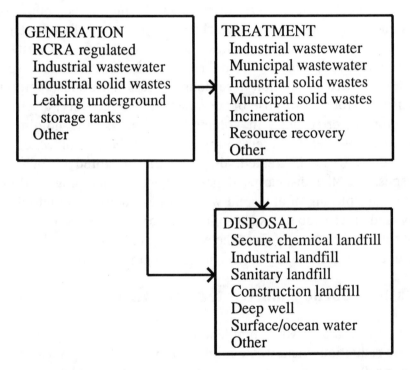

Figure 1.3. System of generation, treatment, and disposal of hazardous wastes.

The **effectiveness** of a hazardous waste system is a measure of how well it reduces the quantities and hazards of wastes, ideally approaching zero for both.[21] In decreasing order of effectiveness the options for handling hazardous wastes are the following:

- Measures that prevent generation of wastes
- Recovery and recycle of waste constituents
- Destruction and treatment, conversion to non-hazardous waste forms
- Disposal (storage, landfill)

Treatment, Storage, and Disposal Facilities

A crucial part of the regulation of hazardous wastes in the United States deals with **treatment, storage, and disposal facilities** (TSDF). In this system, the following definitions apply:[22] Treatment is "any method, technique, or process, including neutralization, designed to change the physical, chemical, or biological character or composition of any hazardous waste so as to neutralize such waste, or so as to recover energy or material resources from the waste, or so as to render such waste non-hazardous, or less hazardous; safer to transport, store, or dispose of; or amenable for recovery, amenable for storage, or reduced in volume." Storage is "the holding of hazardous wastes for a temporary period, at the end of which the hazardous waste is treated, disposed of, or stored elsewhere." Disposal is "the discharge, deposit, injection, dumping, spilling, leaking or placing of any solid waste or hazardous waste into or on any land or water so that such solid waste or hazardous waste or any constituent thereof may enter the environment or be emitted into the air or discharged into any waters, including ground waters."

Waste Reduction and Waste Minimization

Many hazardous waste problems can be avoided at early stages by **waste reduction**[23] and **waste minimization**. As these terms are most commonly used, waste reduction refers to source reduction—less waste-producing materials in, less waste out. Waste minimization can include treatment processes, such as incineration, which reduce the quantities of wastes for which ultimate disposal is required. Reference is sometimes made to waste abatement in terms of the four Rs—reduction, reuse, reclamation, and recycling.[24] These concepts are discussed in more detail in Chapter 11.

Waste Treatment

Under the category of treatment it is necessary to consider both municipal wastewater and municipal solid wastes along with hazardous wastes. The goal of many industrial wastewater and sludge treatment processes is to produce an effluent that meets standards for release to a municipal wastewater treatment plant (publicly owned treatment works, POTW, see Section 3.7 and Figure 13.4) and in some cases to produce solids that can be codisposed with municipal solid wastes. Incineration of muncipal solid wastes may produce some solids that have to be treated as hazardous.

An overall scheme for the treatment of hazardous wastes is shown in Figure 1.4, which may serve as a frame of reference for subsequent discussions of waste treatment.

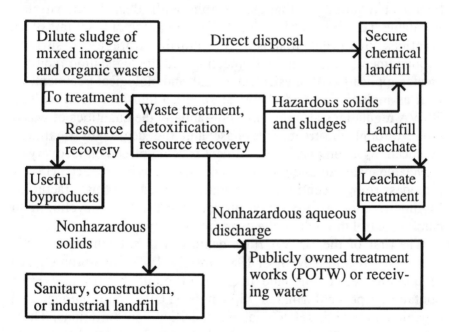

Figure 1.4. Treatment options for mixed hazardous wastes.

There are many options for the treatment of hazardous wastes and they are discussed in detail later in this work. These include industrial wastewater treatment, hazardous waste incinerators, industrial furnaces and boilers, and resource recovery, such as solvent reclamation. The ideal treatment process reduces the quantity of hazardous waste material to a small fraction of the original amount and converts it to a non-hazardous form. However, most treatment processes yield material, such as sludge from wastewater treatment or incinerator ash, which requires disposal and which may be hazardous to some extent.

Direct disposal of minimally treated hazardous wastes is becoming more severely limited with new regulations coming from the Hazardous and Solid Waste Amendments of 1984 (HSWA). Under its "land-ban" rules, this act bans the land disposal of more than 400 chemicals and waste streams unless they are treated or can be shown not to migrate during the time that they remain hazardous. The ultimate objective of these rules is to reduce the amounts of hazardous wastes generated, although quantities are expected to increase during the next decade.[25] Under HSWA, the level of waste treatment required is that achievable with Best Demonstrated Available Technology (BDAT). HSWA requirements curtail older practices of direct disposal to landfill, storage in surface impoundments, and deep well injection and the legislation requires greater application of physical, chemical, and biological treatment, as well as incineration and chemical stabilization/solidification. The HSWA regulations also increase substantially the quantities of waste going through the treatment phase and the degree of treatment practiced. More emphasis in treatment is being placed on recovery of recyclable materials and production of innocuous byproducts. HSWA provides strong incentives to generate less wastes in manufacturing by modification of processes, product substitution, recycling, and careful control throughout the manufacturing system.

The first of the HSWA rules dealing with solvents and dioxin went into effect in November 1986. In July, 1987, EPA promulgated regulations restricting land disposal of liquid hazardous wastes containing polychlorinated biphenyls (PCBs) and halogenated organic compounds (HOCs) above specified concentrations. These wastes are from the "California list" of wastes restricted from land disposal by the State of California and subsequently incorporated by Congress into the 1984 Amendments to RCRA.[26] In addition to the wastes mentioned above, the California list contains liquid hazardous wastes including free cyanides at concentrations greater than or equal to 1,000 mg/L; liquid hazardous wastes containing in excess of specified levels of arsenic, cadmium, chromium, lead, mercury, nickel, selenium, or thallium; and liquid hazardous wastes having a pH less than or equal to 2.0.

Effective June 8, 1989, EPA finalized restrictions on the land disposal of 67 wastes, including (F007) spent cyanide plating bath solutions from electroplating operations, (K009) distillation bottoms from the production of acetaldehyde from ethylene, and (K011, K013, K014) three kinds of bottom streams from acrylonitrile production.[27] In the same action EPA extended for two years land disposal of four wastes that are usually injected underground and some soil and debris wastes.[28]

1.5. APPROPRIATE TREATMENT TECHNOLO-GIES

The appropriate treatment technology for hazardous wastes obviously depends upon the nature of the wastes. An important factor to be considered is that wastewater comprises more than 90 percent by weight of the average RCRA hazardous waste, which presents a challenge in terms of quantity of material handled and the diluted state of hazardous components. However, when the water can be removed and purified sufficiently for discharge, the quantity of waste can be greatly reduced.

Hazardous wastes can be divided between the major categories of organic and inorganic materials and among the physical states of liquids, solids, and sludges. On that basis RCRA hazardous wastes can be classified as the following:[29] Inorganic liquids (91%), inorganic solids (1%), inorganic sludges (5%), organic liquids (2%) and organic sludges (1%).

Treatment Technologies

Major treatment technologies are mentioned here and are covered in greater detail in later chapters. **Incineration** is used primarily to destroy liquid organic hazardous wastes, as well as some solids and sludges. Some aqueous hazardous wastes are subjected to **chemical wastewater treatment** that involves processes such as neutralization, precipitation, and solid sorption. Some wastewaters, usually those containing organic contaminants, are purified by **biological wastewater treatment**, in which microorganisms degrade the wastes. Both chemical and biological wastewater treatment processes produce sludges requiring disposal or additional treatment. **Steam stripping** is used to remove hazardous constituents from wastewater. **Solidification** is used predominantly on inorganic sludges to produce a solid material with minimum permeability that is suitable for disposal.

1.6. LAND DISPOSAL OF WASTES

Land disposal of wastes consists of placing the wastes on, or below the surface of the land. In some cases the wastes are mixed with the soil; in others, efforts are made to keep the wastes from contact with the soil. As the full force of HSWA becomes effective, land disposal is becoming severely curtailed and is allowed only for wastes or treated wastes that meet specific standards. However, wastes and the byproducts of their treatment must go somewhere, so some form of land disposal will still be needed and consideration will still need to be given to unfortunate legacies of improper land disposal from years past.

Hazardous waste **landfills** are disposal pits, usually below ground level. Pits constructed under more modern regulations require a synthetic lining material, an appropriate cap when closed, and a leachate collection system. Uses of such pits include disposal of incinerator ash or sludge from surface impoundments. Because of higher costs and more stringent regulation, there is an increasing tendency toward solidification of inorganic solids and sludges before disposal in surface impoundments.

Land treatment, in some cases called **land farming**, makes use of biological processes by microorganisms to degrade wastes spread on the land and often mixed with soil. It has found greatest application in the petroleum industry for the biodegradation of waste petroleum solids and sludges.

Aqueous wastes are commonly placed in **surface impoundments**, which usually have a large surface area relative to depth. In such impoundments wastes are exposed to atmospheric air, sometimes with the aid of mechanical aerators floating on the surface. Chemical reactions and biological processes occur over a period of time, along with precipitation and settling of solids. Chemicals may be added to the aqueous wastes. Overall, a major objective is stabilization of the wastes. In more arid climates net loss of water by evaporation may occur. In the most favorable cases the water remaining in a surface impoundment is safe to discharge and the wastes have become concentrated in a relatively small quantity of sediment. New regulations will likely result in a shift from surface impoundments to tanks for treatment of aqueous wastes.

Large quantities of aqueous hazardous wastes have been disposed of by **deep well injection**. As the name implies, such a system pumps aqueous wastes deep underground at depths where contamination of useful aquifers is not likely to occur.

1.7. HAZARDOUS SUBSTANCES AND HEALTH

In recent years the health aspects of hazardous substances have received increased attention by the public and by legislative bodies. A basic question is the linkage between the health of people living near Superfund sites and the chemicals found in the sites. This concern gained increased recognition in the U.S. with passage of the 1986 SARA act, greatly expanding the health authorities sections of the 1980 CERCLA act.[30] A specific agency, the Agency for Toxic Substances and Disease Registry (ATSDR) under the Public Health Service of the Department of Health and Human Services, is responsible for the health aspects of toxic substances release. ATSDR was authorized by the 1980 CERCLA act. Originally its limited responsibilities included maintaining files of information and data on the health effects and diseases potentially caused by toxic substances, keeping records of exposure to toxic substances, and listing areas where public access was restricted because of contamination by toxic substances.

The SARA legislation gave ATSDR much more substantial responsibilities and charged it with being the major conduit of information on the health effects of hazardous substances, as well as playing an active role in the response and remediation activities at Superfund waste sites. The agency was required to publish a prior-itized list of the most important hazardous substances which are found at Superfund sites and which pose the greatest threat to human health because of their toxicities and potential for human exposure.[31] The risk to human health of these substances was ranked according to their toxicities, frequency of occurrence at sites, and potential for human exposure. In addition, ATSDR was charged with preparing extensive Toxicological Profile documents pertaining to specific hazardous substances encountered at Superfund sites. Toxicological Profiles for 275 substances are to be prepared by October, 1991. The materials that are the subjects of these profiles are those both commonly encountered at NPL facilities and likely to pose substantial health hazards. For each substance, the Toxicological Profile contains a self-contained health effects statement for the general public, information about manufacture and uses, physical and chemical properties, toxicological data, potential for human exposure, and methods of analysis in biological and environmental materials.

Selection of Substances for ATSDR List

Toxicities of substances to be included in the Toxicological Profiles were based upon a scoring scheme derived for **Reportable Quantities (RQ)** of hazardous substances (1, 10, 100, 1,000, and 5,000 pounds in decreasing order of hazard) based upon primary criteria of acute and chronic toxicity and carcinogenicity. These values were adjusted for environmental persistence, considering biodegradation, hydrolysis, and photolysis. The frequency of occurrence at National Priorities List (NPL) sites was determined from data taken from samples gathered at 358 sites and including 3,000 waste samples from 1980 to 1984, with additional data on volatile organic compounds from 1981 to 1987. The potential for human exposure at or near hazardous waste sites should ideally involve consideration of detailed, site-specific information about the substances, pathways to human exposure, expected levels and duration of exposure, and characterization of populations potentially exposed. In practice, the estimate of human exposure was based upon the three following criteria:

- Average concentration of the substance in groundwater and surface water analyzed at the 385 sites as part of the survey
- Frequency of detection of the substance in groundwater and surface water
- Whether or not the substance is an indicator substance (that is, one that has been selected for detailed exposure and risk assessment at Superfund Remedial sites)

ATSDR has been charged with performing health assessments at sites placed by the EPA on the National Priorities List. These health assessments seek to evaluate information about the release of hazardous substances at particular sites, taking into account the type and degree of contamination, the mechanisms of human exposure, and the probability of human exposure, including estimates of numbers of potentially exposed people. Such assessments should enable possible health impacts to be foreseen and appropriate preventive actions to be taken. In some cases health assessments may result in additional health-related investigations, such as epidemiological studies. Examples of such studies include testing of blood for polychlorinated biphenyls (PCBs) in residents near a site where PCBs were processed and monitoring of lead levels in children in lead mining areas.

1.8. SHORT-TERM ACTIONS AND LONG-TERM REMEDIES

Hazardous wastes can pose acute problems requiring a rapid response and an often temporary remedy, or chronic problems that demand less hasty, but more permanent remedial actions. Superfund provides for both types of measures.

Emergency or Removal Actions

In the parlance of CERCLA an emergency response to a release or threatened release of hazardous wastes to the environment is also known as a "removal action." Such actions are normally short-term, relatively low cost measures. Some of the conditions that can trigger a removal action are probable fire or explosion; actual or potential exposure of people, wildlife, drinking water, or food; presence of hazardous substances in containers (drums, tanks) from which they may be released or in near-surface soils from which they may migrate; and ambient conditions, such as those of weather, that may result in migration or release.

Removal actions begin with monitoring and assessment of the problem and proceed to cleanup, removal, and disposal of the hazardous material. Measures may be taken to reduce damage to public health or the environment from the hazardous substance.

Remedial Actions

Remedial actions to prevent or minimize hazardous substance release are designed to provide permanent remedies to hazardous waste threats, and may be taken subsequent to removal actions. A wide variety of remedial actions may be taken, such as pumping and treating water from the site, lining the site with barriers to prevent hazardous substance migration, and installation of clay caps to prevent release of chemicals to the surface and to minimize surface water inflow. The tendency now is definitely toward remedial responses that will permanently eliminate threats from hazardous substances from a Superfund site. Such measures include, for example, excavation and incineration of hazardous substances.

Remedial Investigation/Feasibility Study

Remedial actions are usually based upon the Remedial Investigation/Feasibility Study (RI/FS) process. The major phases of the RI/FS process are shown in Figure 1.5.

Remedial Design and Remedial Action

Design and construction required for remedial action is undertaken as part of the **Remedial Design/Remedial Action (RD/RA)** phase under the supervision of the U.S. Army Corps of Engineers. This phase is concerned with how well the design can be constructed, whether or not potential contractors can submit realistic bids based upon the information available, and the responsibilities and liabilities of parties involved in design and construction.

Superfund Accomplishments

As of mid-1989, Superfund had become a relatively large, complex program.[32] It had spent $4.5 billion of a total of $10.1 billion authorized. The number of Superfund sites stood at 1,173, with another 30,844 potential Superfund sites. It had accomplished 1,837 emergency cleanups, 103 remedial actions had been completed, and 174 remedial actions were underway. Given the number of potential Superfund sites and the limited number of cleanups started and finished, it was apparent that renewals of Superfund would be necessary to accomplish its original goals.

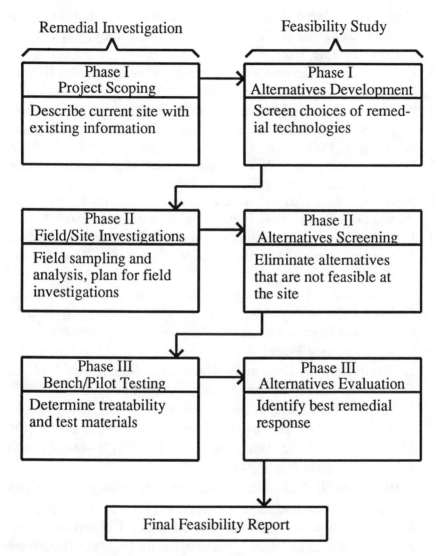

Figure 1.5. Remedial Investigation/Feasibility Study process

LITERATURE CITED

1. Ember, Lois, "Occidental Agrees to Store, Treat Love Canal Wastes," *Chemical and Engineering News*, June 19, 1989, pp. 20-21.
2. Nott, Sam, Caren Arnstein, Stephen Ramsey, and Maureen Crough, *Superfund Handbook*, 2nd ed., Sidley and Austin, Chicago, Illinois, 1987.
3. *Community Right-to-Know Manual*, Thompson Publishing Group, Salisbury, Maryland, 1989.
4. Adkins, Janis M., *Formula for Safety: Making Communities Safer from Chemical Hazards*, Atlantic Information Services, Washington, D.C., 1988.
5. "Ocean Dumping," *SFI Bulletin*, No. 403, Sport Fishing Institute, Washington, D.C., April, 1989, pp. 1-3.
6. Weiss, G., Ed., *Hazardous Chemicals Data Book*, 2nd ed., Noyes Data Corporation, Park Ridge, New Jersey, 1986.
7. U. S. Environmental Protection Agency, *Extremely Hazardous Substances*, Noyes Data Corporation, Park Ridge, New Jersey, 1988.
8. "Identification and Listing of Hazardous Waste," *Code of Federal Regulations*, **40**, July 1, 1986, Part 261, U.S. Government Printing Office, Washington, DC, pp. 359-408.
9. "The Emergency Planning and Community Right to Know Act of 1986 List of Extremely Hazardous Substances," U. S. Environmental Protection Agency, Washington, D.C., 1988. (Also published as Appendices A and B to the final rule (40 CFR 355) in the *Federal Register* on April 22, 1987, (FR 13376), revised on December 17, 1987 (FR 48072) and February 25, 1988 (FR 5574)).
10. Hanson, David, "Chemical Emergency Handling Study Cites Needs," *Chemical and Engineering News*, June 27, 1988, p. 24.
11. Reinhardt, John R., "Summary of Resource Conservation and Recovery Act Legislation," Section 1.2 in *Standard Handbook of Hazardous Waste Treatment and Disposal*, Freeman, Harry M., Ed., McGraw-Hill Book Company, New York, 1989, pp. 1.9-1.27.

12. Wolbeck, Bernd, "Political Dimensions and Implications of Hazardous Waste Disposal," in *Hazardous Waste Disposal*, Lehman, John P., Ed., Plenum Press, New York, 1982, pp. 7-18.

13. Lehman, John P., "Hazardous Waste Definition and Recommended Procedures," in *Hazardous Waste Disposal*, Lehman, John P., Ed., Plenum Press, New York, 1982, pp. 45-68.

14. Monenco Limited, "OWMC Treatment Facility: Status Report, Facility Design," Ontario Waste Management Corporation, Toronto, Ontario, Canada, 1985.

15. "Summary Report of ASME Mixed Waste Workshop," The 1989 Incineration Conference, Knoxville, Tennessee, May 1, 1989.

16. Manahan, Stanley E., "Humic Substances and the Fates of Hazardous Waste Chemicals," Chapter 6 in *Influence of Aquatic Humic Substances on Fate and Treatment of Pollutants*, Advances in Chemistry Series No. 219, American Chemical Society, Washington, DC, 1989, pp. 83-92.

17. Hoye, Robert L., and S. Jackson Hubbard, "Mining Waste," Section 4.5 in *Standard Handbook of Hazardous Waste Treatment and Disposal*, Harry M. Freeman, Ed., McGraw Hill, New York, 1989, pp. 4.47-4.51.

18. Forester, William S., "Solid Waste: There's a Lot More Coming," *EPA Journal*, Washington, D.C., 1988, pp. 3-4.

19. Dorian, Gerry, "Household Hazardous Waste Management," Section 4.6 in *Standard Handbook of Hazardous Waste Treatment and Disposal*, Freeman, Harry M., Ed., McGraw Hill Book Company, New York, 1989, pp. 4.53-4.60.

20. "The Hazardous Waste System," U.S. Environmental Protection Agency Office of Solid Waste and Emergency Response, Washington, D.C., 1987.

21. Andrews, Richard N. L., and Francis M. Lynn, "Siting of Hazardous Waste Facilities," Section 3.1 in *Standard Handbook of Hazardous Waste Treatment and Disposal*, Harry M. Freeman, Ed., McGraw Hill, New York, 1989, pp. 3.3-3.16.

22. "Hazardous Waste Management System: General," *Code of Federal Regulations*, **40**, July 1, 1986, Part 260, U.S. Government Printing Office, Washington, DC, pp. 339-358.

23. Bishop, Jim, "Waste Reduction," *Hazmat World*, October, 1988, pp. 56-61

24. Brandt, Andrew S., "Canadian Perspectives," *Proceedings of the Thirty-First Ontario Waste Conference*, Ontario Ministry of the Environment, Toronto, Ontario, Canada, 1984, pp. 3-20.

25. Hanson, David J., "Hazardous Waste Management: Planning to Avoid Future Problems," *Chemical and Engineering News*, July 31, 1989, pp. 9-18.

26. "Land Disposal Restrictions for Certain 'California List' Hazardous Wastes and Modifications to the Framework," *Federal Register*, **52**(130), July 8, 1987, 25760-25792.

27. "EPA Bans Land Disposal of 67 Waste Streams," *Chemical and Engineering News*, June 19, 1989, p. 18

28. "72 Wastes are Added to Land-Ban List," *Hazmat World* February, 1989, pp. 16-17.

29. "National Screening Survey of Hazardous Waste Treatment, Storage, Disposal, and Recycling Facilities," U.S. Environmental Protection Agency Office of Policy, Planning, and Information, Washington, D.C., 1986.

30. Siegel, Martin R., "Agency for Toxic Substances and Disease Registry Health-Related Activities," in *Hazardous Wastes, Superfund, and Toxic Substances*, American Law Institute American Bar Association Committee on Continuing Professional Education, Philadelphia, PA, 1987, pp. 19-27.

31. "Notice of the First Priority List of Hazardous Substances that will be Subject to Toxicological Profiles," *Federal Register*, **52**(74), April 17 1987, pp. 12866-12874.

32. Ember, Lois R., "Review to Call for Revamp of Superfund's Implementation," *Chemical and Engineering News*, May 29, 1989, pp. 17-20.

General Chemistry and Organic Chemistry

2.1. INTRODUCTION

Although all hazardous substances are inorganic or organic chemicals, many of which are dangerous because of their biochemical effects, individuals who must deal with hazardous wastes often have only minimal backgrounds in general chemistry or organic chemistry. Therefore, the purpose of this chapter is to outline the major and most basic concepts of these disciplines that readers with little knowledge of chemistry need to know in order to understand subject matter in the remainder of the book. More detailed coverage of these areas can be obtained in a text on applied chemistry[1] or other basic books dealing with general chemistry or organic chemistry. Biochemistry, the chemistry of life, is covered along with toxicology in Chapter 4.

The general chemistry section of this chapter defines a number of key chemical terms such as elements, compounds, and chemical bonds. It also discusses fundamental ideas required to understand chemistry, including the periodic table, atomic structure, and chemical formulas, with emphasis upon concepts covered later in the book. Organic chemistry, which includes most carbon-containing compounds, is briefly summarized. Areas of organic chemistry discussed include the unique features of the molecular structure of carbon-containing compounds that make organic chemicals so diverse, functional groups, and basic principles of organic chemical nomenclature.

Because of space limitations, several other areas of chemistry that are often pertinent to hazardous wastes are not covered here. For a basic summary of analytical chemistry, including both the classical and instrumental chemical analysis techniques used to detect and monitor hazardous substances, the reader may want to refer to the book *Quantitative Chemical Analysis*,[2] other undergraduate text-books on quantitative analysis, books on instrumental analysis, and works covering areas such as spectroscopic analysis of elements, chromatographic methods, and mass spectrometry. A recent work dealing with the assessment of human exposure to hazardous substances is also recommended[3]. Other specialized areas of chemistry to which the reader may want to refer include soil chemistry,[4] soil chemistry of hazardous materials,[5] and photochemistry.[6]

The basic theme of this book is environmental chemistry of hazardous wastes. Environmental chemistry is covered in Chapter 3 and is discussed in detail in a book on that subject.[7]

2.2. ELEMENTS

Subatomic Particles and Atoms

Consideration of general chemistry may begin with the three major subatomic particles: positively charged **protons**, negatively charged **electrons**, and uncharged (neutral) **neutrons**. These particles are contained in **atoms**, which may be considered to be the basic building blocks of matter. Protons and neutrons have relatively high masses compared to electrons and are contained in the positively charged **nucleus** of the atom, which has essentially all of the mass, but occupies virtually none of the volume, of the atom. An uncharged atom has the same number of electrons as protons. The electrons in an atom are contained in a cloud of negative charge around the nucleus that occupies most of the volume of the atom.

Atoms and Elements

All of the literally millions of different substances are composed of only around 100 **elements**. Each atom of a particular element is chemically identical to every other atom and contains the same number of protons in its nucleus. This number of protons in the nucleus of each atom of an element is the **atomic number** of the element. Atomic numbers are integers ranging from 1 to more than 100, each of which denotes a particular element. In addition to atomic numbers, each element has a name and a **chemical symbol**, such as carbon, C; potassium, K (for its Latin name kalium); or cadmium, Cd.

Although atoms of the same element are chemically identical, atoms of most elements consist of two or more **isotopes** that have different numbers of neutrons in their nuclei. Some isotopes are **radioactive isotopes** or **radionuclides**, which have unstable nuclei that give off charged particles and gamma rays in the form of **radioactivity**. Radioactivity may have detrimental, or even fatal, health effects; a number of hazardous substances are radioactive.

In addition to atomic number, name, and chemical symbol, each element has an **atomic mass** (atomic weight). The atomic mass of each element is the average mass of all atoms of the element, including the various isotopes of which it consists. The **atomic mass unit, u** (also called the **dalton**), is used to express masses of individual atoms and molecules (aggregates of atoms). The atomic mass unit is defined as a mass equal to exactly 1/12 that of an atom of carbon-12, the isotope of carbon that contains 6 protons and 6 neutrons in its nucleus.

The Periodic Table

When elements are considered in order of increasing atomic number, it is observed that their properties are repeated in a periodic manner. For example, elements with atomic numbers 2, 10, and 18 are gases that do not undergo chemical reactions and consist of individual molecules, whereas those with atomic numbers larger by 1 – 3, 11, and 19 – are unstable, highly reactive metals. An arrangement of the elements in a manner that reflects this recurring behavior is known as the **periodic table** (Figure 2.1). The periodic table is extremely useful in understanding chemistry and predicting chemical behavior. As shown in Figure 2.1, the entry for each element in the periodic table gives the element's atomic number, name, symbol, and atomic mass. More detailed versions of the table include other information as well.

Active metals

Transition metals

Nonmetals

1A 1	2A 2		3B 3	4B 4	5B 5	6B 6	7B 7	8B 8	8B 9	10	1B 11	2B 12	3A 13	4A 14	5A 15	6A 16	7A 17	8A 18
1 H 1.0079																		2 He 4.00260
3 Li 6.941	4 Be 9.01218												5 B 10.81	6 C 12.011	7 N 14.0067	8 O 15.9994	9 F 18.998403	10 Ne 20.179
11 Na 22.98977	12 Mg 24.305												13 Al 26.98154	14 Si 28.0855	15 P 30.97376	16 S 32.06	17 Cl 35.453	18 Ar 39.948
19 K 39.0983	20 Ca 40.078		21 Sc 44.9559	22 Ti 47.88	23 V 50.9415	24 Cr 51.996	25 Mn 54.9380	26 Fe 55.847	27 Co 58.9332	28 Ni 58.69	29 Cu 63.546	30 Zn 65.38	31 Ga 69.72	32 Ge 72.61	33 As 74.9216	34 Se 78.96	35 Br 79.904	36 Kr 83.80
37 Rb 85.4678	38 Sr 87.62		39 Y 88.9059	40 Zr 91.22	41 Nb 92.9064	42 Mo 95.94	43 Tc (98)	44 Ru 101.07	45 Rh 102.9055	46 Pd 106.42	47 Ag 107.8682	48 Cd 112.41	49 In 114.82	50 Sn 118.69	51 Sb 121.75	52 Te 127.60	53 I 126.9045	54 Xe 131.29
55 Cs 132.9054	56 Ba 137.33		57 *La 138.9055	72 Hf 178.49	73 Ta 180.9479	74 W 183.85	75 Re 186.207	76 Os 190.2	77 Ir 192.22	78 Pt 195.08	79 Au 196.9665	80 Hg 200.59	81 Tl 204.383	82 Pb 207.2	83 Bi 208.9804	84 Po (209)	85 At (210)	86 Rn (222)
87 Fr (223)	88 Ra 226.0254		89 +Ac 227.0278	104 Rf (261)	105 Ha (262)	106 Unh (263)	107 Uns (262)		109 Une (266)									

*Lanthanide series

58 Ce 140.12	59 Pr 140.9077	60 Nd 144.24	61 Pm (145)	62 Sm 150.36	63 Eu 151.96	64 Gd 157.25	65 Tb 158.9254	66 Dy 162.50	67 Ho 164.9304	68 Er 167.26	69 Tm 168.9342	70 Yb 173.04	71 Lu 174.967

+Actinide series

90 Th 232.0381	91 Pa 231.0359	92 U 238.0289	93 Np (237)	94 Pu (244)	95 Am (243)	96 Cm (247)	97 Bk (247)	98 Cf (251)	99 Es (252)	100 Fm (257)	101 Md (258)	102 No (259)	103 Lr (260)

The larger labels are common American usage. The smaller labels have recently been proposed and are presently under consideration.

Figure 2.1. The periodic table of the elements.

Features of the Periodic Table

Groups of elements having similar chemical behavior are contained in vertical columns in the periodic table. **Main group** elements may be designated as A groups (1A and 2A on the left, 3A through 8A on the right). **Transition elements** are those between main groups 2A and 3A. **Noble gases** (group 8A), a group of gaseous elements that are virtually chemically unreactive, are in the far right column. The chemical similarities of elements in the same group is especially pronounced for groups 1A, 2A, 7A, and 8A.

Horizontal rows of elements in the periodic table are called **periods,** the first of which consists of only hydrogen (H) and helium (He). The second period begins with atomic number 3 (lithium) and terminates with atomic number 10 (neon), whereas the third goes from atomic number 11 (sodium) through 18 (argon). The fourth period includes the first row of transition elements, whereas lanthanides and actinides are listed separately at the bottom of the table.

Electrons in Atoms

Although a detailed discussion of the placement of electrons in atoms determines how the atoms behave chemically and, therefore, the chemical properties of each element, it is beyond the scope of this book to discuss electronic structure in detail. Several key points pertaining to this subject are mentioned here.

Electrons in atoms occupy **orbitals** in which electrons have different energies, orientations in space, and average distances from the nucleus. Each orbital may contain a maximum of 2 electrons. The placement of electrons in its orbitals determines the chemical behavior of an atom; in this respect the outermost orbitals and the electrons contained in them are the most important. This is because these outermost electrons, which are the ones beyond those of the immediately preceding noble gas in the periodic table, become involved in the sharing and transfer of electrons through which chemical bonding occurs that results in the formation of huge numbers of different substances from only a few elements.

Lewis Symbols of Atoms

Outer electrons are called **valence electrons** and are represented by dots in **Lewis symbols**, as shown for carbon and argon in Figure 2.2, below:

$$\cdot \overset{\displaystyle \cdot}{\underset{\displaystyle \cdot}{C}} \cdot \qquad\qquad \overset{\displaystyle \cdot\cdot}{\underset{\displaystyle \cdot\cdot}{:Ar:}}$$

Lewis symbol of carbon   Lewis symbol of argon

Figure 2.2. Lewis symbols of carbon and argon.

The four electrons shown for the carbon atom are those added beyond those possessed by the noble gas that immediately precedes carbon in the periodic table (helium, atomic number 2). Eight electrons are shown around the symbol of argon. This is an especially stable electron configuration for noble gases known as an **octet**. (Helium is the exception among noble gases in that it has a stable shell of only two electrons.) When atoms interact through the sharing, loss, or gain of electrons to form molecules and chemical compounds (see Section 2.3) many attain an octet of outer shell electrons. This tendency is the basis of the **octet rule** of chemical bonding. (Two or three of the lightest elements, most notably hydrogen, attain stable helium-like electron configurations containing two electrons when they become chemically bonded.)

2.3. CHEMICAL BONDING

Only a few elements, particularly the noble gases, exist as individual atoms; most atoms are joined by chemical bonds to other atoms. This can be illustrated very simply by elemental hydrogen, which exists as **molecules**, each consisting of 2 H atoms linked by a **chemical bond** as shown in Figure 2.3. Because hydrogen molecules contain 2 H atoms, they are said to be diatomic and are denoted by the **chemical formula**, H_2. The H atoms in the H_2 molecule are held together by a **covalent bond** made up of 2 electrons, each contributed by one of the H atoms, and shared between the atoms. (Bonds formed by transferring electrons between atoms are described

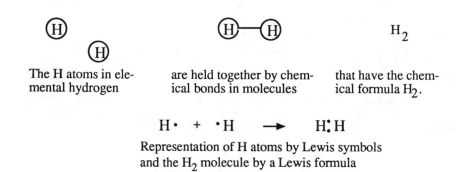

The H atoms in ele-
mental hydrogen

are held together by chemical bonds in molecules

that have the chem-
ical formula H_2.

$$H\cdot \ + \ \cdot H \ \longrightarrow \ H\overset{\cdot}{\cdot}H$$

Representation of H atoms by Lewis symbols
and the H_2 molecule by a Lewis formula

Figure 2.3. Molecule and Lewis formula of H_2.

later in this section.) The shared electrons in the covalent bonds
holding the H_2 molecule together are represented by two dots
between the H atoms in Figure 2.3. By analogy with Lewis symbols
defined in the preceding section, such a representation of molecules
showing outer-shell and bonding electrons as dots is called a **Lewis
formula**.

Chemical Compounds

Most substances consist of two or more elements joined by chemical bonds. As an example consider the chemical combination of the
elements hydrogen and oxygen shown in Figure 2.4. Oxygen, chemical symbol O, has an atomic number of 8 and an atomic mass of
16.00 and exists in the elemental form as diatomic molecules of O_2.
Hydrogen atoms combine with oxygen atoms to form molecules
in which 2 H atoms are bonded to 1 O atom in a substance with a
chemical formula of H_2O (water). A substance such as H_2O that
consists of a chemically bonded combination of two or more
elements is called a **chemical compound**. In the chemical formula
for water the letters H and O are the chemical symbols of the two
elements in the compound and the subscript 2 indicates that there are
2 H atoms per O atom. (The absence of a subscript after the O
denotes the presence of just 1 O atom in the molecule.) As shown in
Figure 2.4, each of the hydrogen atoms in the water molecule is
connected to the oxygen atom by a chemical bond composed of two
electrons shared between the hydrogen and oxygen atoms. For each
bond one electron is contributed by the hydrogen and one by oxygen.

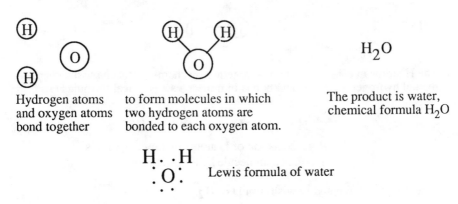

Hydrogen atoms and oxygen atoms bond together

to form molecules in which two hydrogen atoms are bonded to each oxygen atom.

The product is water, chemical formula H_2O

Lewis formula of water

Figure 2.4. Formation and Lewis formula of a chemical compound, water.

The two dots located between each H and O in the Lewis formula of H_2O represent the two electrons in the covalent bond joining these atoms. Four of the electrons in the octet of electrons surrounding O are involved in H-O bonds and are called bonding electrons. The other four electrons shown around the oxygen that are not shared with H are non-bonding outer electrons.

Molecular Structure

As implied by the representations of the water molecule in Figure 2.4, the atoms and bonds in H_2O form an angle somewhat greater than 90 degrees. The shapes of molecules are referred to as their **molecular geometry**, which is crucial in determining the chemical and toxicological activity of a compound and structure-activity relationships.

Ionic Bonds

As shown in Figure 2.5, the transfer of electrons from one atom to another produces charged species called **ions**. Positively charged ions are called **cations** and negatively charged ions are called **anions**. Ions that make up a solid compound are held together by **ionic bonds** in a **crystalline lattice** consisting of an ordered arrangement of the ions in which each cation is largely surrounded by anions and each anion by cations. The attracting forces of the oppositely charged ions in the crystalline lattice constitute ionic bonds in the compound.

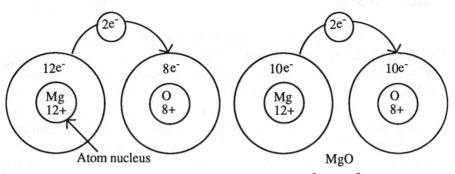

The transfer of an electron from a magnesium atom to an oxygen atom

yields Mg^{2+}and O^{2-}ions that are bonded together by ionic bonds in the compound MgO.

Formation of ionic MgO as shown by Lewis formulas and symbols. In MgO, Mg has lost 2 electrons and is in the +2 oxidation state {Mg(II)} and O has gained 2 electrons and is in the -2 oxidation state.

Figure 2.5. Ionic bonds are formed by the transfer of electrons and the mutual attraction of oppositely charged ions in a crystalline lattice.

The formation of magnesium oxide is shown in Figure 2.5. In naming this compound, the cation is simply given the name of the element from which it was formed, magnesium. However, the ending of the name of the anion, ox*ide*, is different from that of the element from which it was formed, ox*ygen*.

Rather than individual atoms that have lost or gained electrons, many ions are groups of atoms bonded together covalently and having a net charge. A common example of such an ion is the ammonium ion, NH_4^+,

$$
\begin{array}{c}
\text{H} \\
\vdots \\
\text{H} \colon \text{N} \colon \text{H} \quad +\\
\vdots \\
\text{H}
\end{array}
\quad \text{Lewis formula of the ammonium ion}
$$

which consists of 4 hydrogen atoms covalently bonded to a single nitrogen (N) atom and having a net electrical charge of +1 for the whole cation, as shown by its Lewis formula above.

Oxidation State

The loss of two electrons from the magnesium atom as shown in Figure 2.5 is an example of **oxidation**, and the Mg^{2+} ion product is said to be in the +2 **oxidation state**. (A positive oxidation state or oxidation number is conventionally denoted by a Roman numeral in parentheses following the name or symbol of an element as in magnesium(II) and Mg(II)). In gaining 2 negatively charged electrons in the reaction that produces magnesium oxide, the oxygen atom is **reduced** and is in the -2 oxidation state. (Unlike positive oxidation numbers, negative ones are not conventionally shown by Roman numerals in paretheses.) In chemical terms an **oxidizer** is a species that takes electrons from a reducing agent in a chemical reaction. Many hazardous waste substances are oxidizers or strong reducers and oxidation/reduction is a strong driving force behind many dangerous chemical reactions. For example, the reducing tendencies of the carbon and hydrogen atoms in propane cause it to burn violently or explode in the presence of oxidizing oxygen in air. The oxidizing ability of concentrated nitric acid, HNO_3, enables it to react destructively with organic matter, such as cellulose or skin.

Covalently bonded atoms that have not actually lost or gained electrons to produce ions are also assigned oxidation states. This can be done because in covalent compounds electrons are not shared equally. Therefore, an atom of an element with a greater tendency to attract electrons is assigned a negative oxidation state compared to the positive oxidation state assigned to an element with a lesser tendency to attract electrons. For example, Cl atoms attract electrons more strongly than do H atoms so that in hydrogen chloride gas, HCl, the Cl atom is in the -1 oxidation state and the H atoms are in the +1 oxidation state. **Electronegativity** values are assigned to elements on the basis of their tendencies to attract electrons.

The oxidation state (oxidation number) of an element in a compound may have a strong influence on the hazards posed by the compound. For example, chromium from which each atom has lost 3 electrons to form a chemical compound, designated as chromium(III) or Cr(III), is not toxic, whereas chromium in the +6 oxidation state (Cr(VI), chromate) is regarded as a cancer-causing chemical when inhaled.

Molecular Mass

The average mass of all molecules of a compound is its **molecular mass** (formerly called molecular weight). The molecular mass of a compound is calculated by multiplying the atomic mass of each element by the relative number of atoms of the element, then adding all the values obtained for each element in the compound. For example, the molecular mass of NH_3 is $14.0 + 3 \times 1.0 = 17.0$.

2.4. CHEMICAL REACTIONS AND EQUATIONS

Chemical reactions occur when substances are changed to other substances through the breaking and formation of chemical bonds. For example, water is produced by the chemical reaction of hydrogen and oxygen:

Hydrogen plus oxygen yields water

Chemical reactions are written as **chemical equations**. The chemical reaction between hydrogen and water is written as the **balanced chemical equation**

$$2 H_2 + O_2 \longrightarrow 2 H_2O \tag{2.4.1}$$

in which the arrow is read as "yields" and separates the hydrogen and oxygen **reactants** from the water **product**. The equation is balanced because it has the same number of each kind of atom on both sides of the arrow, in this case 4 H and 2 O.

Reaction Rates

Most chemical reactions give off heat and are classified as exothermic reactions. The rate of a reaction may be calculated by the Arrhenius equation, which contains absolute temperature in an exponential term. As a general rule the speed of a reaction doubles for each 10°C increase in temperature. Reaction rate factors are important factors in fires or explosions involving hazardous chemicals.

2.5. SOLUTIONS

A **solution** is formed when a substance in contact with a liquid becomes dispersed homogeneously throughout the liquid in a molecular form. The substance, called a **solute**, is said to **dissolve**. The liquid is called a **solvent**. There may be no readily visible evidence that a solute is present in the solvent; for example a deadly poisonous solution of sodium cyanide in water looks like pure water! The solution may have a strong color, as is the case for intensely purple solutions of potassium permanganate, $KMnO_4$. It may have a strong odor, such as that of ammonia, NH_3, dissolved in water. Solutions may consist of solids, liquids, or gases dissolved in a solvent. Technically, it is possible to have solutions in which a solid is a solvent, although such solutions are not discussed in this book.

Solution Concentration

The quantity of solute relative to that of solvent or solution is called the **solution concentration**. Concentrations are expressed in numerous ways. Very high concentrations are often given as percent by weight. For example commercial concentrated hydrochloric acid is 36% HCl, meaning that 36% of the weight has come from dissolved HCl and 64% from water solvent. Concentrations of very dilute solutions, such as those of hazardous waste leachate containing low levels of contaminants, are expressed as weight of solute per unit volume of solution. Common units are milligrams per liter (mg/L) or micrograms per liter (μg/L). Since a liter of water weighs essentially 1,000 grams, a concentration of 1 mg/L is equal to 1 part per million (ppm) and a concentration of 1 μg/L is equal to 1 part per billion (ppb).

Chemists often express concentrations in moles per liter, or **molarity**, M. Molarity is given by the relationship

$$M = \frac{\text{Number of moles of solute}}{\text{Number of liters of solution}} \qquad (2.5.1)$$

The number of moles of a substance is its mass in grams divided by its molar mass. For example, the molecular mass of ammonia, NH_3, is $14 + 1 + 1 + 1$, so a mole of ammonia has a mass of 17 g.

Therefore, 17 g of NH_3 in 1 L of solution has a value of M equal to 1 mole/L.

Concentration of H^+ Ion and pH

Acids (such as HCl, above) and sulfuric acid (H_2SO_4) produce H^+ ion, whereas bases, such as sodium hydroxide and calcium hydroxide (NaOH and $Ca(OH)_2$, respectively), produce hydroxide ion, OH^-. Molar concentrations of hydrogen ion, $[H^+]$, range over many orders of magnitude and are conveniently expressed by pH defined as

$$pH = -\log[H^+] \tag{2.5.2}$$

In absolutely pure water the value of $[H^+]$ is exactly 1×10^{-7} mole/L, the pH is 7.00, and the solution is **neutral** (neither acidic nor basic). **Acidic** solutions have pH values of less than 7 and **basic** solutions have pH values of greater than 7.

Strong acids and strong bases are **corrosive** substances that exhibit extremes of pH. They are destructive to materials and flesh. Strong acids can react with cyanide and sulfide compounds to release highly toxic hydrogen cyanide (HCN) or hydrogen sulfide (H_2S) gases, respectively. Bases liberate noxious ammonia gas (NH_3) from solid ammonium compounds.

Acid-Base Reactions

The reaction between H^+ ion from an acid and OH^- ion from a base is a **neutralization** reaction. As a specific example consider the reaction of H^+ from a solution of sulfuric acid, H_2SO_4, and OH^- from a solution of calcium hydroxide:

$$\begin{array}{c} \overbrace{} \\ \text{Acid, source of } H^+ \text{ ion} \qquad \text{Water} \\ H_2SO_4 \ + \ Ca(OH)_2 \longrightarrow 2H_2O \ + \ CaSO_4 \\ \underbrace{} \\ \text{Base, source of } OH^- \text{ ion} \qquad \text{Salt} \end{array} \tag{2.5.3}$$

In addition to water, which is always the product of a neutralization reaction, the other product is calcium sulfate, $CaSO_4$. This compound is a **salt** composed of Ca^{2+} ions and SO_4^{2-} ions held together by ionic bonds. A salt, consisting of a cation other than H^+ and an anion other than OH^-, is the other product in addition to water produced when an acid and base react. Some salts that are hazardous substances include the following:

- Ammonium perchlorate, NH_4ClO_4,(reactive oxidant)
- (P013) Barium cyanide, $Ba(CN)_2$(toxic)
- (U144) Lead acetate, $Pb(C_2H_3O_2)_2$ (toxic)
- (U215) Thallium(I) carbonate, Tl_2CO_3 (toxic)

Colloidal Suspensions

Very small particles of the order of 1 micrometer or less in size, called **colloidal particles**, may stay suspended in a liquid for an indefinite period of time. Such a mixture is a **colloidal suspension** and it behaves in many respects like a solution. Colloidal suspensions are used in many industrial applications. Many waste materials are colloidal and are often emulsions consisting of colloidal liquid droplets suspended in another liquid, usually wastewater. One of the challenges in dealing with colloidal wastes is to remove a relatively small quantity of colloidal material from a large quantity of water by precipitating the colloid. This process is called **coagulation** or **flocculation** and is often brought about by the addition of chemical agents.

Water as a Solvent

Most liquid wastes are solutions or suspensions of waste materials in water. Water has some unique properties as a solvent which arise from its molecular structure as represented by the Lewis formula of water below:

$$\text{(+)} \quad \overset{\displaystyle H}{\underset{\displaystyle H}{:O:}} \quad \text{(-)}$$

The H atoms are not on opposite sides of the O atom and the two H–O bonds form an angle of 105°. Furthermore, the O atom (-2 oxidation state) is able to attract electrons more strongly than the 2 H atoms (each in the +1 oxidation state) so that the molecule is **polar**, with the O atom having a partial negative charge and the end of the molecule with the 2 H atoms having a partial positive charge. This means that water molecules can cluster around ions with the positive ends of the water molecules attracted to negatively charged anions and the negative end to positively charged cations. This kind of interaction is part of the general phenomenon of **solvation**, (specifically called **hydration** when water is the solvent) and is partially responsible for water's excellent ability to dissolve ionic compounds, including acids, bases, and salts.

Water molecules form a special kind of bond called a **hydrogen bond** with each other and with solute molecules that contain O, N, or S atoms. As its name implies, a hydrogen bond involves a hydrogen atom held between two other atoms of O, N, or S. Hydrogen bonding is partly responsible for water's ability to solvate and dissolve chemical compounds capable of hydrogen bonding.

As noted above, the water molecule is a polar species, which affects its ability to act as a solvent. Solutes may likewise have polar character. In general, solutes with polar molecules are more soluble in water than nonpolar ones. The polarity of an impurity solute in wastewater is a factor in determining how it may be removed from water. Nonpolar organic solutes are easier to take out of water by an adsorbent species such as activated carbon than are more polar solutes.

Metal Ions Dissolved in Water

Metal ions dissolved in water have some unique properties that influence their properties as natural water constituents, heavy metal pollutants, and hazardous wastes. The formulas of metal ions are usually represented by the symbol for the ion followed by its charge. For example, iron(II) ion (from a compound such as iron(II) sulfate, $FeSO_4$) dissolved in water is represented as Fe^{2+}. Actually, in water solution each iron(II) ion is strongly solvated and bonded to water molecules, so that the formula is more correctly shown as $Fe(H_2O)_6^{2+}$. Many metal ions have a tendency to lose hydrogen ions from the solvating water molecules as shown by the following:

$$Fe(H_2O)_6^{2+} \longrightarrow Fe(OH)(H_2O)_5^{2+} + H^+ \qquad (2.5.4)$$

Ions of the next higher oxidation state, iron(III) have such a tendency to lose H^+ ion in aqueous solution that, except in rather highly acidic solutions, they precipitate out as solid iron(III) hydroxide, $Fe(OH)_3$:

$$Fe(H_2O)_6^{3+} \longrightarrow Fe(OH)_3(s) + 3H_2O + 3H^+ \qquad (2.5.5)$$

Complex Ions Dissolved in Water

It was noted above that metal ions are solvated (hydrated) by binding to water molecules in aqueous solution. Some species in solution have a stronger tendency than water to bond to metal ions. An example of such a species is cyanide ion, CN^-, which displaces water molecules from some metal ions in solution as shown below

$$Ni(H_2O)_4^{2+} + 4CN^- \rightleftharpoons Ni(CN)_4^{2-} + 4H_2O \qquad (2.5.6)$$

The species that bonds to the metal ion, cyanide in this case, is called a ligand and the product of the reaction is a complex ion or metal complex. The overall process is called complexation.

2.6. ORGANIC CHEMISTRY

Most carbon-containing compounds are **organic chemicals** and are discussed under the heading of **organic chemistry**. Organic chemistry is a vast, diverse, discipline because of the enormous number of organic compounds that exist as a consequence of the versatile bonding capabilities of carbon. Such diversity is due to the ability of carbon atoms to bond to each other in a limitless variety of straight chains, branched chains, and rings as illustrated in Figure 2.6. This figure shows several **alkanes**, which are **hydrocarbon compounds** containing only carbon and hydrogen and no double bonds (see alkenes later in this section), triple bonds, or aromatic entities (see Section 2.8). The four hydrocarbon molecules in Figure 2.6 contain 8 carbon atoms each. In one of the molecules, all of the carbon atoms are in a straight chain and in two they are in branched chains, whereas in a fourth 6 of the carbon atoms are in a ring.

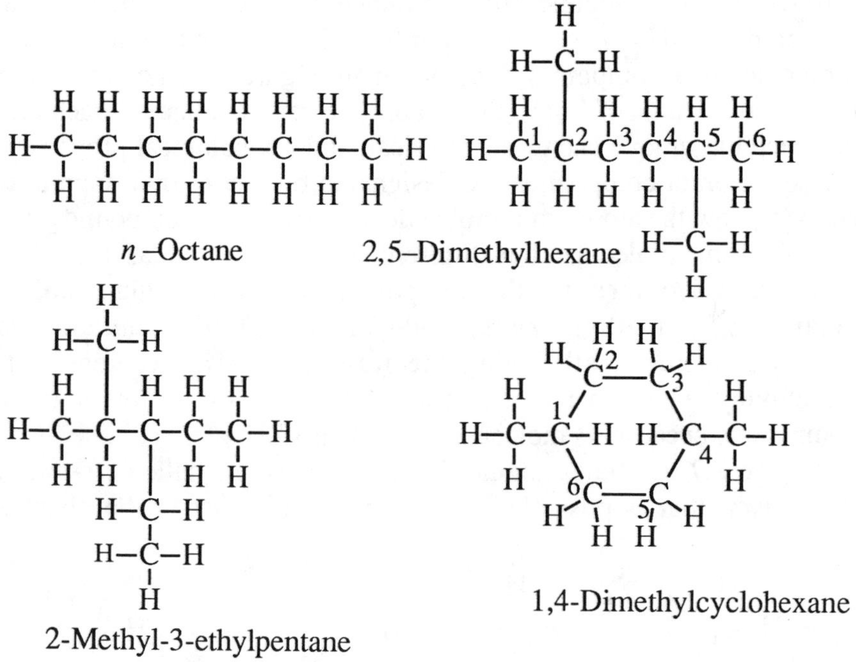

Figure 2.6. Structural formulas of four hydrocarbons, each containing 8 carbon atoms, that illustrate the structural diversity possible with organic compounds. Numbers used to denote locations of atoms for purposes of naming are shown on two of the compounds.

Among organic chemicals are included the majority of important industrial compounds, synthetic polymers, agricultural chemicals, biological materials, and most substances that are of concern because of their toxicities and other hazards. These few pages present only a few of the most fundamental concepts and definitions of organic chemistry needed to understand hazardous organic substances and do not pretend to teach this vast discipline to a reader unfamiliar with it. The reader needing a more complete background in the subject is referred to one of the briefer organic chemistry texts, such as those by Atkins and Carey[8] or Fessenden and Fessenden.[9]

2.7. ORGANIC FORMULAS, STRUCTURES AND NAMES

Formulas of organic compounds present information at several different levels of sophistication. **Molecular formulas**, such as that of octane (C_8H_{18}), give the number of each kind of atom in a molecule of a compound. As shown in Figure 2.6, however, the molecular formula of C_8H_{18}, may apply to several alkanes, each one of which has unique chemical, physical, and toxicological properties. These different compounds are designated by **structural formulas** showing how the atoms in a molecule are arranged. Compounds that have the same molecular, but different structural, formulas are called **structural isomers**. Of the compounds shown in Figure 2.6, *n*-octane, 2,5-dimethylhexane, and 2-methyl-3-ethylpentane are structural isomers, all having the formula C_8H_{18}, whereas 1,4-dimethylcyclohexane is not a structural isomer of the other three compounds because its molecular formula is C_8H_{16}.

Figure 2.7 illustrates another kind of isomerism, called *cis-trans*, isomerism, that is possible for alkenes. In contrast to the alkanes

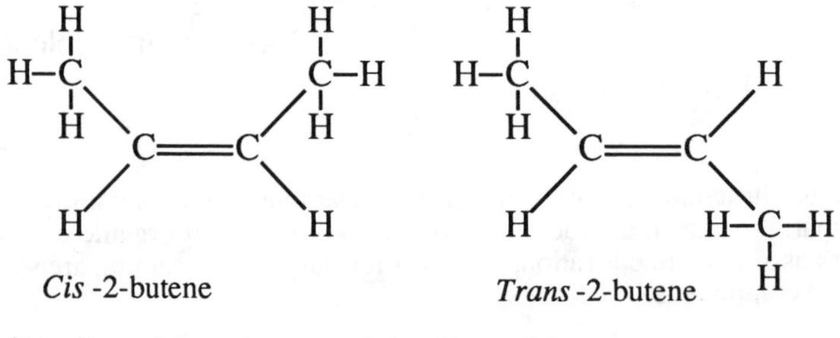

Figure 2.7. *Cis* and *trans* isomers of the alkene, 2-butene.

shown in Figure 2.6, **alkenes**, sometimes still called **olefins**, are hydrocarbons with a **double bond** composed of four shared electrons. The double bond is represented in structural formulas by a double line, =. The two carbon atoms connected by the double bond cannot rotate relative to each other, so that some alkenes are *cis-trans* isomers that have different parts of the molecule oriented differently in space, although these parts occur in the same order. Both alkenes illustrated in Figure 2.7 have a molecular formula of C_4H_8. In the case of *cis*-2-butene, the two CH_3 groups attached to the C=C carbon atoms are on the same side of the molecule, whereas in *trans*-2-butene they are on opposite sides.

Condensed Structural Formulas

To save space, structural formulas are conveniently abbreviated as **condensed structural formulas**. The condensed structural formula of 2-methyl-3-ethylpentane is $CH_3CH(CH_3)CH(C_2H_5)CH_2CH_3$ where the CH_3 (methyl) and C_2H_5 (ethyl) groups are placed in parentheses to show that they are branches attached to the longest continuous chain of carbon atoms, which contains 5 carbon atoms. It is understood that each of the methyl and ethyl groups is attached to the carbon immediately preceding it in the condensed structural formula (methyl attached to the second carbon atom, ethyl to the third).

As illustrated by the examples in Figure 2.8, the structural formulas of organic molecules may be represented in a very compact form by lines and by figures such as hexagons. The ends and inter-sections of straight line segments in these formulas indicate the locations of carbon atoms. Carbon atoms at the terminal ends of lines are understood to have three H atoms attached, C atoms at the intersections of two lines are understood to have *two* H atoms attached to each, *one* H atom is attached to a carbon represented by the intersection of three lines, and *no* hydrogen atoms are bonded to C atoms where four lines intersect. Other atoms or groups of atoms, such as the Cl atom or OH group, that are substituted for H atoms are shown by their symbols attached to a C atom with a line.

Organic Nomenclature

Systematic names, from which the structures of organic molecules can be deduced, have been assigned to all known organic compounds. The more common organic compounds, including hazardous organic sustances, likewise have **common names** that have no structural implications. Although it is not possible to cover organic nomenclature in any detail in this chapter, the basic approach to nomenclature is presented here along with some pertinent examples.

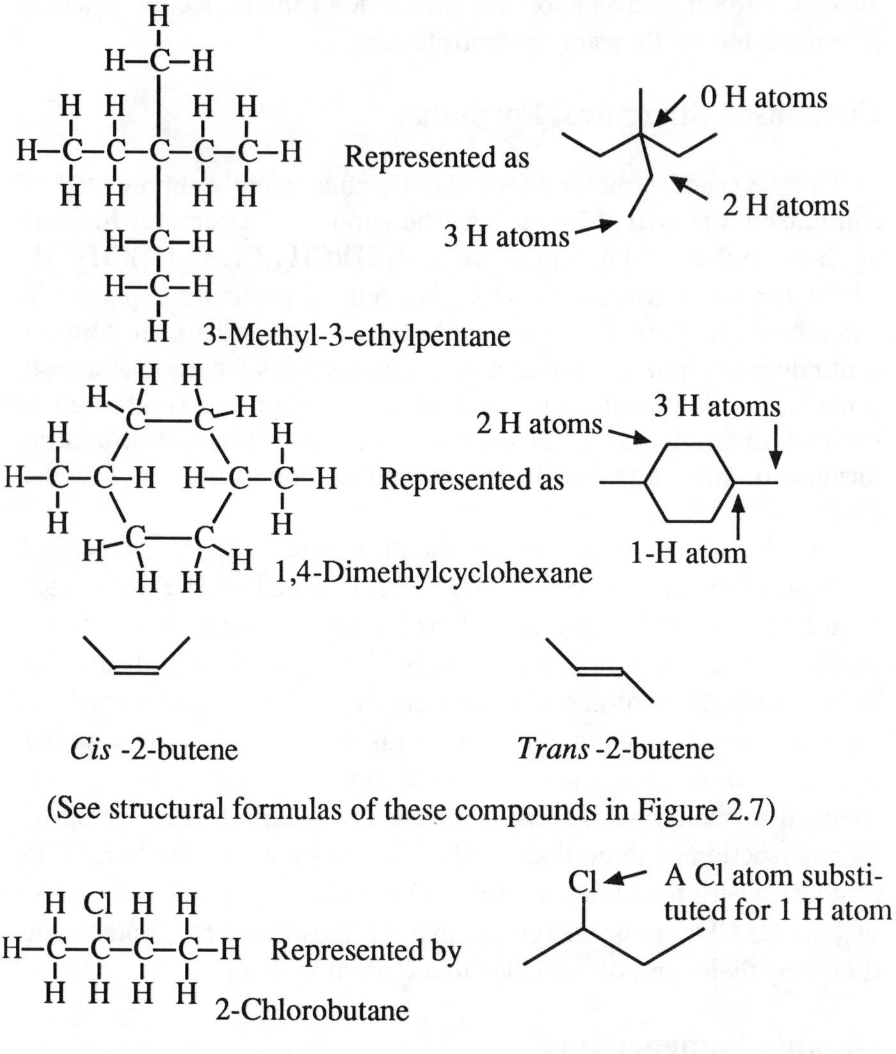

3-Methyl-3-ethylpentane

1,4-Dimethylcyclohexane

Cis -2-butene

Trans -2-butene

(See structural formulas of these compounds in Figure 2.7)

2-Chlorobutane

Figure 2.8. Representation of structural formulas with lines. A carbon atom is understood to be at each corner and at the end of each line. The numbers of hydrogen atoms attached to carbons at several specific locations are shown by arrows.

Consider the alkanes shown in Figure 2.6. The fact that *n*-octane has no side chains is denoted by "*n*", that it has 8 carbon atoms is denoted by "oct," and that it is an alkane is indicated by "ane." The names of compounds with branched chains or atoms other than H or C attached make use of numbers that stand for positions on the longest continuous chain of carbon atoms in the molecule. This convention is illustrated by the second compound in Figure 2.6. It gets the hexane part of the name from the fact that it is an alkane with 6 carbon atoms in its longest continuous chain ("hex" stands for 6). However, it has a methyl group (CH_3) attached on the second carbon atom of the chain and another on the fifth. Hence the full systematic name of the compound is 2,5-dimethylhexane, where "di" indicates two methyl groups. In the case of 2-methyl-3-ethylpentane, the longest continuous chain of carbon atoms contains 5 carbon atoms, denoted by pentane, a methyl group is attached to the second carbon atom, and an ethyl group, C_2H_5, on the third carbon atom The last compound shown in the figure has 6 carbon atoms in a ring, indicated by the prefix "cyclo," so it is a cyclohexane compound. Furthermore, the carbon in the ring to which one of the methyl groups is attached is designated by "1" and another methyl group is attached to the fourth carbon atom around the ring. Therefore, the full name of the compound is 1,4-dimethylcyclohexane.

Molecular Geometry

The three-dimensional shape of a molecule, that is, its molecular geometry, determines in part its properties, particularly its inter-actions with biological systems. Shapes of molecules are represented in drawings by lines of normal, uniform thickness for bonds in the plane of the paper, broken lines for bonds extending away from the viewer, and heavy lines for bonds extending toward the viewer. These conventions are shown by the example of dichloromethane (methylene chloride, Hazardous Waste No. U080), CH_2Cl_2, an important organochloride solvent and extractant, is illustrated in Figure 2.9.

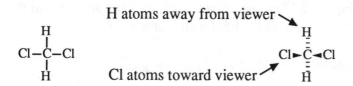

Structural formula of dichloromethane in two dimensions

Structural formula of dichloro-methane represented in three dimensions

Figure 2.9. Structural formulas of dichloromethane, CH_2Cl_2; the formula on the right provides a three-dimensional representation.

2.8. AROMATIC ORGANIC COMPOUNDS

Aromatic compounds (arenes, or **aryl compounds)** constitute an important special class of organic substances. They are unique because of the nature of their bonds, a subject that is beyond the scope of this chapter. Most aromatic compounds discussed in this book contain 6-carbon-atom benzene rings as shown for benzene,

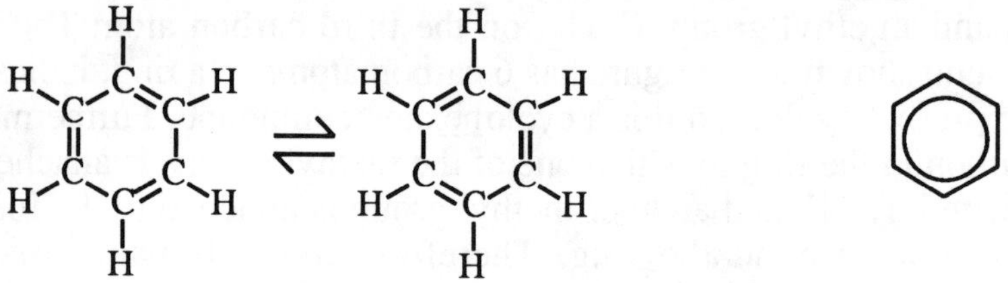

Figure 2.10. Representation of the aromatic benzene molecule with two resonance structures (left) and, more accurately, as a hexagon with a circle in it (right). Unless shown by symbols of other atoms, it is understood that a C atom is at each corner and that one H atom is bonded to each C atom.

C_6H_6, in Figure 2.10. Aromatic compounds such as benzene have ring structures and are held together in part by particularly stable bonds that contain delocalized clouds of so-called π (pi, pronounced "pie") electrons. In an oversimplified sense, the structure of benzene can be visualized as resonating between the two equivalent structures shown on the left in Figure 2.10 by the shifting of electrons in chemical bonds. This structure can be shown more simply and accurately by a hexagon with a circle in it.

Many hazardous waste compounds, such as benzene, toluene, naphthalene, and chlorinated phenols, are arenes (see Figure 2.11). As shown in Figure 2.11, some arenes, such as naphthalene and the polycyclic aromatic compound, benzo(a)pyrene, contain fused rings.

Naming Aromatic Compounds

The chlorinated derivative of phenol, 3-chlorophenol, shown in Figure 2.11, may be used to illustrate the numbering of positions in a 6-membered aromatic ring for the purpose of naming of arenes. Phenol has an -OH group attached to the benzene ring, the carbon to which this group is attached is assigned the number 1, and the other carbon atoms are numbered clockwise around the ring. When chlorine is substituted on the third carbon atom from the -OH group, the compound is 3-chlorophenol. This example also illustrates an older system of nomenclature in which the prefixes, *ortho*, *meta*, and *para*, are employed to show substituent groups on carbons 2 (or 6), 3 (or 5), and 4, respectively.

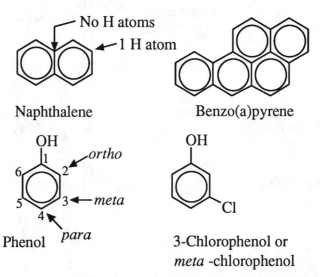

Figure 2.11. Aromatic compounds containing fused rings (top) and showing the numbering of carbon atoms in an aromatic ring for purposes of nomenclature.

2.9. ORGANIC FUNCTIONAL GROUPS

The discussion of organic chemistry so far in this chapter has emphasized hydrocarbon compounds that contain only hydrogen and carbon and the influence of molecular structure upon chemical behavior. It has been shown that hydrocarbons may exist as alkanes, alkenes, and arenes, depending upon the kinds of bonds between carbon atoms. The presence of elements other than hydrogen and carbon in organic molecules greatly increases the diversity of their chemical behavior. **Functional groups** consist of specific bonding configurations of atoms in organic molecules. Most functional groups contain at least one element other than carbon or hydrogen, although two carbon atoms joined by a double bond (alkenes) or triple bond (alkynes) are likewise considered to be functional groups. Table 2.1 shows some of the major functional groups that determine the nature of organic compounds.

Table 2.1. Examples of some important functional groups

Type of functional group	Example compound	Structural formula of functional group[1]
Alkene (olefin)	Propene (propylene)	$H_2C=CH-CH_3$ (with $=CH-$ outlined)
Alkyne	Acetylene	$H-C\equiv C-H$
Alcohol (-OH attached to alkyl group)	2-Propanol	$CH_3-CH(OH)-CH_3$
Phenol (-OH attached to aryl group)	Phenol	C_6H_5-OH
Ketone	Acetone	$CH_3-CO-CH_3$

$$\text{(When } -\overset{\displaystyle O}{\underset{\displaystyle \|}{C}}-H \text{ group is on end carbon,}$$
compound is an aldehyde)

Amine	Methylamine	CH_3-NH_2
Nitro compounds	Nitromethane	CH_3-NO_2
Sulfonic acids	Benzenesulfonic acid	$C_6H_5-SO_2-OH$
Organohalides	1,1-Dichloro-ethane	$Cl-CH_2-CHCl$

[1] Functional group outlined by dashed line

Organic Acids, Alcohols, and Esters

Various functional groups in organic molecules may undergo reactions to form other kinds of functional groups. Among the more important of these in life processes (see the discussion of biochemistry in Chapter 4) are **esters** produced by the reaction of alcohols and carboxylic acids with the loss of water. The formation of methyl acetate ester is shown in Figure 2.12. An ester can also undergo the reverse reaction, hydrolysis with the addition of water, to regenerate the original alcohol and acid. In addition to their occurrence as fats, oils, and waxes in biological systems, some esters are important industrial chemicals and occur in hazardous wastes.

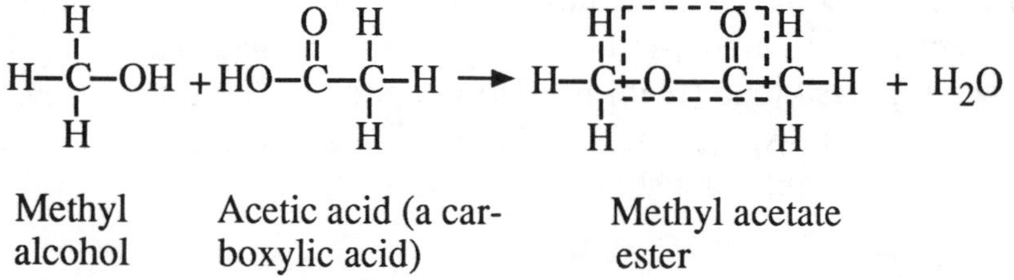

Methyl alcohol

Acetic acid (a carboxylic acid)

Methyl acetate ester

Figure 2.12. Formation of a simple ester with the loss of water. The functional groups characteristic of carboxylic acids and esters formed from alcohols and carboxylic acids are outlined by dashed rectangles.

2.10. SYNTHETIC POLYMERS

A large fraction of the chemical industry worldwide is devoted to polymer manufacture, which is very important in the area of hazardous substances and hazardous wastes. Synthetic **polymers** are produced when small molecules called **monomers** bond together to form a much smaller number of very large molecules. Many natural products are polymers; for example, cellulose in wood, paper, and many other materials is a polymer of the sugar glucose. Synthetic polymers form the basis of many industries, such as rubber, plastics, and textiles manufacture.

An important example of a polymer is that of polyvinylchloride, shown in Figure 2.13. This polymer is synthesized in large quantities for the manufacture of water and sewer pipe, water-repellant liners,

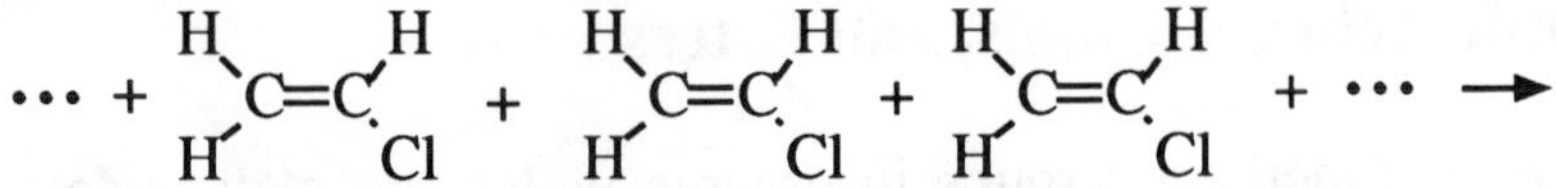

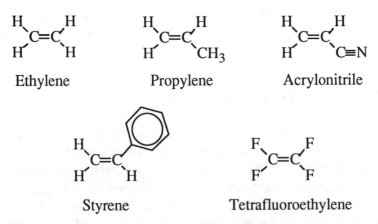

"n" vinyl chloride monomers

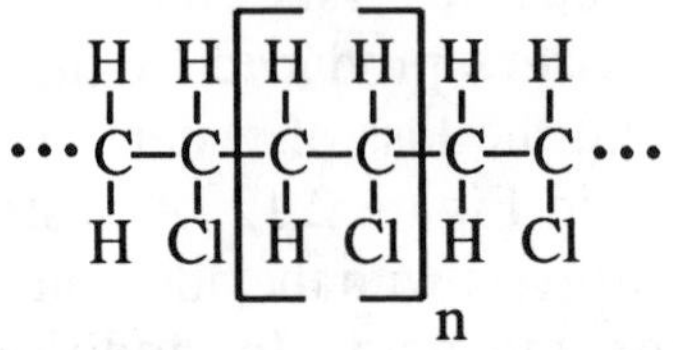

Polyvinylchloride polymer containing a large number, "n," monomer units per molecule

Figure 2.13. Polyvinylchloride polymer.

and other plastic materials. Other major polymers include polyethylene (plastic bags, milk cartons), polypropylene, (impact-resistant plastics, indoor-outdoor carpeting), polyacrylonitrile (Orlon, carpets), polystyrene (foam insulation), and polytetrafluoroethylene (Teflon coatings, bearings); the monomers from which these substances are made are shown in Figure 2.14.

Ethylene   Propylene   Acrylonitrile

Styrene   Tetrafluoroethylene

Figure 2.14. Monomers from which commonly used polymers are synthesized.

Many of the hazards from the polymer industry arise from the monomers used as raw materials. Many monomers are reactive and flammable, with a tendency to form explosive vapor mixtures with air. All have a certain degree of toxicity; vinyl chloride is a known human carcinogen. The combustion of many polymers may result in the evolution of toxic gases, such as hydrogen cyanide (HCN) from polyacrylonitrile or hydrogen chloride (HCl) from polyvinylchloride. Another hazard presented by plastics results from the presence of **plasticizers** added to provide essential properties, such as flexibility. The most widely used plasticizers are phthalates (see Chapter 8), which are environmentally persistent, resistant to treatment processes, and prone to undergo bioaccumulation.

Polymers have a number of applications in waste treatment and disposal. Waste disposal landfill liners are made from synthetic polymers as are the fiber filters which remove particulate pollutants from flue gas in baghouses. Membranes used for ultrafiltration and reverse osmosis treatment of water are composed of very thin sheets of synthetic polymers. Organic solutes can be removed from water by sorption onto hydrophobic (water-repelling) organophilic beads of Amberlite XAD resin. Heavy metal pollutants are removed from wastewater by cation exchange resins made of polymers with anionic functional groups. Typically, these resins exchange harmless sodium ion, Na^+, on the solid resin for toxic heavy metal ions in water. Figure 2.15 shows a segment of the polymeric structure of a cation exchange resin in the sodium form.

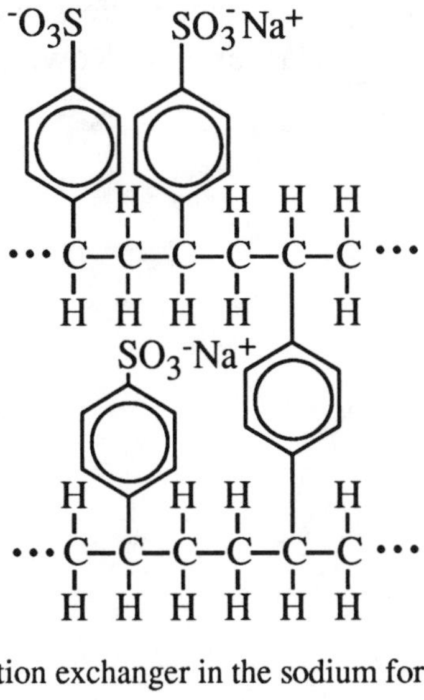

Figure 2.15. Polymeric cation exchanger in the sodium form.

LITERATURE CITED

1. Manahan, Stanley E., *General Applied Chemistry*, 2nd ed., Brooks/Cole Publishing Co., Pacific Grove, CA, 1982.
2. Manahan, Stanley E., *Quantitative Chemical Analysis*, Brooks/Cole Publishing Co., Pacific Grove, CA, 1986.
3. Kneip, Theodore J., and John V. Crable, *Methods for Biological Monitoring*, American Public Health Association, Washington, DC, 1988.
4. Sposito, Garrison, *The Chemistry of Soils*, Oxford University Press, New York, 1989.
5. Dragun, James, *The Soil Chemistry of Hazardous Materials*, Hazardous Materials Control Research Institute, Silver Spring, Maryland, 1988.
6. Wayne, Richard P., *Principles and Applications of Photochemistry*, Oxford University Press, New York, 1988.
7. Manahan, Stanley E., *Environmental Chemistry*, 4th ed., Lewis Publishers, Chelsea, Michigan, 1984.
8. Atkins, Robert C., and Francis A. Carey *Organic Chemistry — A Brief Course*, McGraw-Hill Publishing Company, New York, 1990.
9. Fessenden, Ralph J., and Joan S. Fessenden, *Fundamentals of Organic Chemistry*, Harper and Row, Publishers, Scranton, Pennsylvania, 1990.

3

Environmental Chemical Processes

3.1. WHAT IS ENVIRONMENTAL CHEMISTRY?

Environmental chemistry has been defined as *the study of the sources, reactions, transport, effects, and fates of chemical species in the water, air, terrestrial and living environments.*[1] As an illustration of this definition, consider sulfur dioxide produced by the oxidation of pyrite, FeS_2, when coal is burned. The sulfur dioxide, SO_2, enters the atmosphere through the power plant stack, is transported through the atmosphere by atmospheric processes and is oxidized in part to sulfuric acid, H_2SO_4, by oxidants in the atmosphere. The sulfuric acid is washed from the atmosphere as "acid rain," which may harm plants or fish or dissolve limestone from building walls. Eventually the sulfuric acid reaches a semipermanent resting place, such as a body of water or the soil, known as a **sink**. This simple example illustrates the production, chemical transitions, and fate of a chemical substance in the environment and how it may be interchanged among the various "spheres" of the environment.

Hazardous wastes and their products may be encountered in water, the atmosphere, soil, and living systems. Obviously, environmental chemistry must be strongly involved in efforts to understand and deal with hazardous waste problems. That is why environmental chemistry is summarized in this chapter.

As its definition implies, environmental chemistry can be divided into the four broad areas of aquatic chemistry, atmospheric

55

chemistry, "terrestrial chemistry" (including geochemistry and soil chemistry), and environmental biochemistry, including toxicological chemistry.

One of the most important aspects of environmental chemistry consists of environmental chemical processes by which pollutant species are interchanged between the "spheres" mentioned above. These interactions are illustrated in Figure 3.1.

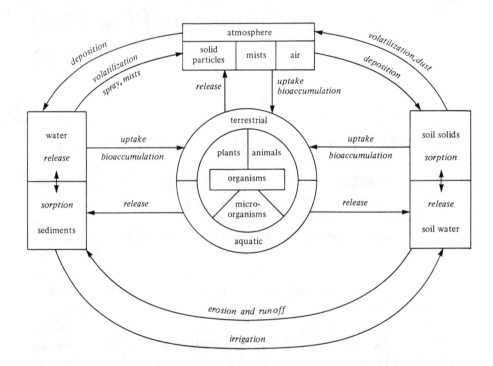

Figure 3.1. Interchange of pollutant species among the atmosphere, hydrosphere, geosphere, and biosphere.

Although analytical chemistry is an important component of environmental chemistry, the two disciplines should not be confused with each other. As applied to environmental chemistry, analytical chemistry is used to determine the nature and quantities of contaminants in the environment. Environmental chemistry has a broader role in that it seeks to predict the behavior and effects of chemical species in the environment and to avoid problems of chemical pollution before they arise.

3.2. AQUATIC CHEMISTRY

Figure 3.2 summarizes the more important aspects of **aquatic chemistry** as it applies to environmental chemistry. As shown in

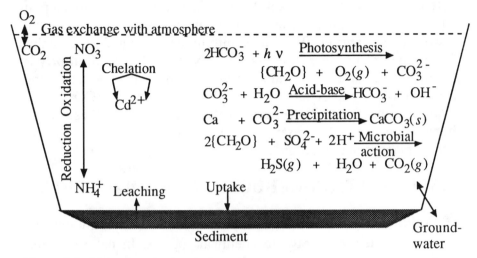

Figure 3.2. Major aquatic chemical processes.

this figure, a number of chemical phenomena occur in water. Many aquatic chemical processes are influenced by the action of algae and bacteria in water. For example, it is shown that algal photosynthesis fixes inorganic carbon from HCO_3^- ion in the form of biomass (represented as $\{CH_2O\}$), in a process that also produces carbonate ion, CO_3^{2-}. Carbonate undergoes an acid-base reaction to produce OH^- ion and raise the pH, or it reacts with Ca^{2+} ion to precipitate solid $CaCO_3$. Most of the many oxidation-reduction reactions that occur in water are mediated (catalyzed) by bacteria. For example, bacteria convert inorganic nitrogen largely to ammonium ion, NH_4^+, in the oxygen-deficient (anaerobic) lower layers of a body of water. Near the surface, where O_2 is available, bacteria convert inorganic nitrogen to nitrate ion, NO_3^-. Metals in water may be bound to organic chelating agents, such as pollutant nitrilotriacetic acid (NTA) or naturally-occurring fulvic acids. Gases are exchanged with the atmosphere, and various solutes are exchanged between water and sediments in bodies of water.

Groundwater is of the utmost importance insofar as hazardous wastes are concerned. This is because of groundwater's susceptibility to contamination by improperly disposed wastes, such as those in poorly constructed landfills. Once a groundwater source is contaminated, it is extremely difficult to restore to its original condition.

Several important characteristics of unpolluted water should be noted. One of these is **gas solubility**. Since it is required to support aquatic life and maintain water quality, oxygen is the most important dissolved gas in water. Water in equilibrium with air at 25°C contains 8.3 milligrams per liter (mg/L) of dissolved O_2. Water **alkalinity** is defined as the ability of solutes in water to neutralize added strong acid. Alkalinity is generally due to bicarbonate ion, HCO_3^-, with contributions from OH^- and CO_3^{2-} at higher pH values. Water **hardness** is due to the presence of calcium ion, Ca^{2+}, and, to a lesser extent, magnesium ion, Mg^{2+}.

3.3. OXIDATION-REDUCTION AND BACTERIA

Oxidation-reduction (redox) reactions in water involve the transfer of electrons between chemical species. In natural water, wastewater, and soil, most significant oxidation-reduction reactions are carried out by bacteria, so they are considered in this section as well.

The relative oxidation-reduction tendencies of a chemical system depend upon the activity of the electron e^-. When the electron activity is relatively high, chemical species (even including water) tend to accept electrons,

$$2H_2O + 2e^- \rightleftharpoons H_2(g) + 2OH^- \qquad (3.3.1)$$

and are said to be reduced. When the electron activity is relatively low, the medium is **oxidizing**, and chemical species such as H_2O may be **oxidized** by the loss of electrons:

$$2H_2O \rightleftharpoons O_2(g) + 4H^+ + 4e^- \qquad (3.3.2)$$

The relative tendency toward oxidation or reduction is based upon the electrode potential, E, which is relatively more positive in an oxidizing medium and negative in a reducing medium. It is defined in terms of the half reaction,

$$2H^+ + 2e^- \rightleftharpoons H_2(g) \tag{3.3.3}$$

for which E is defined as exactly zero when the activity of H^+ is exactly 1 (concentration approximately 1 mole per liter) and the pressure of H_2 gas is exactly 1 atmosphere. Because electron activity in water varies over many orders of magnitude, environmental chemists find it convenient to discuss oxidizing and reducing tendencies in terms of pE, a parameter analogous to pH and defined conceptually as the negative log of the electron activity:

$$pE = -\log(a_{e^-}) \tag{3.3.4}$$

$$pH = -\log(a_{H^+}) \tag{3.3.5}$$

The value of pE is calculated from E by the relationship,

$$pE = \frac{E}{\frac{2.303RT}{F}} \tag{3.3.6}$$

At 25°C for E in volts, $pE = \dfrac{E}{0.0591}$

where R is the gas constant, T is the absolute temperature, and F is the Faraday.

The nature of chemical species in water is usually a function of both pE and pH. A good example of this is shown by a simplified pE-pH diagram for iron in water, assuming that iron is in one of the four forms of Fe^{2+} ion, Fe^{3+} ion, solid $Fe(OH)_3$, or solid $Fe(OH)_2$ as shown in Figure 3.3. Water in which the pE is higher than that shown by the upper dashed line is thermodynamically unstable toward oxidation (Reaction 3.3.2), and below the lower dashed line water is thermodynamically unstable toward reduction (Reaction 3.3.1). It is seen that Fe^{3+} ion is stable only in a very oxidizing, acidic medium such as that encountered in acid mine water, whereas Fe^{2+} ion is stable over a relatively large region as reflected by the common occurrence of soluble iron(II) in oxygen-deficient groundwaters. Highly insoluble $Fe(OH)_3$ is the predominant iron species over a very wide pE-pH range.

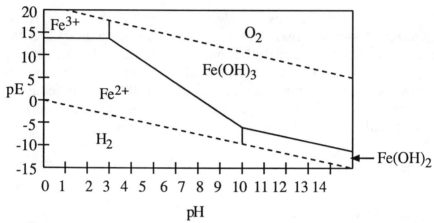

Figure 3.3. A simplified pE-pH diagram for iron in water (maximum total soluble iron concentration 1.0×10^{-5} M).

Bacteria in Oxidation-Reduction Processes

Bacteria in water use oxidation-reduction reactions to extract the energy that they need for their own growth and reproduction. Bacteria are single-celled organisms roughly a micrometer in size. Some bacteria require elemental molecular O_2 for their metabolic needs and are called **aerobic bacteria**. Other **anaerobic bacteria** extract their oxygen from sources such as NO_3^-, SO_4^{2-}, and organic matter, represented as $\{CH_2O\}$.

One of the most common bacterially mediated reactions in oxidation-reduction processes in water and soil is the oxidation of organic matter,

$$O_2 + \{CH_2O\} \longrightarrow CO_2 + H_2O \qquad (3.3.7)$$

a process called **aerobic respiration**. It provides the means for degrading organic wastes, such as sewage. If this reaction occurs at a rate faster than processes for replenishing oxygen in a body of water, the level of dissolved oxygen may become so diminished that the water no longer supports fish life. Some other important oxidation-reduction reactions carried out by bacteria in water are given in Table 3.1.

Table 3.1. Some Oxidation-Reduction Reactions Mediated by Bacteria

Reaction	Significance
$2\{CH_2O\} \rightarrow CH_4 + CO_2$	Fermentation reaction, atmospheric methane source
$SO_4^{2-} + 2\{CH_2O\} + 2H^+$ $\rightarrow H_2S(g) + 2CO_2$ $+ 2H_2O$	Sulfate reduction, source of atmospheric H_2S
$2FeS_2 + 2H_2O + 7O_2 \rightarrow$ $4H^+ + 4SO_4^{2-} + 2Fe^{2+}$	Production of acid mine water
$4NO_3^- + 5\{CH_2O\} + 4H^+$ $\rightarrow 2N_2 + 5CO_2 + 7H_2O$	Denitrification (conversion of fixed nitrogen back to atmospheric N_2)

Bacteria interact with hazardous wastes in a number of ways. Some very toxic materials impede, or even totally stop, bacterial action. A number of organic wastes are partially or completely degraded by bacteria. In a few cases, the organic products are even more toxic than the original wastes. Bacteria can modify the soil and water medium in which hazardous wastes are contained, thus affecting the mobility of the wastes. Prominently, bacterial degradation of organic matter (Reaction 3.3.7) can consume all the oxygen in a landfill and greatly lower the pE. Under these conditions, anaerobic bacteria can reduce insoluble iron(III) oxide (hydroxide or hydrated oxide) to soluble Fe^{2+} ion. This process may release heavy metal ions incorporated with the insoluble iron(III) oxide or soluble iron(II) salts can be mobilized and seep from the site, resulting in iron contamination of the surrounding area.

3.4. COMPLEXATION AND CHELATION

Metal ions in water are always bonded to water molecules in the form of hydrated ions represented by the general formula, $M(H_2O)_x^{n+}$, from which the H_2O is often omitted for simplicity. Other species may be present that bond to the metal ion more strongly than does water. For example, cadmium ion, Cd^{2+}, reacts with cyanide ion, CN^-, as follows:

$$Cd^{2+} + CN^- \rightarrow CdCN^+ \qquad (3.4.1)$$

The product of the reaction is called a **complex** (or complex ion) and the cyanide ion is called a **ligand**. Some (particularly organic) ligands can bond with a metal ion in two or more places, forming particularly stable complexes. One such ligand is the nitrilotriacetate (NTA) ligand, which has the following formula:

This ion has four binding sites, each marked with an asterisk in the preceding illustration, which may simultaneously bond to a metal ion, forming a structure with three rings. Such a species is known as a **chelate**, and NTA is a **chelating** agent.

Complexation and chelation have strong effects upon hazardous wastes. For example, complexation with negatively charged ligands may convert a soluble metal species from a cation, which is readily bound and immobilized by ion exchange processes in soil, to an anion, such as $Ni(CN)_4^{2-}$, that is not strongly held by soil. Thus, codisposal of metal salts and chelating agents can result in increased hazards from heavy metals. On the other hand, some chelating agents are used for the treatment of heavy metal poisoning and insoluble chelating agents, such as chelating resins, can be used to remove metals from waste streams. Metal ions chelated by hazardous waste chelating agents, such as NTA from metal plating bath solutions, may be especially mobile in water. Naturally-occurring humic substance chelating agents may either increase or decrease the solubility of heavy metal wastes. [2]

Complexation may affect the fates of hazardous waste ligand species. For example, whereas cyanide ion is very toxic, but easily oxidized in water, complexed cyanide ion is relatively less toxic, but more difficult to destroy.

Another major type of metal species important in hazardous wastes consists of **organometallic compounds**, which differ from complexes and chelates in that the organic portion is bonded to the metal by a carbon-metal bond and the organic ligand is frequently not capable of existing as a stable separate species. Typical examples of organometallic compound species are monomethylmercury ion and dimethylmercury:

$$Hg^{2+} \qquad\qquad Hg(CH_2)^+ \qquad\qquad Hg(CH_2)^+$$

Mercury(II) ion   Monomethylmercury ion   Dimethylmercury

Organometallic compounds may enter the environment directly as pollutant industrial chemicals and some, including organometallic mercury, tin, selenium, and arsenic compounds, are synthesized biologically by bacteria. Some of these compounds are particularly toxic because of their mobilities in living systems and abilities to cross cell membranes.

3.5. WATER INTERACTIONS WITH OTHER PHASES

Most of the important chemical phenomena associated with water do not occur in solution, but rather through interaction of solutes in water with other phases. For example, the oxidation-reduction reactions catalyzed by bacteria occur in bacterial cells. Many organic hazardous wastes are carried through water as emulsions of very small particles suspended in water. Some hazardous wastes are deposited in sediments in bodies of water, from which they may later enter the water through chemical or physical processes and cause severe pollution effects.

Figure 3.4 summarizes some of the most significant types of interactions between water and other phases, including solids, immisicible liquids, and gases. Films of organic compounds, such as hydrocarbon liquids, may be present on the surface of water. Exposed to sunlight, these compounds are subject to photochemical reactions (see Section 3.9). Gases such as O_2 , CO_2, CH_4, and H_2S are exchanged with the atmosphere. Photosynthesis occurs in suspended cells of algae, and other biological processes, such as biodegradation of organic wastes, occur in bacterial cells. Particles contributing to the turbidity of water may be introduced by physical processes, including the erosion of streams or sloughing of water impoundment banks. Chemical processes, such as the formation of solid $CaCO_3$ illustrated in Figure 3.4, may also form particles in water.

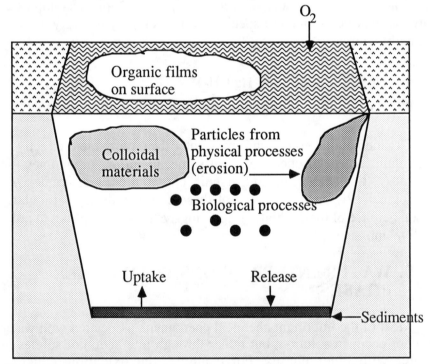

Figure 3.4. Aquatic chemical processes at interfaces between water and gases, solids, or other liquids.

Sediments are repositories of a wide variety of chemical species and the site of many chemical and biochemical processes. Anaerobic fermentation of organic matter by bacteria (see Table 3.1) produces methane gas that is evolved from sediments. Similar bacteria produce mobile $HgCH_3^+$ and $Hg(CH_3)_2$ from insoluble, relatively harmless inorganic mercury compounds. Sediments are sinks for many hazardous organic compounds and heavy metal salts that have gotten into water.

Colloidal particles are very small particles ranging from 0.001 micrometer (μm) to 1 μm in diameter. Colloids have a strong influence on aquatic chemistry. Because of their extremely small size, these particles have a very high surface-to-volume ratio. They are small enough to remain suspended in water, enabling maximum exposure to the water and solutes dissolved in it. Toxic substances in colloidal form are much more available to organisms in water than are such substances in bulk form. Special measures are required to remove colloidal particles from water. Usually, chemical treatment measures are applied to cause colloidal particles to aggregate together (processes called **coagulation** or **flocculation**), and the solids are removed by filtration.

The ability to undergo **ion exchange** processes is an important characteristic of some solids in contact with water. These processes are usually cation exchange reactions represented by

$$\{Solid\}^-M^+ + Z \leftrightarrow \{Solid\}^-Z^+ + M^+ \qquad (3.5.1)$$

where M^+ and Z^+ represent two different cations. Cationic hazardous wastes may be retained on sediments by cation exchange.

3.6. WATER POLLUTANTS

Natural waters are afflicted with a wide variety of inorganic, organic, and biological pollutants, a significant fraction of which come from the improper disposal of hazardous wastes. In some cases, such as that of highly toxic cadmium, a pollutant is directly toxic at a relatively low level. In other cases the pollutant, itself, is not toxic, but its presence results in conditions detrimental to water quality. For example, biodegradable organic matter in water is often not toxic, but the consumption of oxygen during its degradation prevents the water from supporting fish life. Some contaminants, such as NaCl, are normal constituents of water at low levels, but harmful pollutants at higher levels. Table 3.2 summarizes the major water pollutants and their significance.

Table 3.2. General Categories of Water Pollutants

Pollutant category	Examples and significance
Asbestos	Fibrous minerals known to cause lung cancer when inhaled. Effects in water are unknown.
Alkalinity, acidity, salinity	HCO_3^-, H_2SO_4, and NaCl, respectively. Harmful pollutants if in excess.
Biochemical oxygen demand, BOD	Organic matter that consumes oxygen when degraded. Affects water quality, reduces dissolved O_2 levels.
Carcinogens	Aflatoxins, nitrosamines, polycyclic aromatic hydrocarbons (PAHs), others. May cause cancer.
Detergents	Surface active agents and their builders. Harm wildlife, affect esthetics.
Fertilizers (algal nutrients)	Phosphates, K^+, NO_3^-. Cause excessive algal growth. Subsequent decay of algal biomass consumes oxygen and causes eutrophication of water.
Inorganic pollutants	CN^-, NH_3, SO_2. Toxicity, esthetics, general detrimental effects on water quality.
Metal species with organic entities	Heavy metal chelates, organometallic compounds. Toxicity, metal transport.
Pathogens	Disease-causing bacteria and viruses. Cause illness. Impose constraints on water reuse and increase difficulty of water treatment.
Pesticides	Parathion, carbaryl. Toxic to humans and wildlife.
Petroleum wastes, "oil spills"	Petroleum and petroleum products. Harmful to wildlife, esthetics.
Radionuclides	Radium, plutonium. Allowable only at very low levels in drinking water.
Sediments	Detrimental to wildlife and esthetics. Fill streams and impoundments.
Sewage	Sanitary and other wastes discharged into sewer systems. Toxic, contains pathogens, consumes oxygen during biodegradation.
Trace elements and heavy metals	Cadmium, lead, arsenic. Directly toxic to wildlife and humans.
Trace organics	Aromatics, organohalides. Toxic, poorly biodegradable.

3.7. WATER TREATMENT

The treatment of water can be considered under the two major categories of (1) treatment before use and (2) treatment of contaminated water after it has passed through a municipal water system or industrial process. In both cases, it is becoming increasingly evident that consideration must be given to contamination by hazardous wastes. In many areas, contamination by such wastes is threatening sources of municipal water supply and putting extra demands upon water treatment processes. A major objective of many hazardous waste treatment processes is the removal of water that can be safely discharged to municipal waste treatment processes (publicly owned treatment works, POTW).

Several operations may be employed to treat water prior to use. Aeration is employed to drive off odorous gases, such as H_2S^+, and to oxidize soluble Fe^{2+} and Mn^{2+} ions to insoluble forms. Lime is added to remove dissolved calcium (water hardness) as solid $CaCO_3$. Aluminum sulfate coagulant may be added to form a sticky precipitate of $Al(OH)_3$, which pulls very fine particles from suspension. Various filtration and settling processes are employed. Chlorine, Cl_2, is added to kill bacteria. However, chlorine may react with humic substances in the water to produce toxic halogenated byproducts. These can be removed by activated carbon filtration, although it is preferable to eliminate the organic materials from which the halogenated byproducts are formed prior to chlorination.

The treatment of municipal wastewater consists of primary, secondary, and advanced water treatment. **Primary** water treatment consists of settling and skimming operations that remove grit, grease, and physical objects from water. **Secondary** water treatment is designed to take out biochemical oxygen demand, BOD. This is normally accomplished by introducing air and microorganisms such that biomass $\{CH_2O\}$ undergoes aerobic respiration (see Section 3.3) to remove biodegradable biomass:

$$\{CH_2O\} + O_2 \rightarrow CO_2 + H_2O \tag{3.7.1}$$

This process is most commonly carried out in an activated sludge unit in which biodegradation occurs in a tank through which water is pumped. Water from this tank is transferred to a settling basin where the microorganisms responsible for the degradation settle out as a sludge. The sludge is then pumped back to the tank to provide a high concentration of bacteria for the degradation process. Excess sewage sludge tends to collect hazardous waste substances, particularly heavy metals and insoluble organic species, suggesting caution in its disposal or use as fertilizer.

Advanced waste treatment is used when it is necessary to upgrade secondary sewage effluent for release to receiving waters or for further use. Special filtration techniques can be employed to remove small particles left from secondary wastewater treatment. Filtration over activated carbon removes particulate matter and even some dissolved organic substances. Special measures are employed to take out nitrate and phosphate "fertilizers," which tend to cause eutrophication in receiving waters (see Table 3.2). Excess salts can be removed by several processes, the most successful of which is reverse osmosis, in which water is forced under high pressure through a membrane that allows H_2O molecules to pass, but not dissolved salts.

3.8. SOIL

Soil consists of a large variety of material composing the uppermost layer of the earth's crust upon which plants grow. In addition to its essential role as a medium for plant growth, soil is often the repository for a number of substances produced by human activities, including pesticides, petroleum, and hazardous wastes. Typically, soil consists of about 95% mineral matter and 5% organic material, although it may vary from almost pure mineral matter to almost pure organic material. The bulk of finely divided mineral matter in soil consists of weathered products of bedrock.

Some of the main environmental chemical aspects of soil are illustrated in Figure 3.5. The most active and important part of soil is **topsoil**, the layer in which plants are rooted and in which most biological activity occurs. In addition to finely divided mineral matter, topsoil contains water and air. During the growing season substantial amounts of water are lost to the atmosphere by sorption through plant roots followed by evaporation from leaves, a process called **transpiration**. Soil is a very active medium for biochemical processes, including bacterial fixation of atmospheric nitrogen, biodegradation of degradable organics, and oxidation-reduction transitions of inorganic substances, such as the bacterially mediated oxidation of fertilizer ammonia to nitrate, which can be utilized by plants:

$$NH_3 + 2O_2 \text{ (bacteria)} \rightarrow H^+ + NO_3^- + H_2O \qquad (3.8.1)$$

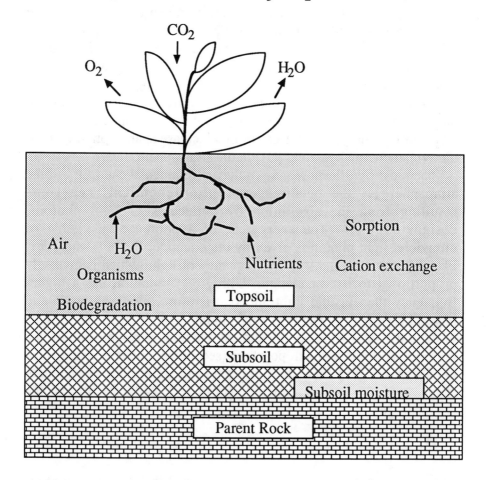

Figure 3.5. Layers of soil.

3.9. THE ATMOSPHERE

The **atmosphere** consists of the thin layer of mixed gases covering the earth's surface. Exclusive of water, atmospheric air is 78.1% (by volume) nitrogen, 21.0% oxygen, 0.9% argon, and 0.03% carbon dioxide. Normally air contains 1-3% water vapor by volume.

In addition, air contains a large variety of trace level gases at levels below 0.002%, including neon, helium, methane, krypton, nitrous oxide, hydrogen, xenon, sulfur dioxide, ozone, nitrogen dioxide, ammonia, and carbon monoxide.

The atmosphere is divided into several layers on the basis of temperature. Of these, the most significant are the troposphere extending in altitude from the earth's surface to approximately 11 kilometers (km) and the stratosphere from about 11 km to approximately 50 km. The temperature of the troposphere ranges from an average of 15°C at sea level to an average of -56°C at its upper boundary. The average temperature of the stratosphere increases from -56°C at its boundary with the troposphere to -2°C at its upper boundary. The reason for this increase is absorption of solar ultraviolet energy by ozone (O_3) in the stratosphere.

The most significant feature of the environmental chemistry of the atmosphere is the occurrence of **photochemical reactions** resulting from the absorption by molecules of light photons, designated hv.[3] (The energy, E, of a photon of a visible or ultraviolet light photon is given by the equation, $E = hv$, where h is Planck's constant and v is the frequency of light, which is inversely proportional to its wavelength. Ultraviolet radiation has a higher frequency than visible light and is, therefore, more energetic and more likely to break chemical bonds in molecules that absorb it.) One of the most significant photochemical reactions is responsible for the presence of ozone in the troposphere (see above) and is initiated when O_2 absorbs highly energetic ultraviolet radiation in the wavelength ranges of 135-176 nanometers (nm) and 240-260 nm in the stratosphere:

$$O_2 + hv \rightarrow O + O \qquad\qquad (3.9.1)$$

The oxygen atoms produced by the photochemical dissociation of O_2 react with oxygen molecules to produce ozone, O_3,

$$O + O_2 + M \rightarrow O_3 + M \qquad\qquad (3.9.2)$$

where M is a third body, such as a molecule of N_2, which absorbs excess energy from the reaction. The ozone that is formed is very effective in absorbing ultraviolet radiation in the 220-330 nm wavelength range in the stratosphere, which causes the temperature increase observed in the stratosphere. The ozone serves as a very valuable filter to remove ultraviolet radiation from the sun's rays. If this radiation reached the earth's surface, it would cause skin cancer and other damage to living organisms.

Physical phenomena in the atmosphere, which are studied under the science of **meteorology**, have a strong influence on atmospheric chemistry. **Temperature inversions**, in which a layer of cold air is trapped beneath hot air, are particularly influental because they limit vertical circulation air and trap bodies of air for long periods of time. In the presence of pollutant nitrogen oxides and hydrocarbons and sunlight, which initiates photochemical reactions, photochemical smog (see Section 3.11) develops in air masses held stationary by temperature inversions.

3.10. OXIDES IN THE ATMOSPHERE

Oxides of carbon, sulfur, and nitrogen are important constituents of the atmosphere and pollutants at higher levels. Of these, the most abundant is carbon dioxide, O_2, which is required for plant growth. As of now, carbon dioxide is present at a level of about 340 parts per million (ppm) by volume in the global atmosphere and is increasing at a rate of about 1 ppm per year. This gas reabsorbs some of the infrared radiation by which the earth's surface re-radiates energy absorbed from the sun. Increasing CO_2 levels resulting from accelerated use of fossil fuels and destruction of CO_2-fixing forests may cause an increase in the earth's temperature (greenhouse effect) with disastrous effects on climate.

Carbon monoxide, CO, is a serious air pollutant in areas where automobile exhaust emissions result in excessive levels. Carbon monoxide combines with the blood's hemoglobin (see Section 7.3) preventing it from transporting oxygen to body tissues. Illness and death can result.

Several oxides of nitrogen are significant air constituents and pollutants. Nitrous oxide, N_2O, is a non-toxic atmospheric gas formed by bacterial action on nitrate salts. Although it participates in atmospheric photochemical reactions, N_2O is not generally regarded as a pollutant. Nitric oxide, NO, and nitrogen dioxide, NO_2, are collectively denoted as "NO_x." Both are regarded as pollutants in the atmosphere. Most pollutant sources of NO_x produce predominantly

NO. Internal combustion engines produce this gas by the high-temperature, high-pressure reaction of N_2 and O_2:

$$N_2 + O_2 \rightarrow 2NO \qquad\qquad (3.10.1)$$

Much of the chemically-bound nitrogen in coal and residual fuel oils is converted to NO_x during combustion. Photochemical processes in the atmosphere tend to convert NO to NO_2. Further reactions can result in the formation of corrosive nitrate salts or nitric acid, HNO_3. Nitric acid is a constituent of **acid precipitation**, consisting of rainwater, snow, or fog having a very low pH.

Nitrogen dioxide is particularly significant in atmospheric chemistry because of its photochemical dissociation by light with a wavelength less than 430 nm,

$$NO_2 + h\nu \rightarrow NO + O \qquad\qquad (3.10.2)$$

to produce highly reactive O atoms. This is the first step in the formation of photochemical smog, discussed in the next section.

Sulfur dioxide, SO_2, is a reaction product of the combustion of sulfur-containing fuels, such as high-sulfur coal. Part of this sulfur dioxide is converted by the following overall atmospheric chemical process to sulfuric acid, H_2SO_4:

$$2SO_2 + O_2 + 2H_2O \rightarrow 2H_2SO_4 \qquad\qquad (3.10.3)$$

In most localities, sulfuric acid is the predominant contributor to acid precipitation.

3.11. HYDROCARBONS AND PHOTOCHEMICAL SMOG

The most abundant hydrocarbon in the atmosphere is methane, CH_4, released from underground sources as natural gas and produced by the fermentation of organic matter as shown in Table 3.1. Methane is one of the least reactive atmospheric hydrocarbons and is produced by diffuse sources, so that its participation in the formation of pollutant photochemical reaction products is minimal. The most significant atmospheric pollutant hydrocarbons are the reactive ones produced as automobile exhaust emissions. In the presence of NO, under conditions of temperature inversion (see Section 3.10), low humidity, and sunlight, these hydrocarbons produce undesirable **photochemical smog** manifested by the presence of visibility-obscuring particulate matter, oxidants such as ozone, and noxious organic species such as aldehydes. Figure 3.6 shows the variation through the day of the main types of chemical species involved in photochemical smog. Examination of the figure shows that shortly after dawn, hydrocarbons and NO increase as a result of direct emissions from automobiles. This is followed by an abrupt drop in NO and a sharp increase in NO_2 as the intensity of sunlight increases. During the afternoon, levels of oxidants and aldehydes, which give smog its irritating qualities, reach peak values.

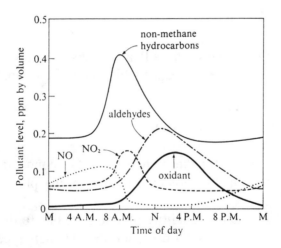

Figure 3.6. Trends in concentrations of major atmospheric species associated with photochemical smog formation.

The smog-forming process is initiated by the photodissociation of nitrogen dioxide as shown by the following reaction:

$$NO_2 + h\nu \rightarrow NO + O \tag{3.11.1}$$

The oxygen atoms react with hydrocarbons, designated as R-H,

$$\text{R-H} + \text{O} \rightarrow \text{R·} + \text{HO·} \tag{3.11.2}$$

to produce hydrocarbon radicals. (For example, if R-H were methane, CH_4, R· would be the methyl radical, $H_3C·$) In keeping with the representation of electrons in Lewis formulas by dots (see Section 2.3), the dots, ·, in the preceding reaction represent electrons. Single (unpaired) electrons are characteristic of usually highly reactive and unstable species called **free radicals**. The hydroxyl radical, HO·, is a key intermediate in the smog-forming process, in that it reacts with a number of relatively stable molecules, sustaining chain reactions in which it is generated. (Hydroxyl radical is a key participant in many atmospheric chemical processes, including those in which methane and carbon monoxide are eliminated from the atmosphere by oxidation to CO_2.) The hydrocarbon free radicals, R·, react with molecules of oxygen,

$$\text{R·} + \text{O}_2 \rightarrow \text{ROO·} \tag{3.11.3}$$

to produce organic peroxyl radicals. These, in turn, undergo a number of additional reactions, of which one of the most significant is,

$$\text{NO} + \text{ROO·} \rightarrow \text{NO}_2 + \text{RO·} \tag{3.11.4}$$

which is responsible for the increase in NO_2 concentration under conditions such that this molecule is undergoing photodissociation back to NO.

Many of the irritating qualities of smog are due to the production of aldehydes and oxidants. Of the latter, the two main types are ozone and organic oxidants, such as peroxyacetyl nitrate, PAN (see below):

$$\overset{\displaystyle O}{\underset{\displaystyle}{\overset{\|}{\text{H--C--H}}}} \qquad \text{O}_3 \qquad \overset{\displaystyle O}{\underset{\displaystyle}{\overset{\|}{\text{H}_3\text{C--C--OO--NO}_2}}}$$

Formaldehyde   Ozone   PAN

These compounds cause eye and respiratory tract irritation. The oxidants are quite toxic to animals and plants and cause damage to rubber and other materials.

Although the automobile is the major culprit contributing to atmospheric hydrocarbon pollution leading to photochemical smog formation, other sources of hydrocarbons are likewise important. Even pine and citrus trees contribute highly reactive terpene hydrocarbons to the atmosphere, which contribute to smog formation. Volatile hydrocarbons emitted from improperly disposed hydrocarbon hazardous wastes can also add to photochemical smog.

3.12. PARTICULATE MATTER

Particles ranging from aggregates of a few molecules to pieces of dust readily visible to the naked eye are commonly found in the atmosphere. Some of these particles, such as sea salt formed by the evaporation of water from droplets of sea spray, are natural and even beneficial atmospheric constituents. Very small particles called **condensation nuclei** serve as bodies for atmospheric water vapor to condense upon and are essential for the formation of precipitation.

Colloidal-sized particles in the atmosphere are called **aerosols**. Those formed by grinding up bulk matter are known as **dispersion aerosols**, of which the latter tend to be smaller. Smaller particles are in general the most harmful because they have a greater tendency to scatter light and are the most respirable (tendency to be inhaled into the lungs).

Particles undergo a number of processes in the atmosphere as summarized in Figure 3.7. The aggregation of small particles into

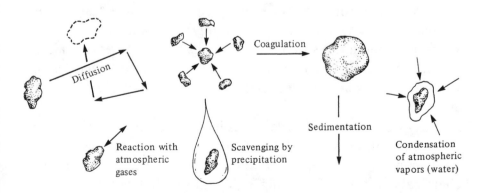

Figure 3.7. Major processes that particles undergo in the atmosphere.

larger ones is called **coagulation** and often precedes removal of particles from the atmosphere. Particles come out of the atmosphere by **sedimentation**, scavenging by precipitation, and **dry deposition** in which they stick upon the surface of soil or plant leaves. Particles are required for the condensation of water vapor from the atmosphere, so they are involved in rainfall. Particles serve as reaction sites for atmospheric chemical processes. These heterogeneous reactions may occur on the surfaces of particulate oxides or carbon or inside water droplets.

3.13. INDICATOR CHEMICALS

The release of chemical species into the environment is one of the most common and troublesome aspects of the hazardous waste problem. A large number of chemical species may be released from waste sites, and it is difficult to monitor them all. Instead, **indicator chemicals** are chosen for monitoring on the basis of their toxicities, mobilities, persistence, quantities present, and amenability to analysis.[4] Indicator chemicals are selected for a site on the basis of their **indicator score**, IS_i, a unitless number calculated by the formula,

$$IS_i = \Sigma (C_{ij}T_{ij}) \tag{3.13.1}$$

where C_{ij} is the concentration of species i in medium j and T_{ij} is a toxicity value for the same chemical in the medium under consideration. The units of C_{ij} are in mg/L, mg/kg, and mg/m^3 for water, soil, and air, respectively, and units of T_{ij} are reciprocals of the units of C_{ij}. Values of IS_i may also be adjusted for frequencies at which chemicals are detected, and time trends for specific chemicals at the site.

The species most often selected as indicator chemicals are organic compounds and heavy metals. As rated by the frequency at which they are encountered at Superfund sites, the most common indicator chemicals are the following: Trichloroethylene, benzene > polychlorinated biphenyls (PCBs), tetrachloroethylene > lead > vinyl chloride, arsenic > cadmium, chromium > 1,1-dichloroethylene >

xylene, 1,1,2-trichloroethylene,1,2-dichloroethane, polynuclear aromatic hydrocarbons (PAH) > 1,1,2,2-tetrachloroethane, chloroform, ethylbenzene, toluene, zinc > dichloromethane, benzo-(a)pyrene (a PAH compound) > copper, nickel, mercury, 2,3,7,8-TCDD (commonly called dioxin), phenol, pentachlorophenol, N-nitrosodiphenylamine. The chemical and toxicological properties of these species are discussed in Chapters 6-9.

LITERATURE CITED

1. Manahan, Stanley E., *Environmental Chemistry*, 4th ed., Lewis Publishers, Chelsea, Michigan, 1984.
2. Manahan, Stanley E., "Humic Substances and the Fates of Hazardous Waste Chemicals," Chapter 6 in *Influence of Aquatic Humic Substances on Fate and Treatment of Pollutants*, Advances in Chemistry Series **219**, American Chemical Society, Washington, DC, 1989, pp. 83-92.
3. Wayne, Richard P., *Principles and Applications of Photochemistry*, Oxford University Press, New York, 1988.
4. Zamuda, Craig, "Superfund Risk Assessments: The Process and Past Experience at Uncontrolled Hazardous Waste Sites," Chapter 6 in *The Risk Assessment of Environmental and Human Health Hazards: A Textbook of Case Studies*, Dennis J. Paustenbach, Ed., John Wiley and Sons, New York, 1989, pp. 266-295.

Biochemistry and Toxicology

4.1. BIOCHEMISTRY AND THE CELL

Ultimately, most hazardous substances are of concern because of their effects upon living organisms. The study of the adverse effects of substances on life processes forms the basis of toxicology, which is discussed in the latter parts of this chapter. An understanding of toxicology, however, requires some basic knowledge of biochemistry, the chemistry of life processes.[1] **Biochemistry** is the science that deals with chemical processes and materials in living systems and is summarized very briefly here, with emphasis upon aspects that are especially pertinent to hazardous toxic substances, including cell membranes, DNA, and enzymes.

The Cell

The focal point of biochemistry and biochemical aspects of toxicants is the **cell**, the basic building block of living systems where most life processes are carried out. The type of cell in higher forms of animal and plant life (multicelled organisms) is the **eukaryotic cell**[2] (Figure 4.1).

Eukaryotic cells are filled with **cytoplasm** and contain a number of different structures. Energy conversion and utilization are mediated by cell **mitochondria**. **Ribosomes** participate in protein synthesis. Some toxicants are metabolized by enzymatic processes that involve the **endoplasmic reticulum**. The cell membrane and nucleus are of particular importance in respect to toxic substances and are discussed below.

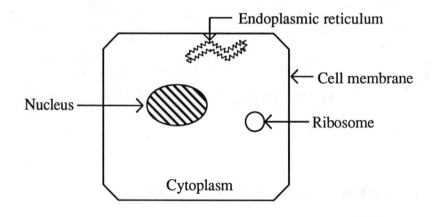

Figure 4.1. Some major features of the eukaryotic cell.

Cell Membrane

The **cell membrane** encloses the cell and regulates the passage of ions, nutrients, metabolic products, and lipid-soluble ("fat-soluble") substances into and out of it. Cell membranes are composed in part of phospholipids that are arranged with their hydrophilic ("water-seeking") heads on the cell membrane surfaces and their hydrophobic ("water-repelling") tails inside the membrane. Cell membranes contain bodies of proteins that are involved in the transport of some substances through the membrane. One reason the cell membrane is very important in toxicology is because it regulates the passage of toxicants and their products into and out of the cell interior. Furthermore, when its membrane is damaged by toxic sustances, a cell may not function properly and the organism may be harmed.

Cell Nucleus

The cell **nucleus** contains **deoxyribonucleic acid, DNA**, a basic unit of which is shown in Figure 4.2. Through its DNA the nucleus is the "control center" of the cell that regulates cell division and protein synthesis. DNA is a polymeric substance with a molecular mass in the range of 6 to 16 million, the basic repeating unit of which is a **nucleotide**. As shown in Figure 4.2, a nucleotide

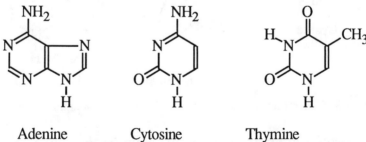

*The other three organic bases in DNA are the following:

Adenine   Cytosine   Thymine

Figure 4.2. Unit of nucleotide in DNA consisting of phosphoric acid, one of four organic bases, and deoxyribose.

is composed of phosphoric acid, deoxyribose (a sugar), and one of four organic bases (adenine, guanine, cytosine, and thymine). DNA determines heredity, cell reproduction, and protein synthesis in cells. When DNA is damaged by foreign substances, various toxic effects, including mutations, cancer, birth defects, and defective immune system function may occur.

4.2. PROTEINS

Proteins are the basic building blocks of biological material. Proteins have essential functions in living systems; they constitute enzymes and most of the cytoplasm inside the cell. Proteins consist of huge molecules with molecular masses of several tens of thousands that are composed primarily of biopolymers of **amino acids**. There are 20 naturally-occurring amino acids having the general formula shown by the first structure in Figure 4.3, where R

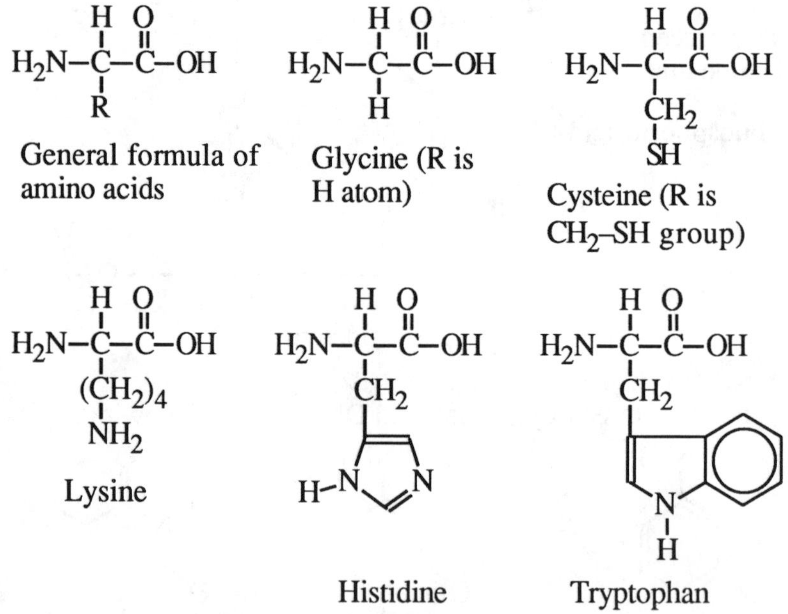

$$
\begin{array}{ccc}
\text{H O} & \text{H O} & \text{H O} \\
\text{H}_2\text{N}-\overset{|}{\underset{|}{\text{C}}}-\overset{\|}{\text{C}}-\text{OH} & \text{H}_2\text{N}-\overset{|}{\underset{|}{\text{C}}}-\overset{\|}{\text{C}}-\text{OH} & \text{H}_2\text{N}-\overset{|}{\underset{|}{\text{C}}}-\overset{\|}{\text{C}}-\text{OH} \\
\text{R} & \text{H} & \text{CH}_2 \\
& & \text{SH}
\end{array}
$$

General formula of amino acids

Glycine (R is H atom)

Cysteine (R is CH$_2$–SH group)

$$
\begin{array}{ccc}
\text{H O} & \text{H O} & \text{H O} \\
\text{H}_2\text{N}-\overset{|}{\underset{|}{\text{C}}}-\overset{\|}{\text{C}}-\text{OH} & \text{H}_2\text{N}-\overset{|}{\underset{|}{\text{C}}}-\overset{\|}{\text{C}}-\text{OH} & \text{H}_2\text{N}-\overset{|}{\underset{|}{\text{C}}}-\overset{\|}{\text{C}}-\text{OH} \\
\text{(CH}_2)_4 & \text{CH}_2 & \text{CH}_2 \\
\text{NH}_2 & &
\end{array}
$$

Lysine

Histidine   Tryptophan

Figure 4.3. General formula and five specific examples of amino acids.

represents a group ranging from H atom (in the amino acid glycine) to moderately complex structures. Five examples of the 20 amino acids found in most proteins are shown in Figure 4.3.

Figure 4.4, which illustrates a tripeptide consisting of only three amino acids, shows how amino acids form biopolymers; protein molecules contain polypeptides formed from many amino acid molecules. The amino acids in proteins are joined at **peptide linkages** outlined by dashed lines in Figure 4.4.

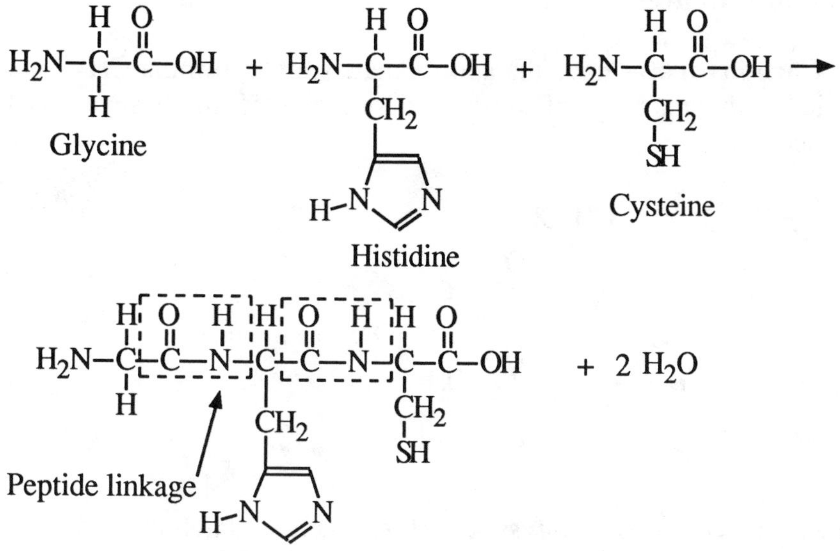

Figure 4.4. A tripeptide formed by the linking of three amino acids. The peptide linkages with which the amino acids are joined are outlined by dashed lines.

The **structures** of protein molecules determine the behavior of proteins in crucial areas such as the processes by which the body's immune system recognizes substances that are foreign to the body. Proteinaceous enzymes depend upon their structures for the very specific functions of the enzymes. The order of amino acids in the protein molecule determines its primary structure. Secondary and tertiary protein structures depend upon the ways in which the polypeptide molecules are bent and folded and by hydrogen bonding (see Section 2.5). The loss of a protein's secondary and tertiary structure is called **denaturation** and may be caused by heat or the action of foreign chemicals. Some corrosive poisons act by denaturing proteins.

4.3. CARBOHYDRATES AND LIPIDS

In addition to proteins, two other major kinds of substances that are synthesized and broken down by metabolic processes are carbohydrates and lipids. The nature and importance of these materials are discussed briefly in this section.

Carbohydrates

Carbohydrates have the approximate simple formula CH_2O and include a diverse range of substances composed of simple sugars such as glucose:

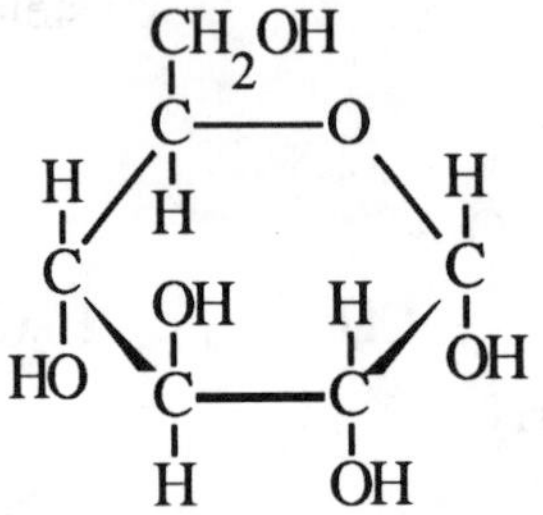

Glucose molecule

High-molecular-mass **polysaccharides**, such as starch and glycogen ("animal starch"), are biopolymers of simple sugars. The major functions of carbohydrates are to store and transfer energy, processes with which toxic substances may interfere.

Lipids

Lipids are substances that can be extracted from plant or animal matter by organic solvents, such as chloroform, diethyl ether, or toluene. The most common lipids are fats and oils. In addition to fats and oils, numerous other biological materials, including waxes, cholesterol, and some vitamins and hormones, are classified as lipids. The general structure of fats and oils is shown in Figure 4.5. These compounds, called **triglycerides**, are esters (see Section 2.9)

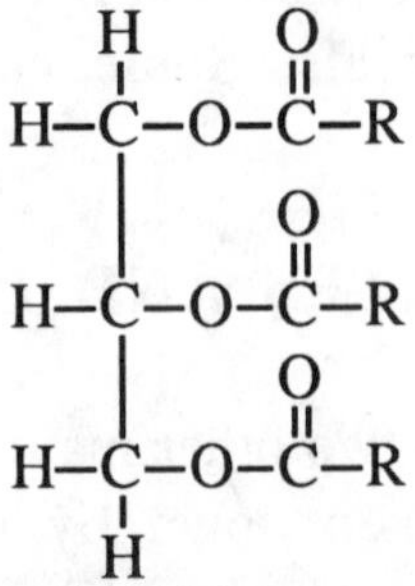

Figure 4.5. General formula of fats and oils. The R group is from a fatty acid and is a hydrocarbon chain, such as $-(CH_2)_{16}CH_3$.

formed from the alcohol glycerol, $CH_2(OH)CH(OH)CH_2(OH)$ and a long-chain fatty acid such as stearic acid, $CH_3(CH_2)_{16}C(O)OH$.

Lipids are toxicologically important for several reasons. Some toxic substances interfere with lipid metabolism, leading to detrimental accumulation of lipids. Some triglyceride lipids, for example, those formed from the unsaturated fatty acid, oleic acid, $CH_3(CH_2)_7(CH=CH)(CH_2)_7C(O)OH$, contain alkenyl carbon-carbon bonds, which are prone to oxidation through the action of toxic substances, such as carbon tetrachloride, CCl_4. Many toxic organic compounds are poorly soluble in water, but are lipid-soluble, so that bodies of lipids in organisms serve to dissolve and store toxicants.

4.4. ENZYMES

Enzymes are proteinaceous substances with highly specific structures that enable biochemical reactions to occur. After participating in a chemical reaction, an enzyme is regenerated intact and can take part in additional reactions. A substance that behaves in such a manner to enable chemical reactions to occur is called a **catalyst**. Enzymes are very specific in their ability to interact with particular substances or class of substances called **substrates**. This specificity is based upon the unique shapes of enzymes as illustrated in Figure 4.6. As shown in the figure, an enzyme "recognizes" a particular substrate by its molecular structure and binds to it to produce an enzyme-substrate complex. This complex then breaks apart to form one or more products different from the original enzyme, regenerating the unchanged enzyme, which is then available to catalyze additional reactions.

Enzymes are named for what they do. For example, **lipase** enzymes cause lipid triglycerides to dissociate and form glycerol and fatty acids. Since these enzymes catalyze the breakdown of biopolymers by the addition of water, they belong to a major system of enzymes called **hydrolyzing enzymes**. Other major systems of enzymes are **oxidoreductase** enzymes that catalyze biological oxidation-reduction reactions; **transferase** enzymes that move chemical groups from one molecule to another; **lyase** enzymes that

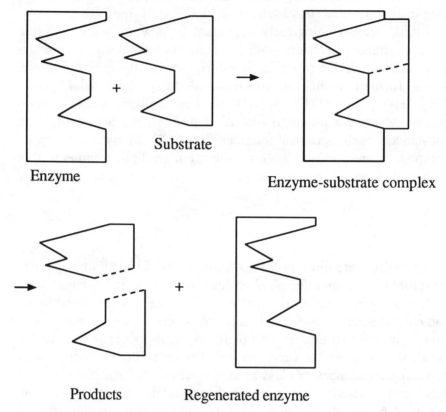

Figure 4.6. Enzyme structure in enzyme-catalyzed biochemical reactions.

remove chemical groups without hydrolysis and participate in the formation of C=C bonds or addition of species to such bonds; **isomerase** enzymes, which enable formation of isomers without loss or gain of atoms; and **ligase** enzymes that work in conjunction with ATP (adenosine triphosphate, a high-energy molecule that plays a crucial role in energy-yielding, glucose-oxidizing metabolic processes) to link molecules together with the formation of bonds such as carbon-carbon or carbon-sulfur bonds.

A major mechanism of toxicity is the alteration or destruction of enzymes by agents such as cyanide, heavy metals, or insecticidal parathion. An enzyme that has been destroyed obviously cannot perform its designated function, whereas one that has been altered may either not function at all or may act improperly. Toxicants can affect enzymes in several ways. Parathion, for example, bonds covalently to the nerve enzyme acetylcholinesterase, which can then no longer serve to stop nerve impulses. Heavy metals tend to bind to sulfur atoms in enzymes (such as sulfur from the amino acid cysteine shown in Figure 4.3) thereby altering the shape and function of the enzyme. Enzymes are denatured by some poisons so that the enzyme no longer has its crucial specific shape.

4.5. ENZYME-CATALYZED REACTIONS OF XENOBIOTIC SUBSTANCES

A **xenobiotic** substance or compound is one that is foreign to living systems and is usually made synthetically. Enzymes catalyze reactions of xenobiotic substances. This usually occurs in a way that is beneficial to an organism, although in some cases the products are more toxic than their precursor compounds. The metabolic processes by which organisms metabolize xenobiotic species are enzyme-catalyzed phase I and phase II reactions,[3] which are described briefly here.

Phase I Reactions

Toxic xenobiotic compounds often consist of lipophilic ("fat-seeking") toxicant molecules that tend to pass through lipid-containing cell membranes and may be bound to, and transported through the body by, lipoproteins. The body can deal with xenobiotic species more readily if they can be rendered more water-soluble and reactive by the attachment of polar functional groups (see Table 2.1). A **phase I reaction** introduces such groups onto the xenobiotic molecule, making it more water-soluble and providing chemically reactive sites for the attachment of conjugating groups in additional phase II reactions.

Phase I reactions are illustrated by Figure 4.7. Most of these are "microsomal mixed-function oxidase" reactions catalyzed by the cytochrome P-450 enzyme system. These enzymes are associated with the **endoplasmic reticulum** of the cell (see Figure 4.1) and occur most abundantly in the livers of vertebrates. The enzyme active sites of cytochrome P-450 enzymes contain iron atoms that can change oxidation state from +2 to +3 and back. In oxidizing a xenobiotic species the enzyme binds to the substrate and molecular O_2, enabling the transfer of an oxygen atom to the xenobiotic molecule.

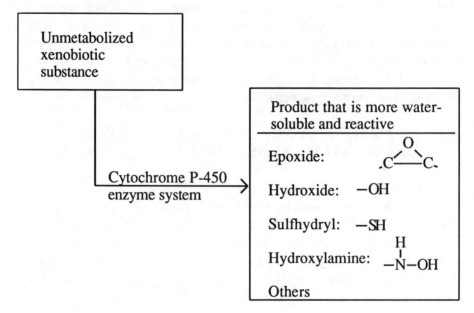

Figure 4.7. Illustration of phase I reactions.

Phase II Reactions

Through **phase II** reactions enzymes enable the attachment of **conjugating agents** to xenobiotics, their phase I reaction products, and non-xenobiotic compounds. The **conjugation product** of such a reaction is usually less toxic than the original xenobiotic compound, less lipid-soluble, more water-soluble, and more readily eliminated from the body. This kind of reaction is known as a **conjugation reaction** because it involves the joining together of a substrate compound with another species that is endogenous to (produced by) the body. Figure 4.8 illustrates the overall process of phase II reactions.

The major conjugating agents and the enzymes that catalyze their phase II reactions are the following:

- Glucuronide (UDP glucuronyltransferase enzyme)
- Glutathione (glutathionetransferase enzyme)
- Sulfate (sulfotransferase enzyme)
- Acetyl (acetylation by acetyltransferase enzymes)

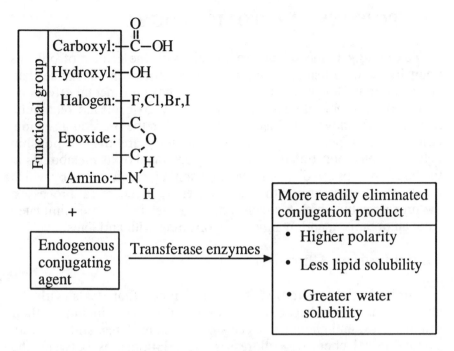

Figure 4.8. Illustration of phase II reactions.

It is beyond the scope of this work to go into any detail regarding the processes by which these conjugates are formed. The most abundant are glucuronides. A glucuronide conjugate is illustrated in Figure 4.9, where -X-R represents a xenobiotic species conjugated to glucuronide and R is an organic moiety. For example, if the xenobiotic compound conjugated is phenol, HXR is HOC_6H_5, X is the O atom, and R represents the phenyl group, C_6H_5.

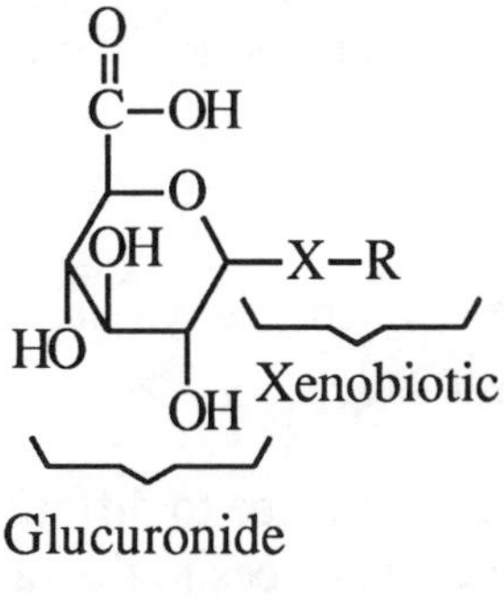

Glucuronide

Figure 4.9. Glucuronide conjugate formed from a xenobiotic, HX-R.

4.6. POISONS AND TOXICOLOGY

Toxicology is the science that deals with the effects of poisons upon living organisms. A **poison** or **toxicant**, is a substance that, above a certain level of exposure or dose, has detrimental effects on tissues, organs, or biological processes. Many toxicants are xenobiotic materials, which were defined and discussed in the preceding section. Among the important aspects of toxicology are the relationship between the demonstrated presence of a chemical or its metabolites in the body and observed symptoms of poisoning, mechanisms by which toxicants are transformed to other species by biochemical processes, the processes by which toxicants and their metabolites are eliminated from an organism, and treatment of poisoning with antidotes.

Toxicological Chemistry

Toxicological chemistry is the science that deals with the chemical nature and reactions of toxic substances, including their origins, uses, and chemical aspects of exposure, fates, and disposal. Toxicological chemistry addresses the relationships between the chemical properties and molecular structures of molecules and their toxicological effects. Figure 4.10 outlines the terms discussed above and the relationships among them.

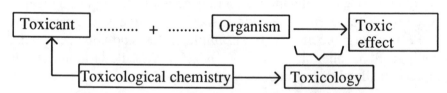

Figure 4.10. Toxicology is the science of poisons. Toxicological chemistry relates toxicology to the chemical nature of toxicants.

4.7. TOXICITIES

The major variables in the ways in which toxicants affect organisms are the following:

- Toxicity of the same substance to different organisms
- Toxicity of different substances to the same organism
- Minimum levels for observable toxic effects

- Sensitivity to small increments of toxicant
- Levels at which most organisms experience the ultimate effect (particularly death)
- Reversibility of toxic effect
- Acute or chronic effect

When a toxicant leaves no permanent effect, either through the action of the organism's natural defense mechanisms or the administration of substances to counteract the toxicant's action (antidotes), it is said to act in a **reversible** manner. Effects that last after the toxicant is eliminated, such as the scar from a sulfuric acid burn on the skin, are termed **irreversible**.

Acute toxicity refers to responses that are observed soon after exposure to a toxic substance. **Chronic toxicity** deals with effects that take a long time to be manifested. Chronic responses to toxicants may have latency periods as long as several decades in humans. Acute effects normally result from brief exposures to relatively high levels of toxicants and are comparatively easy to observe and relate to exposure to a poison. Chronic effects are often obscured by normal background maladies and tend to result from low exposures to a toxicant over relatively long periods of time. Chronic effects are much more difficult to study, but are of greater importance in dealing with hazardous wastes and pollutants.

Dose-Response Relationship

Dose is defined as the degree of exposure of an organism to a toxicant, commonly in units of mass of toxicant per unit of body mass of the organism. The observed effect of the toxicant is the **response**. A plot of the percentage of organisms that exhibit a particular response as a function of dose is a **dose-response** curve.[5] As shown in Figure 4.11 such plots are S-shaped. The statistical estimate of the dose that would cause death in 50 percent of the subjects is the inflection point of the dose-response curve and is designated as LD_{50}. As illustrated in Figure 4.11 the LD_{50} values and the slopes of the dose-response curves may differ substantially.

Most subjects respond to toxicants at doses corresponding to the mid-range of the dose-response curve and are called **normals**. A response to a very low level of toxicant is known as **hyper-**

sensitivity whereas response to only extremely high levels is called **hyposensitivity**. In some cases hypersensitivity develops as an extreme reaction to a toxicant after one or more doses of it. Hypersensitivity usually involves an allergic reaction, which is an exaggerated response of the body's immune system. For example, some individuals develop such a reaction to formaldehyde.

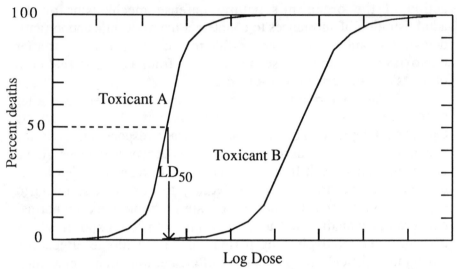

Figure 4.11. Illustration of a dose-response curve in which the response is the death of the organism. The cumulative percentage of deaths of organisms is plotted on the Y axis

Toxicity Ratings

Substances may be assigned **toxicity ratings** that vary from 1 (practically non-toxic) to 6 (supertoxic). These ratings with example substances are summarized below:

- 1. Practically non-toxic, >15 g/kg mass of toxicant per unit body mass (bis(2-ethylhexylphthalate), a plasticizer)
- 2. Slightly toxic, 5–15 g/kg, (ethanol)
- 3. Moderately toxic, 0.5–5 g/kg (malathion insecticide)
- 4. Very toxic, 50–500 mg/kg (heptachlor insecticide)
- 5. Extremely toxic, 5-50 mg/kg (parathion insecticide)
- 6. Supertoxic <5 mg/kg (Sarin, an organophosphorus military poison)

Toxicity ratings are used in this book to express toxicities of various substances.

4.8.   TOXICANTS IN THE BODY

This section discusses what happens to toxicants in an organism, including how they are metabolized, transported, and excreted; their adverse biochemical effects; and manifestations of poisoning. It is convenient to consider these processes as two major phases, a kinetic phase and a dynamic phase. In so doing, it is helpful to keep in mind the major routes and sites of absorption, metabolism, binding, and excretion of toxic substances in the body as illustrated in Figure 4.12.

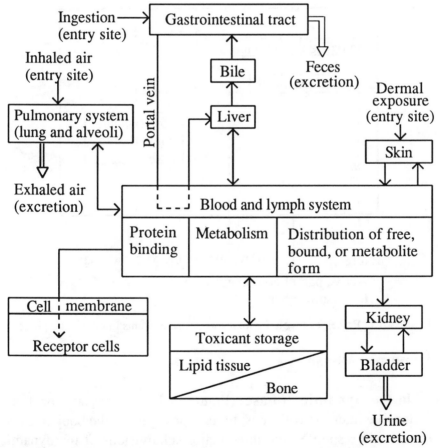

Figure 4.12.   Major routes and sites of absorption, metabolism, binding, and excretion of toxic substances in the body.

Kinetic Phase

In the **kinetic phase** a toxicant or the metabolic precursor of a toxic substance (**protoxicant**) may undergo absorption, metabolism, temporary storage, distribution, and excretion as illustrated in Figure 4.13. A toxicant that is absorbed may be passed through the kinetic phase unchanged as an **active parent compound**, metabolized to a **detoxified metabolite** that is excreted, or converted to a toxic **active metabolite**.

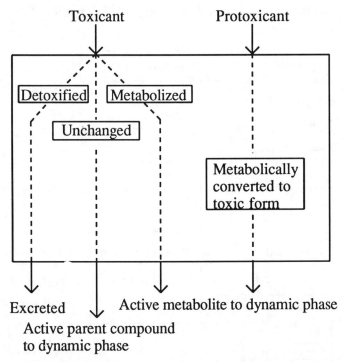

Figure 4.13. Processes involving toxicants or protoxicants in the kinetic phase.

Dynamic Phase

In the **dynamic phase** (Figure 4.14) a toxicant or toxic metabolite interacts with cells, tissues, or organs in the body to cause some toxic response. The three major subdivisions of the dynamic phase are the following:

- **Primary reaction** with a receptor or target organ[6]
- A **biochemical response**
- Observable effects

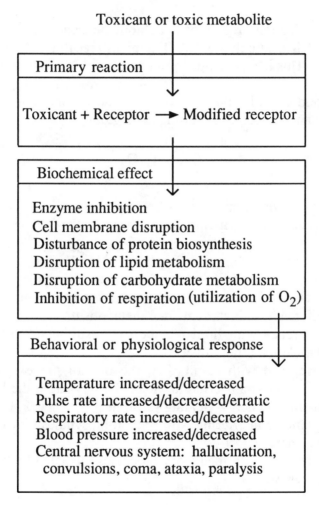

Figure 4.14. The dynamic phase of toxicant action.

Biochemical Effects in the Dynamic Phase

A toxicant or an active metabolite reacts with a receptor to cause a toxic response. Such a reaction occurs, for example, when benzene epoxide,

O   Benzene epoxide

a phase I oxidation product of xenobiotic benzene, forms an adduct with a nucleic acid unit in DNA (Figure 4.2) resulting in alteration of the DNA.[7] This is an example of an **irreversible** reaction between a toxicant and a receptor. A **reversible** reaction that can result in a toxic response is illustrated by the one that occurs between carbon monoxide and oxygen-transporting hemoglobin in blood:

$$O_2Hb \; + \; CO \; \rightleftharpoons \; COHb \; + \; O_2 \qquad\qquad (4.8.1)$$

The binding of a toxicant to a receptor may result in some kind of biochemical effect. The major ones of these are the following:

- Impairment of enzyme function by binding to the enzyme, coenzymes, metal activators of enzymes, or enzyme substrates
- Alteration of cell membrane or carriers in cell membranes
- Interference with carbohydrate metabolism
- Interference with lipid metabolism resulting in excess lipid accumulation ("fatty liver")
- Interference with respiration, the overall process by which electrons are transferred to molecular oxygen in the biological oxidation of energy-yielding substrates
- Stopping or interfering with protein biosynthesis by the action of toxicants on DNA
- Interference with regulatory processes mediated by hormones or enzymes.

Behavioral and Physiological Responses

Prominent among the more chronic responses to toxicant exposure are mutations, cancer, and birth defects (discussed in Section 4.9) and effects on the immune system (discussed in Section 4.10). Other observable effects, some of which may occur soon after exposure, include the following:

- Gastrointestinal illness
- Cardiovascular disease
- Hepatic (liver) disease
- Renal (kidney) malfunction
- Neurologic symptoms (central and peripheral nervous systems)
- Skin abnormalities (rash, dermatitis)

Among the more immediate and readily observed manifestations of poisoning are alterations in the **vital signs** of **temperature, pulse rate, respiratory rate,** and **blood pressure.** The three effects that toxicants may have on pulse rate are **bradycardia** (decreased rate), **tachycardia** (increased rate), and **arrhythmia** (irregular pulse). Some toxicants, such as hexachlorobenzene and sodium fluoroacetate, increase body temperature, whereas others, such as phenobarbital and ethanol, decrease body temperature. Among the toxicants that increase respiratory rate are nitrites (NO_2^- ion), methanol (CH_3OH), and hexachlorobenzene. Cyanide and carbon monoxide may either increase or decrease respiratory rate. Toxic levels of amphetamines, cocaine, phenylcyclidines, and belladonna alkaloids increase blood pressure. Toxic doses of iron, nitrite, cyanide, and mushroom toxins decrease blood pressure.

Poisoning by some substances, including mercury, arsenic, thallium, carbamates, organophosphates, carbon monoxide, and cyanide, may cause an abnormal skin color or excessively moist or dry skin. A blue skin color due to oxygen deficiency in the blood (**cyanotic** appearance) may be evidence of poisoning by cyanide, carbon monoxide, and nitrites. The skin of subjects poisoned by a number of toxicants, including arsenic, arsine gas (AsH_3), iron, aniline dyes, and carbon tetrachloride, may appear **jaundiced,** exhibiting a yellow color resulting from the presence of bile pigments in the blood. Toxic levels of some materials or their metabolites cause the body to have unnatural **odors.** Examples include aromatic odors from hydrocarbons, the odor of violets from the ingestion of turpentine, the bitter almond odor of HCN in tissues of victims of cyanide poisoning, rotten egg odor from hydrogen sulfide (H_2S) or carbon disulfide[8] poisoning, and an extremely potent garlic breath odor from even very slight exposures to some selenium compounds.

Eye **miosis** (excessive or prolonged contraction of the eye pupil) is a toxic response to alcohols, carbamates, organophosphates, and phenycyclidine, whereas excessive pupil dilation (**mydriasis**) is caused by amphetamines, belladonna alkaloids, glutethimide, and tricyclic antidepressants. Exposure to corrosive acids and bases, nitrogen dioxide, hydrogen sulfide, methanol, and formaldehyde causes **conjunctivitis**, a condition marked by inflammation of the mucus membrane that covers the front part of the eyeball and the inner lining of the eyelids. Some poisonings are manifested by **nystagmus**, the involuntary movement of the eyeballs.

Caustic acids, bases, mercury, arsenic, thallium, carbamates, and organophosphates cause a moist condition of the mouth, whereas a dry mouth is symptomatic of poisoning by tricyclic antidepressants, amphetamines, antihistamines, and glutethimide. Gastrointestinal tract effects occur as a result of poisoning by a number of toxic substances. These effects include pain, vomiting, or paralytic ileus (stoppage of the normal paristalsis movement of the intestines).

The central nervous system manifests poisoning by a number of symptoms, particularly **convulsions**, **paralysis**, **hallucinations**, and **ataxia** (lack of coordination of voluntary movements of the body). Abnormal behavior, including agitation, hyperactivity, disorientation, and delirium, is caused by some central nervous system poisons. Severe poisoning by some substances, including alcohols, organophosphates, carbamates, lead, hydrogen sulfide, and hydrocarbons, causes **coma**, the term used to describe a lowered level of consciousness.

Subclinical Effects of Poisoning

Often the effects of toxicant exposure are subclinical in nature and may include the following:

- Some kinds of damage to immune system
- Chromosomal abnormalities
- Modification of functions of liver enzymes
- Slowing of conduction of nerve impulses

Synergism, Potentiation, and Antagonism

When two potentially toxic substances administered together act in a manner that is totally independent of each other, their effects are

additive. Often, however, the response differs substantially from that to any one of the substances alone.[9] This is because chemical interaction between substances may affect their toxicities, both substances may act upon the same physiologic function, or two substances may compete for binding to the same receptor. **Synergistic** interaction occurs when the overall effect is greater than the sum of the separate effects. **Antagonism** results when an active substance decreases the effect of another active one. An inactive substance may enhance (**potentiate**) the effects of an active one.

4.9. TERATOGENESIS, MUTAGENESIS, AND CARINOGENESIS

Teratogenesis

Teratogens are chemical species that cause birth defects. These usually arise from damage to embryonic or fetal cells. However, mutations in germ cells (egg or sperm cells) may cause birth defects, such as Down's syndrome.

The biochemical mechanisms of teratogenesis are varied. These include enzyme inhibition by xenobiotics; deprivation of the fetus of essential substrates, such as vitamins; interference with energy supply; or alteration of the permeability of the placental membrane.

Mutagenesis

Mutagens alter DNA (see Section 4.1) to produce inheritable traits. Although mutation is a natural process that occurs even in the absence of xenobiotic substances, most mutations are harmful. The mechanisms of mutagenicity are similar to those of carcinogenicity and mutagens often cause birth defects as well. Therefore, mutagenic hazardous substances are of major toxicological concern.

Carcinogenesis

Chemical carcinogenesis is the term that applies to the role of substances foreign to the body in causing the uncontrolled cell replication commonly known as cancer. In the public eye chemical carcinogenesis is the aspect of toxicology most commonly associated with hazardous substances.

The two major steps in the overall processes by which xenobiotic chemicals cause cancer are an **initiation stage** followed by a **promotional stage**.[10] Chemical carcinogens usually have the ability to form covalent bonds with macromolecular life molecules, especially DNA.[11] This can alter the DNA in a manner such that the cells replicate uncontrollably and form cancerous tissue. Many chemical carcinogens are **alkylating agents**, which act to attach alkyl groups — such as methyl (CH_3) or ethyl (C_2H_5) — or **arylating agents**, which act to attach aryl moieties, such as the phenyl group, C_6H_5, to DNA. Attachment is made through the N and O atoms in the nitrogenous bases that are contained in DNA.

Primary Carcinogens and Procarcinogens

Chemical substances that cause cancer directly are called **primary** or **direct-acting carcinogens**. Most xenobiotics involved in causing cancer are **precarcinogens** or **procarcinogens**. These species require metabolic activation by phase I or phase II reactions to produce **ultimate carcinogens**, which are actually responsible for carcinogenesis.[12]

Only a few chemicals have definitely been established as human carcinogens, from observations of the occurrence of abnormal levels of specific kinds of cancer in humans who have experienced known exposures to particular substances. A well documented example is provided by vinyl chloride,

$$\begin{array}{ccc} H & & H \\ & \diagdown \; C=C \; \diagup & \\ H & & Cl \end{array}$$

which is known to have caused a rare form of liver cancer (angiosarcoma) in individuals who cleaned autoclaves in the polyvinylchloride fabrication industry. Animals tests are used to infer chemical carcinogenicity in humans.

Bruce Ames Test

Because of their similarities, mutagenicity can be used to infer carcinogenicity. This is the basis of the **Bruce Ames** test, in which observations are made of the reversion of mutant histidine-requiring

Salmonella bacteria back to a form that can synthesize its own histidine.[13] The test makes use of enzymes in homogenized liver tissue to convert potential procarcinogens to ultimate carcinogens. Histidine-requiring *Salmonella* bacteria are inoculated onto a medium that does not contain histidine, and those that mutate back to a form that can synthesize histidine establish visible colonies. These colonies are assayed to provide both a qualitative and quantitative indication of mutagenicity.

4.10. IMMUNE SYSTEM RESPONSE

Adverse effects on the body's immune system are being increasingly recognized as important consequences of exposure to hazardous substances. The **immune system** acts as the body's natural defense system to protect it from xenobiotic chemicals; infectious agents, such as viruses or bacteria; and neoplastic cells, which give rise to cancerous tissue. Toxicants can cause **immunosuppression**, which is the impairment of the body's natural defense mechanisms. Xenobiotics can also cause the immune system to lose its ability to control cell proliferation, resulting in leukemia or lymphoma.

Another major toxic response of the immune system is **allergy** or **hypersensitivity**. This kind of condition results when the immune system overreacts to the presence of a foreign agent or its metabolites in a self-destructive manner. Among the xenobiotic materials that can cause such reactions are beryllium, chromium, nickel, formaldehyde, pesticides, resins, and plasticizers.

4.11. HEALTH HAZARDS

The first toxic effects of hazardous wastes that were documented and dealt with were those that produced readily recognized, usually severe, acute maladies that developed on a short time scale as a result of brief, intense exposure to toxicants. In recent decades attention has shifted to delayed, chronic, often less severe illnesses developing from long-term exposure to low levels of toxicants.[14] Although the total impact of the latter kinds of health effects may be substantial, their assessment is very difficult because of factors such as uncertainties in exposure, low occurrence above background levels of disease, and long latency periods.

Assessment of Potential Exposure

Documentation of exposure to hazardous wastes can be much more difficult than that to occupational exposure to toxicants. The first step is to find out what is present in the waste site. To a certain extent this can be done by examining records of what was placed in the site. Complications include uncertainty in these records, chemical and biological transformations of wastes, selective migration of wastes, and interaction of wastes to produce other products. Some of these uncertainties can be circumvented by chemical analysis of species at the site. However, this approach requires a very extensive sampling and analysis program to obtain a reasonably accurate assessment of chemicals present.

Once an idea is obtained of the waste chemicals present, the next step in assessing the risk posed by them is an evaluation of their chemical, physical, and toxicological properties. This is relatively simple for inorganic wastes compared to organic wastes. Even so, the complexities of waste substances can complicate this phase.

Knowledge of the degree and pathways of chemical migration from the site is important in assessing exposure. Contamination of drinking water supplies is a particular concern. Factors involved in this phase include the nature of the chemicals (particularly water solubility), means of disposal (for example, measures used for immobilization), the nature of soil and other barriers to migration, migration pathways, and time since disposal. Chemical analysis, particularly of potentially contaminated groundwater, can be useful for this purpose.

A final step in exposure assessment is evaluation of potentially exposed populations. The most direct approach to this is to determine chemicals or their metabolic products in organisms. For inorganic species this is most readily done for heavy metals, radionuclides, and some minerals, such as asbestos. Symptoms associated with exposure to particular chemicals may also be evaluated. Examples of such effects include skin rashes or subclinical effects, such as chromosomal damage.

Epidemiologic Evidence

Epidemiologic studies applied to hazardous wastes attempt to correlate observations of particular illnesses with probable exposure to such wastes. There are two major approaches to such studies. One approach is to look for diseases known to be caused by particular agents in areas where exposure is likely from such agents in hazardous wastes. A second approach is to look for **clusters** consisting of an abnormally large number of cases of a particular disease in a limited geographic area, then attempt to locate sources of exposure to hazardous wastes that may be responsible. The most common types of maladies observed in clusters are spontaneous abortions, birth defects, and particular types of cancer.

Epidemiologic studies are complicated by long latency periods from exposure to onset of disease, lack of specificity in the correlation between exposure to a particular waste and the occurrence of a disease, and background levels of a disease in the absence of exposure to a hazardous waste capable of causing the disease.

Estimation of Health Effects Risks

An important part of estimating the risks of adverse health effects from exposure to hazardous wastes involves extrapolation from experimentally observable data. Usually the end result needed is an estimate of a low occurrence of a disease in humans after a long latency period resulting from low-level exposure to a toxicant for a long period of time. The data available are almost always taken from animals exposed at high levels of the substance for a relatively short period of time. Extrapolation is then made using linear or curvilinear projections to estimate the risk to human populations. There are, of course, very substantial uncertainties in this kind of approach.

4.12. TOXICOLOGY AND RISK ASSESSMENT

One of the major ways in which toxicology interfaces with the area of hazardous wastes is in **health risk assessment**, providing guidance for risk management, cleanup, or regulation needed at a hazardous waste site based upon knowledge about the site and the chemical and toxicological properties of wastes in it. Risk assessment includes the factors of site characteristics; substances present, including indicator species; potential receptors; potential exposure pathways; and uncertainty analysis. It may be divided into the following major components:[15]

- Identification of hazard
- Dose-response assessment
- Exposure assessment
- Risk characterization

The ultimate concern over hazardous substances and hazardous wastes is the effects of such materials on living organisms – humans, wildlife, crops. Toxicology is the major tool with which risks from hazardous substances are assessed. Basic knowledge of toxicology, toxicological chemistry, and toxicological biochemistry can be invaluable in estimating risks from toxic substances and in designing appropriate control measures.

LITERATURE CITED

1. Hodgson, Ernest, and Frank E. Guthrie, Eds., *Introduction to Biochemical Toxicology*, Elsevier, New York, 1980.
2. Sears, Curtis C., and Conrad L. Stanitski, *Chemistry for Health-Related Sciences*, 2nd ed., Prentice-Hall, Inc., Englewood Cliffs, NJ, 1983.
3. Hodgson, Ernest, and Walter C. Dauterman, "Metabolism of Toxicants — Phase I Reactions," Chapter 4, and Walter C. Dauterman, "Metabolism of Toxicants," Chapter 5, in *Introduction to Biochemical Toxicology*, Ernest Hodgson and Frank E. Guthrie, Eds., Elsevier, New York, 1980, pp. 67–105.
4. Manahan, Stanley E., *Toxicological Chemistry*, Lewis Publishers, Inc., Chelsea, Michigan, 1989.
5. James, Robert C., "General Principles of Toxicology," Chapter 2 in *Industrial Toxicology*, Philip L. Williams and James L. Burson, Eds., Van Nostrand Reinhold Company, New York, 1985.
6. Cohen, Gerald M., Ed., *Target Organ Toxicity*, CRC Press, Inc., Boca Raton, Florida, 1988.
7. Norpoth, K., W. Stücker, E. Krewet, and G. Müller, "Biomonitoring of Benzene Exposure by Trace Analysis of Phenylguanine," *International Archives of Occupational and Environmental Health*, **60**, 163-168 (1988).

8. Dreisbach, Robert H., and William O. Robertson, *Handbook of Poisoning*, 12th ed., Appleton and Lange, Norwalk, Conn., 1987.

9. James, Robert C., "General Principles of Toxicology," Chapter 2 in *Industrial Toxicology*, Phillip L. Williams and James L. Burson, Eds., Van Nostrand Reinhold, New York, 1985, pp. 7-26.

10. Diamond, Leila, "Tumor Promoters and Cell Transformation," Chapter 3 in *Mechanisms of Cellular Transformation by Carcinogenic Agents*, D. Grunberger and S. P. Goff, Eds., Pergamon Press, New York, 1987, pp. 73–132.

11. Singer, B., and D. Grunberger *Molecular Biology of Mutagens and Carcinogens*, Plenum Press, New York, 1983.

12. Levi, Patricia E., "Toxic Action," Chapter 6 in *Modern Toxicology*, Ernest Hodgson and Patricia E. Levi, Eds., Elsevier, New York, 1987, pp. 133–184.

13. "The Detection of Environmental Mutagens and Potential Carcinogens," Bruce N. Ames, *Cancer*, **53**, 1034-1040 (1984).

14. Heath, Clark W., Jr., "Health Effects of Hazardous Waste," Section 3.2 in *Standard Handbook of Hazardous Waste Treatment and Disposal*, Harry M. Freeman, Ed., McGraw-Hill, New York, 1989, pp. 3.17-3.27.

15. Paustenbach, Dennis J., Ed., *The Risk Assessment of Environmental and Human Health Hazards: A Textbook of Case Studies*, John Wiley and Sons, New York, 1988.

5

Chemical Perspectives of Hazards

5.1. INTRODUCTION

The management of hazards posed by hazardous substances and wastes is a crucial part of the operation of any modern chemical industry.[1] It is a significant and increasing part of the cost of any business dealing with chemical products and processes. Personnel working with such products and processes must have a good understanding of hazardous substances and hazardous wastes.

As defined in Chapter 1, a hazardous substance is a material with a substantial potential to pose a danger to living organisms, materials, structures, or the environment. Having covered the basics of chemistry and toxicological chemistry in Chapters 3-4, it is now possible to discuss hazardous substances and wastes from a more scientific perspective. That is done in this chapter.

5.2. FLAMMABLE AND COMBUSTIBLE SUBSTANCES

In a broad sense a **flammable substance** is something that will burn readily, whereas a **combustible substance** requires relatively more persuasion to burn. Before trying to sort out these definitions it is necessary to define several other terms. Perhaps most chemicals that are likely to burn accidentally are liquids. Liquids form **vapors** which are usually more dense than air, and thus tend to settle. The tendency of a liquid to ignite is measured by a test in which the liquid is heated and periodically exposed to a flame until the mixture of vapor and air ignites at the liquid's surface. The temperature at which this occurs is called the **flash point**.

With these definitions in mind it is possible to divide ignitable materials into four major classes. A **flammable solid** is one that can ignite from friction or from heat remaining from its manufacture, or which may cause a serious hazard if ignited. Explosive materials are not included in this classification. A **flammable liquid** is one having a flash point below 37.8°C (100°F). A **combustible liquid** has a flash point in excess of 37.8°C, but below 93.3°C. Gases are substances that exist entirely in the gaseous phase at 0°C and 1 atm pressure. A **flammable compressed gas** meets specified criteria for lower flammability limit, flammability range (see below), and flame projection. Especially in the case of liquids, there are several subclassifications of flammability and combustibility such as those of the U.S. Department of Transportation and the National Fire Protection Association.[2]

In considering the ignition of vapors, two important concepts are those of flammability limit and flammability range. Values of the vapor/air ratio below which ignition cannot occur because of insufficient fuel define the lower **flammability limit**. Similarly, values of the vapor/air ratio above which ignition cannot occur because of insufficient air define the upper flammability limit. The difference between upper and lower flammability limits at a specified temperature is the **flammability range**. Table 5.1 gives some examples of these values for common liquid chemicals.

The percentage of flammable substance for best combustion (most explosive mixture) is labelled "optimal."[3] In the case of acetone, for example, the optimal flammable mixture is 5.0% acetone.

One of the more disastrous problems that can occur with flammable liquids is a boiling liquid expanding vapor explosion, BLEVE. These are caused by rapid pressure buildup in closed containers of flammable liquids heated by an external source. The explosion occurs when the pressure buildup is sufficient to break the container walls.

Combustion of Finely Divided Particles

Finely divided particles of combustible materials are somewhat analogous to vapors in respect to flammability. One such example is a spray or mist of hydrocarbon liquid in which oxygen has the opportunity for intimate contact with the liquid particles. In this case the liquid may ignite at a temperature below its flash point.

Table 5.1. Flammabilities of Some Common Organic Liquids

| Liquid | Flash point (°C)[a] | Volume percent in air | |
		LFL[b]	UFL[b]
Diethyl ether	-43	1.9	36
Pentane	-40	1.5	7.8
Acetone	-20	2.6	13
Toluene	4	1.27	7.1
Methanol	12	6.0	37
Gasoline (2,2,4-tri-methylpentane)	---	1.4	7.6
Naphthalene	157	0.9	5.9

[a] Closed-cup flash point test

[b] LFL, lower flammability limit; UFL, upper flammability limit at 25°C.

Dust explosions can occur with a large variety of solids that have been ground to a finely divided state. Many metal dusts, particularly those of magnesium and its alloys, zirconium, titanium, and aluminum can burn explosively in air. In the case of aluminum, for example, the reaction is the following:

$$4Al(powder) + 3O_2(from\ air) \rightarrow 2Al_2O_3 \qquad (5.2.1)$$

Coal dust and grain dusts have caused many fatal fires and explosions in coal mines and grain elevators, respectively. Dusts of polymers such as cellulose acetate, polyethylene, and polystyrene can also be explosive.

Oxidizers

Combustible substances are reducing agents that react with **oxidizers** (oxidizing agents or oxidants) to produce heat (see Section 2.3 for a discussion of oxidation and reduction). Diatomic oxygen, O_2, from air is the most common oxidizer. Many oxidizers are chemical compounds that contain oxygen in their formulas. The halogens (periodic table group 7A) and many of their compounds are oxidizers. Some examples of oxidizers are given in Table 5.2.

An example of a reaction of an oxidizer is that of concentrated HNO_3 with copper metal, which gives toxic NO_2 gas as a product:

$$4HNO_3 + Cu \rightarrow Cu(NO_3)_2 + 2H_2O + 2NO_2 \qquad (5.2.1)$$

The toxic effects of some oxidizers are due to their ability to oxidize biomolecules in living systems.

Table 5.2. Examples of Some Oxidizers

Name	Formula	State of matter
Ammonium nitrate	NH_4NO_3	Solid
Ammonium perchlorate	NH_4ClO_4	Solid
Bromine	Br_2	Liquid
Chlorine	Cl_2	Gas (stored as liquid)
Fluorine	F_2	Gas
Hydrogen peroxide	H_2O_2	Solution in water
Nitric acid	HNO_3	Concentrated solution
Nitrous oxide	N_2O	Gas (stored as liquid)
Ozone	O_3	Gas
Perchloric acid	$HClO_4$	Concentrated solution
Potassium permanganate	$KMnO_4$	Solid
Sodium dichromate	$Na_2Cr_2O_7$	Solid

Whether or not a substance acts as an oxidizer depends upon the reducing strength of the material that it contacts. For example, carbon dioxide is a common fire extinguishing material that can be sprayed onto a burning substance to keep air away. However, aluminum is such a strong reducing agent that carbon dioxide in contact with hot, burning aluminum reacts as an oxidizing agent to give off toxic combustible carbon monoxide gas:

$$2Al + 3CO_2 \rightarrow Al_2O_3 \qquad (5.2.2)$$

Oxidizers can contribute strongly to fire hazards because fuels may burn explosively in contact with an oxidizer.

Spontaneous Ignition

Substances that catch fire spontaneously in air without an ignition source are called **pyrophoric**. These include several elements— white phosphorus, the alkali metals (group 1A), and powdered forms of magnesium, calcium, cobalt, manganese, iron, zirconium and aluminum. Also included are some organometallic compounds, such as lithium ethyl (LiC_2H_4) and lithium phenyl (LiC_6H_5), and some metal carbonyl compounds, such as iron pentacarbonyl, $Fe(CO)_5$. Another major class of pyrophoric compounds consists of metal and metalloid hydrides, including lithium hydride, LiH; pentaborane, B_5H_9; and arsine, AsH_3. Moisture in air is often a factor in spontaneous ignition. For example, lithium hydride undergoes the following reaction with water from moist air:

$$LiH + H_2O \rightarrow LiOH + H_2 + heat \qquad (5.2.3)$$

The heat generated from this reaction can be sufficient to ignite the hydride so that it burns in air:

$$2LiH + O_2 \rightarrow Li_2O + H_2O \qquad (5.2.4)$$

Some compounds with organometallic character (see Sections 5.6 and 6.5) are also pyrophoric. An example of such a compound is diethylethoxyaluminum:

$$\begin{matrix} H_5C_2 \diagdown \\ \diagup \end{matrix} Al\text{--}OC_2H_5 \quad \text{Diethylethoxyaluminum}$$
$$H_5C_2$$

Many mixtures of oxidizers and oxidizable chemicals catch fire spontaneously and are called **hypergolic mixtures**. Nitric acid and phenol form such a mixture.

Toxic Products of Combustion

Some of the greater dangers of fires are from toxic products and byproducts of combustion. The most obvious of these is carbon monoxide, CO, which can cause serious illness or death because it forms carboxyhemoglobin with hemoglobin in the blood so that the blood no longer carries oxygen to body tissues. Toxic SO_2, P_4O_{10}, and HCl are formed by the combustion of sulfur, phosphorus, and organochloride compounds, respectively. A large number of noxious organic compounds such as aldehydes are generated as byproducts of combustion. In addition to forming carbon monoxide, combustion under oxygen-deficient conditions produces polycyclic aromatic hydrocarbons consisting of fused ring structures. Some of these compounds, such as benzo(a)pyrene, below, are precarcinogens.

 Benzo(a)pyrene

5.3. REACTIVE SUBSTANCES

Reactive substances are those that tend to undergo rapid or violent reactions under certain conditions. Such substances include those that react violently or form potentially explosive mixtures with water. An example is sodium metal which reacts strongly with water as follows:

$$2Na + 2H_2O \rightarrow 2NaOH + H_2 + heat \qquad (5.3.1)$$

This reaction usually generates enough heat to ignite the sodium. Explosives constitute another class of reactive substances. For regulatory purposes substances are also classified as reactive that react with water, acid, or base to produce toxic fumes, particularly those of hydrogen sulfide or hydrogen cyanide.

Heat and temperature are usually very important factors in reactivity. In order for them to start, many reactions require energy of activation. The rates of most reactions tend to increase sharply with increasing temperature and most chemical reactions give off heat. Therefore, once a reaction is started in a reactive mixture lacking an effective means of heat dissipation, the rate may increase exponentially with time, leading to an uncontrollable event. Other factors that may affect reaction rate include physical form of reactants (for example, a finely divided metal powder may react explosively with oxygen, whereas a single mass of metal barely reacts), rate and degree of mixing of reactants, degree of dilution with nonreactive media (solvent), presence of a catalyst, and pressure.

Some chemical compounds are self-reactive, in that they contain oxidant and reductant in the same compound. Nitroglycerin, a strong explosive with the formula $C_3H_5(ONO_2)_3$ decomposes spontaneously to CO_2, H_2O, O_2, and N_2 with a rapid release of a very high amount of energy. Pure nitroglycerin has such a high inherent instability that only a slight blow may be sufficient to detonate it. Trinitrotoluene (TNT) is also an explosive with a high degree of reactivity. However, it is inherently relatively stable in that some sort of detonating device is required to cause it to explode.

Chemical Structure and Reactivity

As shown in Table 5.3, some chemical structures are associated with high reactivity.[4] In some cases of organic compounds high reactivity results from unsaturated bonds in the carbon skeleton, particularly where multiple bonds are adjacent (allenes, C=C=C) or separated by only one carbon-carbon single bond (dienes, C=C-C=C). Some organic structures involving oxygen are very reactive. Examples are oxiranes, such as ethylene oxide,

hydroperoxides (ROOH), and peroxides (ROOR'), where R and R' stand for hydrocarbon moieties such as the methyl group, $-CH_3$. Many organic compounds containing nitrogen along with carbon and hydrogen are very reactive. Included are triazenes (R-N=N-N), some azo compounds (R-N=N-R'), and some nitriles:

$$R-C\equiv N \quad \text{Nitrile}$$

Functional groups containing both oxygen and nitrogen tend to impart reactivity to an organic compound. Examples are alkyl nitrates ($R-O-NO_2$), alkyl nitrites (R-O-N=O), nitroso compounds (R-N=O), and nitro compounds ($R-NO_2$).

Table 5.3. Examples of Reactive Compounds and Structures

Name	Structure or formula
Organic	
Allenes	$C=C=C$
Dienes	$C=C-C=C$
Azo compounds	$C=N-N=C$
Triazenes	$C-N=N-N$
Hydroperoxides	$R-OOH$
Peroxides	$R-OO-R'$
Alkyl nitrates	$R-O-NO_2$
Nitro compounds	$R-NO_2$
Inorganic	
Nitrous oxide	N_2O
Nitrogen halides	NCl_3, NI_3
Interhalogen compounds	$BrCl$
Halogen oxides	ClO_2
Halogen azides	ClN_3
Hypohalites	$NaClO$

Many different classes of inorganic compounds are reactive. These include some of the halogen compounds of nitrogen (shock-sensitive nitrogen triiodide, NI_3, is an outstanding example), compounds with metal-nitrogen bonds, halogen oxides (ClO_2), and compounds with oxyanions of the halogens. An example of the last group of compounds is ammonium perchlorate, NH_4ClO_4, which was involved in a series of massive explosions that destroyed 8 million lb of the compound and demolished a 40 million lb/year U.S. rocket fuel plant near Henderson, Nevada, in 1988.[4] By late 1989 a new $92 million plant for the manufacture of ammonium perchlorate had been constructed near Cedar City in a remote region of southwest Utah.[5] Prudently, the buildings at the new plant have been placed at large distances from each other!

Explosives such as nitroglycerin or TNT that are single compounds containing both oxidizing and reducing functions in the same molecule are called **redox compounds.** Some redox compounds have more oxygen than is needed for a complete reaction and are said to have a positive balance of oxygen, some have exactly the stoichiometric quantity of oxygen required (zero balance, maximum energy release), and others have a negative balance and require oxygen from outside sources to completely oxidize all components. Trinitrotoluene has a substantial negative balance; ammonium dichromate $((NH_4)_2Cr_2O_7)$ has a zero balance, reacting with exact stoichiometry to H_2O, N_2, and Cr_2O_3; and nitroglycerin has a positive balance as shown by the following reaction:

$$4C_3H_5N_3O_9 \rightarrow 12CO_2 + 10H_2O + 6N_2 + O_2 \qquad (5.3.2)$$

5.4. CORROSIVE SUBSTANCES

Conventionally, **corrosive substances** are regarded as those that dissolve metals or cause oxidized material to form on the surface of metals—rusted iron is a prime example. In a broader sense corrosives cause deterioration of materials, including living tissue, that they contact.[6] Most corrosives belong to at least one of the four following chemical classes: (1) strong acids, (2) strong bases, (3) oxidants, (4) dehydrating agents. Table 5.4 lists some of the major corrosive substances and their effects. For more detailed discussion of corrosive substances the reader is referred to a book on toxicological chemistry.[7]

Sulfuric Acid

Sulfuric acid is a prime example of a corrosive substance. As well as being a strong acid, concentrated sulfuric acid is also a dehydrating agent and oxidant. The tremendous affinity of H_2SO_4 for water is illustrated by the heat generated when water and concentrated sulfuric acid are mixed. If this is done incorrectly by adding water to the acid, localized boiling and spattering can occur that result in personal injury. The major destructive effect of sulfuric acid on skin tissue is removal of water with accompanying release of heat. Sulfuric acid

Table 5.4. Examples of Some Corrosive Substances

Name and formula	Properties and effects
Nitric acid, HNO_3	Strong acid and strong oxidizer, corrodes metal, reacts with protein in tissue to form yellow xanthoproteic acid, lesions are slow to heal
Hydrochloric acid, HCl	Strong acid, corrodes metals, gives off HCl gas vapor, which can damage respiratory tract tissue
Hydrofluoric acid, HF	Corrodes metals, dissolves glass, causes particularly bad burns to flesh
Alkali metal hydroxides, NaOH and KOH	Strong bases, corrode zinc, lead, and aluminum, caustic substances that dissolve tissue and cause severe burns
Hydrogen peroxide, H_2O_2	Oxidizer, all but very dilute solutions cause severe burns
Interhalogen compounds such as ClF, BrF_3	Powerful corrosive irritants that acidify, oxidize, and dehydrate tissue
Halogen oxides such as OF_2, Cl_2O, Cl_2O_7	Powerful corrosive irritants that acidify, oxidize, and dehydrate tissue
Elemental fluorine, chlorine, bromine (F_2, Cl_2, Br_2,)	Very corrosive to mucous membranes and moist tissue, strong irritants

decomposes carbohydrates by removal of water. In contact with sugar, for example, concentrated sulfuric acid reacts to leave a charred mass. The reaction is

$$C_{12}H_{22}O_{11} \xrightarrow{H_2SO_4} 11H_2O(H_2SO_4) + 12C + heat \qquad (5.4.1)$$
Dextrose sugar

Some dehydration reactions of sulfuric acid can be very vigorous. For example, the reaction with perchloric acid produces unstable Cl_2O_7, and a violent explosion can result. Concentrated sulfuric acid produces dangerous or toxic products with a number of other substances, such as toxic carbon monoxide (CO) from reaction with oxalic acid, $H_2C_2O_4$; toxic bromine and sulfur dioxide (Br_2, SO_2) from reaction with sodium bromide, NaBr; and toxic, unstable chlorine dioxide (ClO_2) from reaction with sodium chlorate, $NaClO_3$.

Contact of sulfuric acid with tissue results in tissue destruction at the point of contact. A severe burn results, which may be difficult to heal. Inhalation of sulfuric acid fumes or mists damages tissues in the upper respiratory tract and eyes. Long term exposure to sulfuric acid fumes or mists has caused erosion of teeth!

5.5. TOXIC SUBSTANCES

Toxicity is of the utmost concern in dealing with hazardous substances. This includes both long term chronic effects from continual or periodic exposures to low levels of toxicants and acute effects from a single large exposure. Toxic substances were covered in Chapter 4, so a discussion of the nature of toxic substances and their effects will not be repeated here.

Toxicity Characteristic Leaching Procedure

For regulatory and remediation purposes a standard test is needed to measure the likleihood of toxic substances getting into the environment and causing harm to organisms. The test required by the U.S. EPA is the **Toxicity Characteristic Leaching Procedure** (TCLP) designed to determine the mobility of both organic and inorganic contaminants present in liquid, solid, and multiphasic wastes. For analysis of toxic species a solution is leached from the waste and is designated as the TCLP extract. If no significant solid material is present, the waste is filtered through a 0.6-0.8 μm glass fiber filter and designated as the TCLP extract. In mixed liquid-solid wastes the liquid is separated and analyzed separately. Solid wastes to be extracted are required to have a surface area per gram of material equal to or greater than 3.1 cm^2 or to consist of particles smaller than 1 cm in their most narrow dimension. The kind of extraction fluid used on the solids is determined from the pH of a mixture of 5 g of the solids (reduced to approximately 1 mm in size if necessary) shaken vigorously with 96.5 mL of water. If the pH of the water after mixing is less than 5.0, the extraction fluid used is an acetic acid/sodium acetate buffer of pH 4.93±0.05 and if the pH is greater than 5.0, the extraction fluid is a dilute acetic acid solution with a pH of 2.88±0.05. After the fluid to be used is determined, an amount of extraction fluid equal to 20 times the mass of the solid is used for the extraction which is carried out for 18 hours in a sealed container held on a device that rotates it end-over-end for 18 hours. After the TCLP extract is separated from the solids it is analyzed for 39 specified volatile organic compounds, semivolatile organic compounds, and metals to determine if the waste exceeds specified levels of these contaminants.[8]

5.6. HAZARDOUS SUBSTANCES BY CHEMICAL CLASS

Another way of viewing hazardous substances in the context of their chemical properties is to divide them into classes of chemicals. That is done briefly in this section and more detailed coverage is provided in Chapter 7.

A number of **elements** are used industrially in their elemental forms, in many cases for chemical synthesis. Some of these elements pose hazards of flammability, corrosivity, reactivity, or toxicity. Elemental hydrogen, H_2, is extremely flammable and forms explosive mixtures with air. Three of the halogens, fluorine, chlorine, and bromine, are widely produced as elemental F_2, Cl_2, and Br_2. Fluorine is the strongest elemental oxidant and extremely reactive. It is very corrosive to the skin and inhalation of F_2 can cause severe lung damage. Chlorine, one of the most widely produced industrial chemicals, is a reactive oxidant that forms acid in water and is a corrosive poison to tissue, especially in the respiratory tract. Bromine is a volatile brown liquid which is corrosive to skin in both the liquid and vapor form. Elemental white phosphorus is a reactive substance that may catch fire spontaneously in air. It is a systemic poison. Elemental lithium, sodium, and potassium react with a large number of chemicals and burn readily to give off caustic oxide and hydroxide fumes. Elemental mercury vapor is especially toxic by inhalation. Some metals, commonly known as heavy metals, are particularly toxic in their chemically combined forms. These include lead, cadmium, mercury, beryllium, and arsenic.

Many **inorganic compounds** are hazardous because of reactivity, corrosivity, and toxicity. These compounds are discussed in Chapter 6. Many **organometallic compounds**, which have a metal atom or metalloid atom (such as silicon or arsenic) bonded directly to carbon in a hydrocarbon group or in carbon monoxide, CO, are volatile, reactive, and toxic.

Organic Compounds

There are millions of known **organic compounds,** most of which can be hazardous in some way and to some degree. Most organic compounds can be divided among hydrocarbons, oxygen-containing compounds, nitrogen-containing compounds, organo-halides, sulfur-containing compounds, phosphorus-containing compounds, or combinations thereof. These compounds are discussed further in Chapters 8 and 9.

5.7. ORIGIN AND AMOUNTS OF HAZARDOUS WASTES

Quantities of hazardous wastes produced each year are not known with certainty and depend upon the definitions used for such materials. In 1988 the figure for RCRA-regulated wastes was placed at 290 million tons. However, most of this material is water, with only a few million tons consisting of solids. Some high-water-content wastes are generated directly by processes that require large quantities of water in waste treatment and other aqueous wastes are produced by mixing hazardous wastes with wastewater.

Some wastes that might exhibit a degree of hazard are exempt from RCRA regulation by legislation. These exempt wastes include the following:

- Fuel ash and scrubber sludge from power generation by utilities
- Oil and gas drilling muds
- Byproduct brine from petroleum production
- Cement kiln dust
- Waste and sludge from phosphate mining and beneficiation
- Mining wastes from uranium and other minerals

Eventual reclassification of these kinds of low-hazard wastes could increase the quantities of RCRA-regulated wastes several fold. One problem in dealing with hazardous wastes is the lack of information about these materials. There is in fact a shortage of hard data that effectively quantifies the extent of the hazardous waste problem or that documents what actually happens to a large fraction of hazardous wastes.

Types of Hazardous Wastes

In terms of quantity by weight, more wastes than all others combined are those from categories designated by **hazardous waste numbers** preceded by F and K, respectively. The former are those from nonspecific sources and include the following examples:

- F001 The spent halogenated solvents used in degreasing: tetrachloroethylene, trichloroethylene, methylene chloride, 1,1,1,-trichloroethane, carbon tetrachloride, and the chlorinated fluorocarbons; and sludges from the recovery of these solvents in degreasing operations
- F004 The spent non-halogenated solvents: Cresols, cresylic acid, and nitrobenzene; and still bottoms from the recovery of these solvents
- F007 Spent plating-bath solutions from electroplating operations
- F010 Quenching-bath sludge from oil baths from metal heat treating operations

The "K-type" hazardous wastes are those from specific sources produced by industries such as the manufacture of inorganic pigments, organic chemicals, pesticides, explosives, iron and steel, and nonferrous metals, and from processes such as petroleum refining or wood preservation; some examples are given below:

- K001 Bottoms sediment sludge from the treatment of wastewaters from wood-preserving processes that use creosote and/or pentachlorophenol
- K002 Wastewater treatment sludge from the production of chrome yellow and orange pigments
- K020 Heavy ends (residue) from the distillation of vinyl chloride in vinyl chloride monomer production

- K043 2,6-Dichlorophenol waste from the production of 2,4-D
- K047 Pink/red water from TNT operations
- K049 Slop oil emulsion solids from the petroleum refining industry
- K060 Ammonia lime still sludge from coking operations
- K067 Electrolytic anode slimes/sludges from primary zinc production

The second largest category of wastes generated are reactive wastes, followed by corrosive wastes and toxic wastes. About 1% of wastes are designated as ignitable and another 1% are type "P" wastes (discarded commercial chemical products, off-specification species, containers, and spill residues) or "U" wastes (chemical intermediates from the manufacture of commercial chemicals or chemical formulations). Several percent of wastes are unspecified types.

Hazardous Waste Generators

About 650,000 companies generate hazardous wastes in the U. S., but most of these generators produce only small quantities of wastes. About 99% of hazardous wastes are produced by only about 2% of the generators. Hazardous waste generators are unevenly distributed geographically across the continental U.S., with a relatively large number located in the industrialized upper midwest, including the states of Illinois, Indiana, Ohio, Michigan, and Wisconsin.

Distribution of Quantities of Hazardous Wastes

Industry types of hazardous waste generators can be divided among the 7 following major categories, each containing of the order of 10-20 percent of hazardous waste generators: chemicals and allied products manufacture, petroleum-related industries, fabricated metals, metal-related products, electrical equipment manufacture constitutes another class of generators, "all other manufacturing," and nonmanufacturing and nonspecified generators. About 10% of the generators produce more than 95% of all hazardous wastes. Whereas, as noted above, the number of hazardous waste generators is distributed relatively evenly among several major types of industries, 70-85% of the quantities of hazardous wastes are generated by the chemical and petroleum industries. Of the remainder, about 3/4 comes from metal-related industries and about 1/4 from all other industries.

LITERATURE CITED

1. Kasperson, Roger E., Jeanne X. Kasperson, Christoph Hohenemser, and Robert W. Kates, *Corporate Management of Health and Safety Hazards: A Comparison of Current Practice*, Westview Press, Boulder, Colorado, 1988.
2. Gerlach, Rudolph, "Flammability, Combustibility," Chapter 6 in *Improving Safety in the Chemical Laboratory: A Practical Guide*, Young, Jay A., Ed., John Wiley and Sons, New York, 1987, pp. 59-91.
3. Wray, Tom, "Explosive Limits," *Hazmat World*, June, 1989, p. 52.
4. Bretherick, Leslie, "Chemical Reactivity: Instability and Incompatible Combinations," Chapter 7 in *Improving Safety in the Chemical Laboratory: A Practical Guide*, Young, Jay A., Ed., John Wiley and Sons, New York, 1987, pp. 93-113.
5. Seltzer, Richard, "New Plant Ends Rocket Oxidizer Shortage," *Chemical and Engineering News*, October 8, 1989, p. 5.
6. Meyer, Eugene, *Chemistry of Hazardous Materials*, Prentice-Hall, Englewood Cliffs, New Jersey, 1987.
7. Manahan, Stanley E., *Toxicological Chemistry*, Lewis Publishers, Chelsea, Michigan, 1989.
8. "Hazardous Wastes: EPA adds 25 Organics to RCRA List," *Chemical and Engineering News*, March 12, 1990, p. 4.

Chemistry of Inorganic Hazardous Wastes

6.1. CLASSES OF INORGANIC HAZARDOUS WASTES

This chapter deals with the chemistry of inorganic hazardous wastes and Chapter 7 deals with their toxicities. Chapters 8 and 9 discuss the same aspects of organic wastes. Inorganic substances are widely manufactured and used industrially and numerous inorganic substances occur in waste materials. In this chapter the chemical properties of inorganic wastes are discussed on the basis of chemical type. However, reference is made to waste characteristics and listed wastes designated by the U.S. Environmental Protection Agency in order to relate inorganic substances to regulated hazardous wastes.

Chemical Types of Inorganic Hazardous Wastes

Inorganic hazardous wastes may be subdivided into the four general categories of **elements, hazardous elemental forms, inorganic compounds**, and **organometallic compounds.**[1]

Although any element may be a constituent of hazardous wastes, usually in a combined form, some elements tend to predominate in wastes. These elements are usually classified as hazardous because of the toxicities of their elemental forms or compounds.[2] Most of the elements that are notable for the toxicities of their ions and compounds are the **heavy metals**, such as chemically bound cadmium or lead or elemental mercury which is a systemic poison when inhaled in the vapor form.

Examples of hazardous elemental species include reactive and toxic ozone, O_3 and highly toxic, flammable white phosphorus. These and other hazardous elemental species are discussed in Section 6.3.

Inorganic compounds present a variety of hazards in waste materials. Some are dangerously flammable and others are very reactive, for example, as oxidants. Strong acids and strong bases are highly corrosive. Many inorganic compounds are toxic.[3] A large number of inorganic compounds in wastes are hazardous for more than one reason. For example, extremely toxic hydrogen cyanide is also flammable.

Organometallic compounds tend to have properties of both inorganic and organic compounds. Most exhibit various degrees of toxicity.[4] Many are flammable and others are reactive.

6.2. ELEMENTS IN HAZARDOUS SUBSTANCES

Elements that are mentioned prominently as hazardous waste constituents are summarized in Table 6.1. Elements in this table are hazardous because of their toxicities, although in certain compounds they may be hazardous for other characteristics as well.

Arsenic occurs in both the +3 and +5 oxidation states. Arsenic is directly below phosphorus in the periodic table, similar to phosphorus in some chemical properties, and always encountered as an impurity in phosphorus minerals. The removal of arsenic from phosphoric acid is required for the production of food-grade phosphates and is accomplished by sulfide precipitation:

$$2As^{3+} + 3H_2S \longrightarrow As_2S_3 + 6H^+ \qquad (6.2.1)$$

Inorganic compounds of arsenic specified as hazardous waste constituents include arsenic acid (P010), arsenic(III) oxide (P012), and arsenic(V) oxide (P011).

Beryllium is in Group 2A of the periodic table. Its chemistry is atypical compared to other members of this group, especially in its tendency to form covalent compounds such as nonionic $BeCl_2$. The greatest use for beryllium metal is in the formulation of alloys, particularly with copper. Beryllium oxide, BeO, is used to make ceramics with electronic applications where heat dissipation is required.

Table 6.1. Toxic Elements Often Found in Hazardous Wastes
Element   Characteristics and examples of wastes

Element	Characteristics and examples of wastes
Arsenic (As)[a,b]	Toxic metalloid contained in wastes from arsenic compound processing and wastes from other sources, such as As_2S_3 removed from food-grade phosphoric acid
Barium (Ba)[b]	Toxic
Beryllium (Be)[a]	Toxic light metal contained in beryllium dust (waste No. P015)
Cadmium (Cd)[a,b]	Toxic heavy metal contained in some wastes from steel production
Chromium (Cr)[a,b]	Heavy metal toxic in the Cr(VI) form contained in some wastes from pigment production, steel production and finishing, and electroplating
Lead (Pb)[a,b]	Toxic heavy metal contained in some wastes from pigment production, lead smelting, and other metal processing operations
Mercury (Hg)[b]	Toxic heavy metal contained in brine purification muds and wastewater treatment sludge from the mercury cell process for chlorine manufacture
Nickel (Ni)[a]	Toxic heavy metal contained in some metal plating wastes
Selenium (Se)[b]	Toxic metalloid
Silver (Ag)[b]	Toxic heavy metal
Thallium (Tl)	Toxic heavy metal

[a] In Priority Group 1 of toxic substances found at NPL sites for which Toxicological Profiles have been prepared (see "Notice of the First Priority List of Hazardous Substances that will be the Subject of Toxicological Profiles," *Federal Register* **40**(74), Friday, April 17, 1987, pp. 12866-12874)
[b] Measured as contaminant for characteristic of EP toxicity

Cadmium is directly below zinc in the periodic table and behaves chemically much like zinc. It forms divalent compounds such as $Cd(NO_3)_2$ and exists in aqueous solution as the hydrated ion, $Cd(H_2O)_6^{2+}$. The two major uses of cadmium metal are in electroplating steel parts to protect from corrosion and in NiCd rechargeable batteries. Inorganic cadmium compounds are used for metal finishing, for pigment synthesis, in catalysis, and as plastic stabilizers.

Chromium metal is used to formulate alloys, particularly stainless steel, and as a decorative, corrosion- and wear-resistant plating on other metals. Chromium is a transition metal that exists in the chemically combined form in all oxidation states from +2 through +6. Of these the +3 and +6 states are predominant. Basic chromium(III) sulfate is required for some leather tanning processes; other inorganic chromium compounds are used in metal finishing, pigments, wood preservation, and catalysts. Chromium(VI), commonly called **chromate**, exists in acidic aqueous solution as soluble dichromate, $Cr_2O_7^{2-}$, and in more basic solutions as chromate anion, CrO_4^{2-}. Chromium(VI) is used in plating baths and as a corrosion inhibitor in some cooling tower systems. Chromium(VI) oxide, CrO_3, is used to prepare chrome plating solutions.

Because of the insolubility of the hydroxide, $Cr(OH)_3$, the Cr(III) salts are soluble only in acidic media at pH values below 4, which decreases the environmental and toxicological hazards of inorganic Cr(III) appreciably. However, chromium(VI) species are oxidants that can be reactive under some conditions. Chromium(VI) compounds are regarded as carcinogenic. Some listed hazardous wastes that may contain chromium are the following:

- F006 Wastewater treatment sludges from electroplating operations (may also contain cadmium and nickel).
- F019 Wastewater treatment sludges from the chemical conversion coating of aluminum.
- K002 Wastewater treatment sludge from the production of chrome yellow and orange pigments (also contains lead).
- K003 Wastewater treatment sludge from the production of molybdate orange pigments (also contains lead).
- K004 Wastewater treatment sludge from the production of zinc yellow pigments.

- K005   Wastewater treatment sludge from the production of chrome green pigments (also contains lead).
- K006   Wastewater treatment sludge from the production of chrome oxide green pigments.
- K007   Wastewater treatment sludge from the production of iron blue pigments.
- K008   Oven residue from the production of chrome oxide green pigments.
- K061   Emission control dust/sludge from the primary production of steel in electric furnaces (also contains lead and cadmium).
- K062   Spent pickle liquor generated by steel finishing operations of plants that produce iron or steel (also contains lead).
- K069   Emission control dust/sludge from the primary production of steel in electric furnaces (also contains lead and cadmium).
- K100 Waste leaching solution from acid leaching of emission control dust/sludge from secondary lead smelting (also contains lead and cadmium).

Chromium is a common contaminant at hazardous waste sites. A total of about 100 sites in New Jersey, alone, are thought to be contaminated with chromium.[5] In some cases the chromate wastes are alleged to seep through basement walls forming greenish-yellow crystals of chromate salts.

Lead is used as the metal, as lead(II) compounds, and to a lesser extent as lead(IV) compounds. Metallic lead has a number of applications in which it is usually alloyed with other metals. Lead is alloyed with antimony to make grids, connectors, and terminals in storage batteries. Lead(IV) oxide, PbO_2, is generated when such a battery is charged and lead(II) sulfate, $PbSO_4$, is a product of discharge. Inorganic lead salts are ingredients of some pigments, glass colorizers, and heat stabilizers in plastics. Inorganic compounds of lead specified as hazardous waste constituents include lead acetate (U144), lead subacetate (U146), and lead phosphate (U145).

Mercury is the only metal that is liquid at room temperature. It is a good electrical conductor. It is used for many purposes in the metallic form in barometers, instruments, switches, and seals. Mercury vapor is used in mercury vapor lamps and ultraviolet radiation sources.

Mercury in inorganic compounds is present in either the +1 or +2 oxidation states. Red mercury(II) oxide is mixed with graphite in the Ruben-Mallory dry cell, for which the overall cell reaction is

$$HgO + Zn \longrightarrow Hg + ZnO \qquad (6.2.2)$$

Small quantities of mercury are also used in popular alkali batteries; in 1990 one major manufacturer of these batteries announced that the mercury content was being lowered to 0.025 percent and might be totally eliminated in the future[6]. One of the limitations on the uses of mercury is its high toxicity, whether in the form of vapor, inorganic compounds, or organometallic compounds.

Nickel is a transition metal which occurs in the combined form in all oxidation states from -1 through +4, predominantly as Ni(II). A variety of containers and parts are made from nickel metal, and it is alloyed with a number of metals, including copper, iron, chromium, molybdenum, tungsten, and niobium. Nickel acts as a catalyst for the addition of hydrogen to unsaturated organic compounds, such as in the hydrogenation of unsaturated vegetable oils to make margarine (see Reaction 6.3.1). The most common nickel-based hydrogenation catalyst is Raney nickel. Nickel is an electrode material in alkaline dry cells. Nickel compounds are widely used in electroplating baths, ceramics, and pigments.

Selenium is directly below sulfur in the periodic table and exhibits many of the chemical properties of sulfur. It forms compounds in a variety of oxidation states, including all those from -1 through +2, as well as +4 and +6. Selenium and its compounds are used in chromium plating and in the production of iron, copper, and lead metals and their alloys. Some photoelectric devices and rectifiers contain selenium. Inorganic selenium compounds are also used in photocopying processes and in the manufacture of pigments, ceramics, and glass.

Inorganic thallium exists in the +1 oxidation state (for example, Tl_2SO_4) and in the +3 oxidation state (for example, $TlCl_3$). An alloy containing 8.7% thallium in mercury is liquid down to -60°C and serves as a substitute for mercury metal in low-temperature applications. Some bearing surfaces are made of thallium alloys. The industrial applications of thallium and its compounds have been limited in part by their high toxicities. Thallium(I) compounds resemble those of lead(II) in their toxicological properties. The sulfate can be used as a pesticide to control rodents, moles, and some crawling insects, including cockroaches. Because of thallium's notorious record of human and wildlife poisonings, its pesticidal use is severely limited in the U.S. and many other countries. Inorganic compounds of thallium specified as hazardous waste constituents include thallium(III) oxide (P113), thallium(I) selenite (P114), thallium(I) sulfate (P115), thallium(I) acetate (U214), thallium(I) carbonate (U215), thallium(I) chloride (U216), and thallium(I) nitrate (U217).

6.3.   HAZARDOUS ELEMENTAL FORMS

As mentioned in Section 5.6, some elements are produced and used industrially in their elemental forms, which may be flammable, corrosive, reactive, or toxic. The most common such elements are discussed here.

Elemental hydrogen, H_2, has a number of industrial uses.[7] It exists as a gas and can be transported and handled as a cryogenic liquid. It is used in metallurgy to reduce metal oxides and for welding; in the synthesis of ammonia, hydrochloric acid, aluminum alkyls, alcohols, and aldehydes; as a rocket fuel; in petroleum refining to upgrade hydrocarbon fuels; and in the food industry to hydrogenate oils. The last application is illustrated by the following reaction for the hydrogenation of an unsaturated oil:

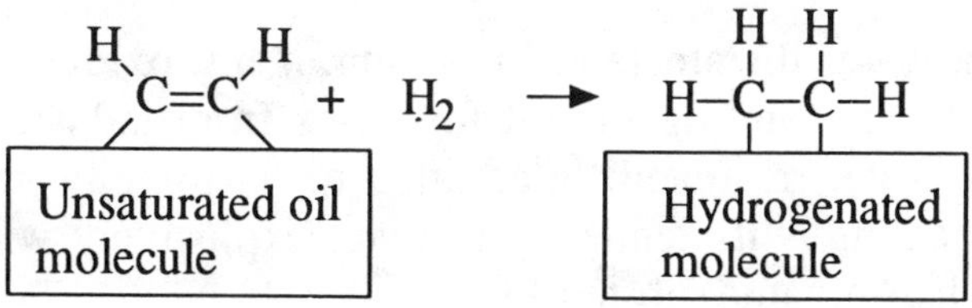

$$\tag{6.3.1}$$

The main danger from elemental hydrogen is its extreme flammability and tendency to form explosive mixtures with air or oxygen. Discarded hydrogenation catalysts can be hazardous because of residual flammable hydrogen.

Carbon black is an amorphous form of elemental carbon produced by the partial combustion or pyrolysis of vapor-phase hydrocarbon fuels as shown by the following general reaction:

$$C_cH_h + \frac{h}{4}O_2 \;\longrightarrow\; cC + \frac{h}{2}H_2O \tag{6.3.2}$$

It is an important industrial chemical manufactured in large quantities for rubber reinforcement, particularly in tires. It is also used in plastics, pigment, inks, and carbon electrodes. Carbon dust or carbon mixed with oxidizing agents such as NH_4NO_3 can be explosive. Elemental carbon in the form of graphite dust has caused fires by shorting out electrical equipment.

Elemental oxygen, O_2, is extracted by the fractional distillation of liquified air and is widely used as an industrial chemical in the forms of gas under pressure or cryogenic liquid. It can be hazardous as an oxidant. Many substances which burn poorly in air do so vigorously in an oxygen atmosphere. Pure oxygen forms violently explosive mixtures when mixed with combustible dusts or vapors, such as those of hydrocarbon vapor. Ozone, O_3, is a reactive, gaseous form of elemental oxygen, second only to fluorine (see below) as an elemental oxidizer. It is generated as needed on-site by passing an electrical discharge or ultraviolet radiation through oxygen as shown by the following reactions:

$$O_2 + \text{electrical discharge} \longrightarrow O + O \qquad (6.3.3)$$

$$O_2 + O \longrightarrow O_3 \qquad (6.3.4)$$

Ozone is used as a water disinfectant, bleaching agent, oxidant in some chemical syntheses, and deodorant for gases from industrial processes and wastewater treatment. In addition to its toxicity, the major hazard from ozone is its tendency to react explosively with reducing agents such as organic substances.

Three of the halogens – fluorine, chlorine, and bromine – are widely produced as elemental F_2, Cl_2, and Br_2, respectively. These elements are toxic, corrosive, and reactive.

Elemental **fluorine**, F_2, is produced by the electrolysis of potassium hydrogen fluoride, KHF_2, containing some additional hydrogen fluoride, HF. It is used to make highly stable sulfur hexafluoride, SF_6, which is employed as a gaseous dielectric in some electrical equipment and electronic devices, and UF_6 gas used for the enrichment of fissionable uranium-235. Fluorine is an ingredient in the synthesis of chlorofluorocarbon compounds. In addition to its high toxicity fluorine is dangerous because it is the most chemically reactive element and the strongest elemental oxidant. Its reactions are often explosive in nature and liberate large amounts of heat. It reacts with a wide variety of substances including hydrocarbons, other oxidizable organic compounds, metals, metal sulfides, and even water. The displacement of hydrogen from organic compounds or water by fluorine produces highly toxic hydrogen fluoride.

Chlorine is one of the most widely produced industrial chemicals, with about 10 million metric tons manufactured in the U.S each year. It is a green gas at room temperature and atmospheric pressure, but is transported and handled primarily in the liquified form. It is used to make organochlorine compounds (which include many important hazardous waste chemicals), for water disinfection, and for bleaching, especially of wood pulp and paper. As a chemical hazard, chlorine is a reactive corrosive oxidant with properties similar to those of fluorine. Fires or explosions may result from contact of Cl_2 with oxidizable substances such as some organic compounds, metals, and inorganic hydrides or sulfides.

Bromine is a volatile brown liquid used to make a large variety of chemical compounds, such as pharmaceuticals, gasoline additives, and agricultural chemicals. Both the liquid and vapor are corrosive to skin. It is a strong oxidizer that can react violently with a variety of substances including some organic compounds, finely divided aluminum, and ammonia.

The most common elemental form of phosphorus is solid **white phosphorus**, a raw material for the manufacture of a large variety of compounds including phosphate fertilizer, food-grade phosphoric acid, and organophosphate pesticides. White phosphorus is very reactive and ignites spontaneously in air. The oxide, P_4O_{10}, is produced as a dense fog of finely divided particles. It is highly deliquescent and very corrosive and it reacts with water or water vapor from the air to give orthophosphoric acid, H_3PO_4. Cases have been reported in which pieces of white phosphorus excavated at waste sites have burned spontaneously and disrupted operations at the site.

Elemental sulfur, S, is usually encountered as a yellow powder. It can be explosive as a dust in air and it burns to produce toxic sulfur dioxide gas, SO_2. It is reactive with either strong oxidants, such as ammonium perchlorate (NH_4ClO_4) or calcium hypochlorite ($Ca(OCl)_2$), or strong reductants, such as tin metal or sodium hydride (NaH). Elemental sulfur that melted as a result of a shipboard fire and then became mixed with ammonium nitrate oxidant caused the massive Texas City, Texas, explosion that killed almost 600 people in 1947.[8]

Lithium, sodium, and **potassium** are alkali metals that are used for various purposes in their elemental forms. They react with a large number of chemicals, including any oxidizing substances, and burn readily to give off caustic oxide and hydroxide fumes. Sodium and potassium can ignite spontaneously in contact with air or water and can cause secondary fires as a result. Fires from these metals are very difficult to extinguish because the metals react with substances commonly used to extinguish fires, including water, nonflammable organohalide compounds (halon fire extinguishers), and carbon dioxide. Burning alkali metals even react with sand, producing alkali silicates, such as Na_2SiO_3. These elements react violently with water to form the hydroxides and evolve explosive hydrogen gas. Their reaction with moisture in skin or tissue can cause thermal and chemical burns.

The specific hazards of the Group IIA metal, beryllium, are discussed in Section 6.2. Other potentially hazardous elemental forms in this group include magnesium and calcium. These metals are strong reducing agents. Although magnesium is used as a light-weight structural metal, it is dangerously flammable in the finely divided form. Magnesium turnings must be handled with special care in magnesium fabrication operations to minimize fire hazards. Calcium metal is less likely to be encountered than magnesium. One of the hazards from calcium metal is its release of explosive hydrogen gas by reaction with water:

$$Ca + 2H_2O \rightarrow Ca(OH)_2 + H_2 \qquad (6.3.5)$$

A liquid at room temperature, elemental **mercury** metal has a number of applications such as in thermometers, switches, and instruments. It has a relatively high vapor pressure and the toxic vapor can build up to dangerous levels in enclosed spaces.

6.4. HAZARDOUS INORGANIC COMPOUNDS

Many inorganic compounds are hazardous because of reactivity, corrosivity, and toxicity. Some of the more significant of these compounds taken from a list developed as part of the Emergency

Planning and Community Right-to-Know Act of 1986[9] are given in Table 6.2. This section discusses simple inorganic compounds and ions that may occur in hazardous wastes. Many of these are listed by the Environmental Protection Agency as **hazardous constituents**.[10]

Table 6.2. Examples of Hazardous Inorganic Compounds

Name and formula	Properties and effects
Aluminum phosphide, AlP	Insecticide and fumigant, releases toxic phosphine (PH_3) in contact with water
Arsenic salts	Toxic
Boron trichloride, BCl_3	Reactive and toxic, releases HCl with water
Carbon disulfide, CS_2	Flammable, explosion hazard, toxic to the central nervous system.
Cyanide salts (CN^-)	Toxic
Hydrazine, N_2H_4	Powerful reductant, explosion hazard, skin sensitizer, systemic poison
Gallium trichloride, $GaCl_3$	Toxic
Hydrocyanic acid, HCN	Flammable, very toxic
Hydrogen sulfide, H_2S	Flammable, very toxic
Mercury(II) chloride, "corrosive sublimate," $HgCl_2$	Deadly poison
Nitric oxide, NO	Toxic
Nitrogen dioxide, NO_2	Toxic, reacts strongly with some chemicals
Phosphine, PH_3	Very toxic flammable gas producing corrosive fumes when burned
Phosphorus oxychloride, $POCl_3$	Toxic irritant, reacts with water to produce heat and corrosive fumes
Sodium azide, NaN_3	Very unstable explosive, toxic
Sulfur tetrafluoride, SF_4	Powerful irritant, reacts with water to produce toxic, corrosive fumes
Thallium salts such as thallous sulfate, Tl_2SO_4	Toxic, thallium sulfate used as rodenticide
Zinc phosphide, Zn_3P_2	Highly toxic, used as rodenticide

Ammonia, NH_3, is manufactured in large quantities as a chemical raw material, fertilizer, and refrigerant fluid. It is a gas that is readily liquified, and it is stored and transported in a liquid form. Ammonia vapor is toxic and NH_3 is reactive with some substances. Improperly discarded or stored ammonia can be a hazardous waste. In 1989 a tank containing 10,000 gallons of liquid ammonia at a closed Nassau County, Long Island, manufacturing plant was drained under emergency conditions by the U.S. Coast Guard to prevent release of ammonia to the surrounding area.[11] The tank was partially under a building which was in danger of collapse, an event that might have ruptured the full tank and caused release of ammonia gas.

Ammonium nitrate is an oxidant that can be hazardous. Mixed with fuel oil, it forms a relatively safe explosive that serves as a substitute for dynamite in quarry blasting and construction. About 45,000 pounds of this mixture that was detonated as the result of a fire was involved in two massive explosions that killed 6 firefighters at a construction site in Kansas City, Missouri, on November 29, 1988.

Cyanides are the most prominently mentioned inorganic hazardous waste constituents. Among the inorganic cyanide compounds listed by the Environmental Protection Agency as hazardous constituents are hydrogen cyanide, hydrocyanic acid, and cyanides of barium, calcium, nickel, potassium, silver, sodium, and zinc. Specific listed inorganic hazardous wastes that may contain cyanide include the following:

- F006 Wastewater treatment sludges from electroplating operations (may contain complexed cyanide).
- F019 Wastewater treatment sludges from the chemical conversion coating of aluminum (may contain complexed cyanide).
- F007 Spent cyanide plating bath solutions from electroplating operations.
- F008 Plating bath residues from the bottom of plating baths from electroplating operations where cyanides are used in the process.
- F009 Spent stripping and cleaning bath solutions from electroplating operations where cyanides are used in the process.

- F010 Quenching bath residues from oil baths from metal heat treating operations where cyanides are used in the process.
- F011 Spent cyanide solutions from salt bath pot cleaning from metal heat treating operations.
- F012 (T) Quenching waste water treatment sludges from metal heat treating operations where cyanides are used in the process (contains complexed cyanide).

Cyanides may exist in hazardous wastes in any of the following forms:

- Hydrogen cyanide gas, HCN
- Unionized hydrocyanic acid dissolved in water, HCN(aq)
- Dissolved cyanide ion, CN^-
- Cyanide salts, such as NaCN or KCN
- Complexed cyanide in ions such as $Fe(CN)_6^{4-}$ or $Ni(CN)_4^{2-}$

The greatest hazard from cyanide is its toxicity in the dissolved or gaseous (HCN) form. Some complexed cyanide species are not very toxic, but complexed cyanide resists measures that are used to destroy free cyanide. Cyanide complex ions of metals, such as $Fe(CN)_6^{4-}$, are negatively charged anions. Unlike hydrated cationic metal species such as $Fe(H_2O)_6^{2+}$, these anionic complex ions are not well retained by ion exchange processes in soils and have a strong tendency to migrate with hazardous waste leachate, which mobilizes both metals and cyanide species.

A major hazard with uncomplexed cyanide waste salts such as the cyanides of sodium, potassium, or calcium is the liberation of hydrogen cyanide gas by strong acids as shown by the reaction

$$NaCN + H^+ \longrightarrow HCN + Na^+ \qquad (6.4.1)$$

Inhalation of relatively small doses of HCN gas can be fatal.

Metal sulfides can react with strong acids to liberate hydrogen sulfide as shown by the reaction

$$CaS + 2H^+ \longrightarrow H_2S + Ca^{2+} \qquad (6.4.2)$$

Hydrogen sulfide is a very toxic odorous gas. In addition to H_2S, itself, sulfides of selenium and strontium are listed by the Environmental Protection Agency as hazardous constituents of wastes. Arsenic sulfide was mentioned as a hazardous waste in Section 6.2. A test method for sulfides in solid wastes is given in the standard EPA manual for solid wastes analysis.[12]

Phosphine, PH_3, is a colorless gas that is very toxic and potentially fatal. It is used in the synthesis of organophosphorus compounds. Sometimes industrial and laboratory processes involving other phosphorus compounds produce phosphine inadvertently, adding to its hazard. Phosphine undergoes autoignition at 100°C, its fire hazard is high, and it burns to produce a choking fog of particulate P_4O_{10} and H_3PO_4. A gas related to phosphine is arsine, AsH_3. Although arsine poses only moderate fire or explosion hazards, it is highly toxic.

Some metal **phosphides**, which are inorganic compounds containing P(-III), are regarded as hazardous wastes. The EPA lists aluminum phosphide, AlP, and zinc phosphide, Zn_3P_2, as hazardous constituents. Phosphides liberate flammable, toxic phosphine in vigorous reactions with water or acid. Phosphides burn to produce noxious fumes of metal and phosphorus oxides.

Several hazardous waste substances are simple compounds of nitrogen. Of these, three are oxides of nitrogen – nitrogen(I) oxide (nitrous oxide, N_2O), nitrogen(II) oxide (nitric oxide, NO), and nitrogen(IV) oxide (nitrogen dioxide, NO_2/N_2O_4). Nitrous oxide ("laughing gas"), is a central nervous system depressant and asphyxiant used as a dental anesthetic and oxidant gas. Because it supports combustion, it is a flammability hazard. It can be very reactive and form explosive mixtures with reductants such as aluminum, hydrazine and phosphine. Nitric oxide is a toxic gas produced in the combustion of nitrogen-containing fuels and by the reaction of N_2 and O_2 under some conditions. It is strongly reactive in contact with a number of reducing substances such as aluminum, carbon disulfide, phosphine, and ethylene. The reaction of nitric oxide with some substances, such as steam, produces toxic fumes of higher nitrogen oxides (NO_2). Nitrogen dioxide, consisting of NO_2 in equilibrium with N_2O_4, is the most toxic of the three nitrogen oxides discussed here. It is violently reactive with some substances, including elemental fluorine and formaldehyde.

Hydrazine, N_2H_4, is a colorless, fuming liquid that is used in rocket fuels. It is a systemic poison, and exposure to the skin can result in hypersensitivity. It is a strong reductant and its autoignition is catalyzed by iron rust at temperatures as low as 23°C. Hydrazine explodes in contact with a number of substances such as liquid oxygen, chromium(VI) salts, sodium, and elemental fluorine and chlorine.

Gaseous cyanogen (NCCN), volatile liquid cyanogen bromide (BrCN), and highly volatile cyanogen chloride (ClCN, b.p. 13.1°C) are reactive compounds and cyanogen reacts strongly with oxidizers. These compounds are toxic irritants to the respiratory tract. One of their major hazards is reaction with steam or water to produce hydrogen cyanide and halogen acids as shown by the reaction,

$$XCN \ + \ H_2O \ \longrightarrow \ HOX \ + \ HCN \qquad\qquad (6.4.3)$$

where X is NC-, Cl-, or Br-. Cyanogen yields cyanic acid (HOCN); cyanogen chloride yields HOCl; and cyanogen bromide yields HOBr.

Carbon disulfide (CS_2) was once applied as an insecticide and fumigant and is used as a solvent in viscose rayon manufacture, cellophane manufacture, and chemical analysis. It is toxic volatile, flammable liquid with a very wide explosive range of 1.3% to 50%. It reacts explosively with a number of substances including aluminum, zinc, potassium, chlorine, and fluorine.

Phosgene (carbon oxychloride, $COCl_2$, b.p. 8.3°C) exists as either a colorless gas or volatile liquid used in the manufacture of polyurethane, polycarbonates, and herbicides. It has been manufactured as a military poison. Phosgene reacts strongly with aluminum, sodium, potassium, and some other substances. Carbon oxyfluoride has properties similar to those of phosgene and hydrolyzes very rapidly.

Inorganic azides have the azide anion, N_3^-, bound to either H (in hydrogen azide or azoic acid, HN_3) or metal cations (such as Na^+ in sodium azide, NaN_3). Azides are explosively unstable compounds. Sodium azide is listed as a hazardous waste (P105).

Osmium tetroxide, OsO_4, is a solid transition metal oxide that is volatile enough to have a detectable odor. It is used as a catalyst for organic reactions. It is dangerous because of its toxicity, particularly to eye tissue, upon which it can form an opaque coating. It is designated by the EPA as a hazardous waste (PO87). Because of the high cost and rarity of osmium, the tetroxide is not likely to be encountered as a waste material.

Acids

Large quantities of hazardous wastes consist of strong mineral acids, such as sulfuric acid (H_2SO_4) and hydrochloric acid (HCl). Strong acids exhibit the characteristic of *corrosivity*[13] because of their low pH and ability to corrode steel. Acids may exacerbate hazards from other substances such as by causing the evolution of HCN or H_2S from cyanide or sulfide salts (see Reactions 6.4.1 and 6.4.2). They are corrosive poisons that can destroy or wound exposed flesh.

Alkaline Wastes

Alkaline wastes or bases are classified as corrosive because they dissolve to produce solutions with a pH greater than 12.5. Bases are corrosive to exposed tissue and can react with some metals, such as aluminum. In contact with ammonium salts bases can liberate toxic ammonia gas as shown by the following reaction:

$$NH_4NO_3 + NaOH \rightarrow NH_3 + NaNO_3 + H_2O \qquad (6.4.4)$$

6.5. HAZARDOUS ORGANOMETALLIC SUBSTANCES

Organometallic compounds and **organometalloid compounds** are those in which metal atoms or metalloid atoms, respectively, are bonded to carbon atoms in organic groups. In subsequent discussion, *organometallic* will be used as a term to designate both organometallic and organometalloid compounds and *metal* will refer to both metals and metalloids, unless otherwise indicated.

Some organometallic compounds, such as organoarsenicals used as drugs, organomercury fungicides, and tetramethyl- and tetraethyllead used as antiknock additives for gasoline, have been known and used for many years. The chemical and toxicological properties of these compounds are rather well known and their applications as human drugs and pesticides have declined sharply as less toxic and environmentally harmful substitutes have been developed. However, in the last two decades potential hazards have arisen from the significant number of new organometallic compounds that have come into use for various "high-tech" applications in catalysis, chemical synthesis, and semiconductor manufacture.

Classification of Organometallic Compounds

For discussion of their hazardous properties organometallic compounds are conveniently classified as (1) **alkyl compounds** such as tetraethyllead, $Pb(C_2H_5)_4$; (2) **carbonyl compounds** in which the organic group is carbon monoxide, CO; (3) compounds in which the organic group is a π **electron donor**, such as benzene; and (4) combinations of these kinds of compounds.[14,15] Structural formulas of some typical organometallic compounds are given in Figure 6.1.

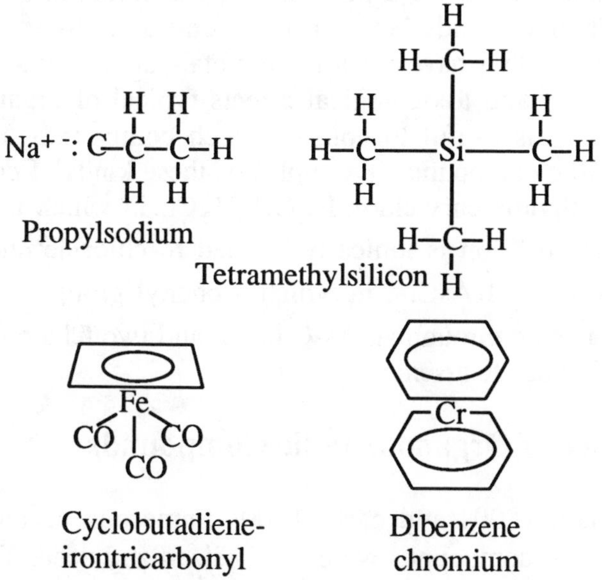

Figure 6.1. Examples of organometallic compounds

Related Compounds

A number of compounds have properties similar to those of organometallic compounds defined above, but have organic groups bonded to a metal atom through atoms other than carbon. Such compounds can be classified as organometallics for the discussion of their occurrence in hazardous wastes. The most notable such compounds are alkoxides having the general formula M^+ ^-OR, in which R is a hydrocarbon group. An alkoxide is formed, for example by the reaction of sodium with methanol

$$2CH_3OH + 2Na \rightarrow 2Na^+ {}^-OCH_3 + H_2 \qquad (6.5.1)$$

to yield sodium methoxide and hydrogen gas. The alkoxide compounds are highly basic and caustic, reacting with water to produce the corresponding hydroxides as illustrated by the following reaction:

$$K^+ {}^-OCH_3 + H_2O \rightarrow KOH + CH_3OH \qquad (6.5.2)$$

A large number of compounds exist that have at least one bond between the metal and a C atom on an organic group, as well as other covalent or ionic bonds between the metal and atoms other than carbon. Because they have at least one metal-carbon bond, as well as properties, uses and toxicological effects typical of organometallic compounds, it is useful to consider such compounds along with organometallic compounds. Examples of these kinds of compounds are monomethylmercury chloride, CH_3HgCl, in which the organometallic CH_3Hg^+ ion is ionically bonded to chloride and phenyldichloroarsine, $C_6H_5AsCl_2$, in which a phenyl group is covalently bonded to arsenic through an As-C bond, and two Cl atoms are also covalently bonded to arsenic.

Occurrence of Organometallic Compounds

In the late 1800s and early 1900s numerous organomercury pharmaceutical compounds were synthesized and used. These have since been replaced by more effective and safe non-mercury substitutes. Organomercury compounds have been widely used as pesticidal fungicides (see Figure 6.2), but these applications are now declining because of the adverse effects of mercury in the environment.

Soluble and volatile dimethylmercury (($CH_3)_2Hg$) and soluble monomethylmercury (CH_3Hg^+) salts are significant environmental pollutants. Biomethylation of inorganic mercury by anaerobic bacteria in sediments produces methylmercury compounds and plays a major role in the mobilization of waste inorganic mercury and mercury metal.

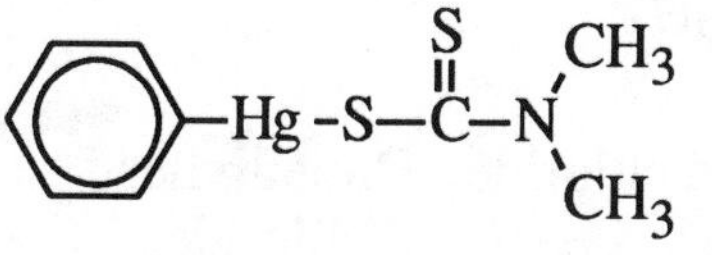

Phenylmercurydimethyldithiocarbamate (mold retardant for paper and slimicide for wood pulp)

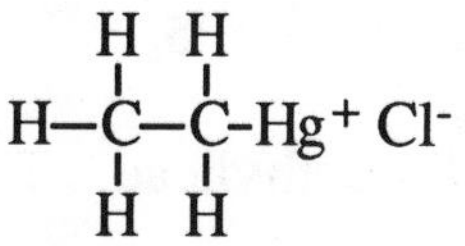

Ethylmercury chloride (seed fungicide)

Figure 6.2. Fungicidal Organomercury Compounds.

Of all the metals, tin has the greatest number of organometallic compounds in commercial use. Major industrial applications of organotin compounds include fungicides, acaricides, disinfectants, antifouling paints, stabilizers to lessen the effects of heat and light in PVC plastics, catalysts, and precursors for the formation of films of SnO_2 on glass. Tributyl tin (TBT) compounds are widely used as industrial biocides because of their bactericidal, fungicidal, and insecticidal properties. Organotin compounds applied as biocides include tributyl tin chloride,

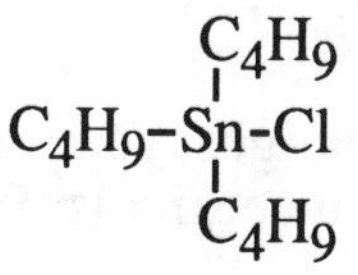

$$C_4H_9-\underset{\underset{C_4H_9}{|}}{\overset{\overset{C_4H_9}{|}}{Sn}}-Cl \qquad \text{Tributyltin chloride}$$

tributyltin hydroxide, the naphthenate, bis(tributyltin) oxide, and tris(tributylstannyl) phosphate. TBT compounds are used to preserve wood, leather, paper, and textiles;[16] in boat and ship hull coatings to prevent the growth of fouling organisms;[17] and as antifungal slimicides in cooling tower water.

The commercially important organolead compounds are predominantly methyl and ethyl lead alkyls and their salts. Some examples of these compounds are shown in Figure 6.3. Tetraethyllead has been very important commercially, toxicologically, and in the environment because of its use for several decades as a gasoline octane booster. Fortunately this use is being largely phased out.

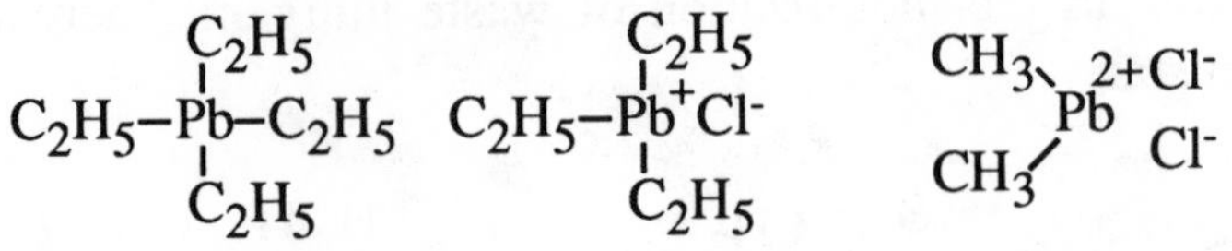

$$\begin{array}{ccc}
\underset{\displaystyle C_2H_5-\overset{\displaystyle\overset{C_2H_5}{|}}{\underset{\displaystyle\underset{C_2H_5}{|}}{Pb}}-C_2H_5}{} &
\underset{\displaystyle C_2H_5-\overset{\displaystyle\overset{C_2H_5}{|}}{\underset{\displaystyle\underset{C_2H_5}{|}}{Pb^+Cl^-}}}{} &
\underset{\displaystyle \overset{CH_3}{\underset{CH_3}{}}Pb^{2+}\,\substack{Cl^-\\Cl^-}}{}
\end{array}$$

Tetraethyllead   Triethyllead chloride   Dimethyllead dichloride

Figure 6.3. Examples of organolead compounds.

Organoarsenic compounds were the first effective synthetic pharmaceutical compounds and were widely used in the early 1900s. For most applications they have been replaced by less toxic pharmaceuticals. However, several organoarsenic compounds such as arsanilic acid,

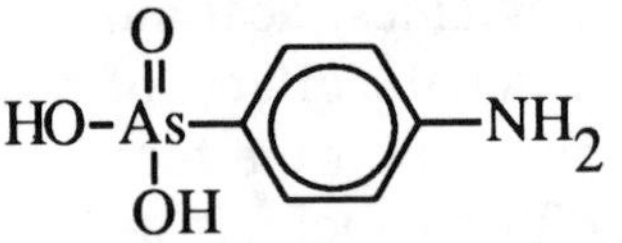

$$\underset{\displaystyle \underset{OH}{|}}{HO-\overset{\displaystyle\overset{O}{\|}}{As}}\!\!-\!\!\bigcirc\!\!-\!NH_2 \qquad \text{Arsanilic acid}$$

are still used as veterinary pharmaceuticals and as animal feed additives. Hazardous wastes from specific sources associated with the manufacture of veterinary pharmaceuticals and that may contain organoarsenic species are the following:

- K084 Wastewater treatment sludges generated during the production of veterinary pharmaceuticals from arsenic or organo-arsenic compounds.
- K101 Distillation tar residues from the distillation of aniline-based compounds in the production of veterinary pharmaceuticals from arsenic or organo-arsenic compounds.
- K102 Residue from the use of activated carbon for decolorization in the production of veterinary pharmaceuticals from arsenic or organo-arsenic compounds.

Grignard reagents such as methylmagnesium iodide,

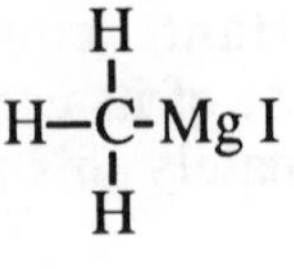

$$\underset{\displaystyle \underset{H}{|}}{H-\overset{\displaystyle\overset{H}{|}}{C}}\!-\!Mg\,I \qquad \text{Methylmagnesium iodide}$$

are widely used in organic chemical synthesis for the attachment of a hydrocarbon group such as methyl, $-CH_3$, to molecules. Grignard reagents are reactive with water and air. Ethyl ether solutions of methylmagnesium bromide (CH_3MgBr) ignite spontaneously in contact with water, which can result in bad ethyl ether fires. Contact with Grignard reagents can damage skin and inhalation can damage pulmonary tissue.

Many transition metal carbonyl compounds are known. The one of these that is the most significant as a hazardous waste because of its widespread occurrence and extremely poisonous nature is the nickel carbonyl compound, $Ni(CO)_4$, hazardous waste No. PO73. In addition to its toxicity, this compound is flammable and reacts violently with oxidizing agents including O_2.

Reactions of Organometallic Compounds

Sigma-covalently bonded organometallic compounds hydrolyze in contact with water to produce metal hydroxides and organic species as shown by the following example of cyclopentadienyl sodium:

$$C_5H_5^- \ Na^+ \ + \ H_2O \ \longrightarrow \ C_5H_6 \ + \ NaOH \qquad (6.5.3)$$

They also oxidize to produce metal oxides, carbon dioxide, and water as shown below for diethylzinc:

$$Zn(C_2H_5)_2 \ + \ 7O_2 \ \longrightarrow ZnO(s) \ + \ 5H_2O(g) \ + \ 4CO_2(g) \ (6.5.4)$$

Because of the high stabilities of the metal oxide, water, and carbon dioxide products of combustion, organometallic compounds have very high heats of combustion. Both dimethylmagnesium, $Mg(CH_3)_2$, and diethylmagnesium, $Mg(C_2H_5)_2$, are pyrophoric compounds that are violently reactive to water and steam and that self-ignite in air. Diethylmagnesium even burns in carbon dioxide (like the elemental form, magnesium in an organometallic compound removes O from CO_2 to form MgO and release elemental carbon).

Liquid trimethylaluminum reacts almost explosively with water or water and air:

$$Al(CH)_3 \xrightarrow[\{O_2\}]{H_2O} Al(OH)_3 + \text{Organic products} \qquad (6.5.5)$$

In addition to the dangers posed by the vigor of the reaction, noxious organic products may be evolved. Accidental exposure to air in the presence of moisture can result in the generation of sufficient heat to cause complete combustion of trimethylaluminum to the oxides of aluminum and carbon and to water.

LITERATURE CITED

1. Büchner, Werner, Reinhard Schleibs, Gerhard Winter, and Karl Heinz Büchel, Eds., *Industrial Inorganic Chemistry*, VCH Publishers, Inc., New York, 1989.
2. "Toxic Elements," Chapter 5 in *Toxicological Chemistry*, Stanley E. Manahan, Lewis Publishers, Inc., Chelsea, Michigan, 1989, pp. 93–116.
3. "Toxic Inorganic Compounds," Chapter 7 in *Toxicological Chemistry*, Stanley E. Manahan, Lewis Publishers, Inc., Chelsea, Michigan, 1989, pp. 143-164.
4. "Organometallics and Organometalloids," Chapter 6 in *Toxicological Chemistry*, Stanley E. Manahan, Lewis Publishers, Inc., Chelsea, Michigan, 1989, pp. 117-141.
5. Hanley, Robert, "Buried Chromium Poses a Threat to New Jersey," *New York Times*, June 30, 1989, p. 12.
6. Stroud, Jerri, "Mercury Cut in Eveready Batteries," *St. Louis Post-Dispatch*, April 6, 1990, p. C1
7. Mandelik, B. G., and David Newsome, "Hydrogen," in *Kirk-Othmer Concise Encyclopedia of Chemical Technology*, John Wiley and Sons, New York, 1985.
8. Meyer, Eugene, *Chemistry of Hazardous Materials*, Prentice Hall, Englewood Cliffs, New Jersey, 1977.
9. "List of Extremely Hazardous Substances," U. S. Environmental Protection Agency, 40 CFR 355 (Sections 302 and 304), Washington, D.C., 1988.
10. "Appendix VIII—Hazardous Constituents," *Code of Federal Regulations: Protection of Environment*, **40**, Part 261, Office of the Federal Register National Archives and Records Administration, Washington, D.C., 1986, pp. 395-400.

11. Hevesi, Dennis, "Coast Guard Orders Draining of Hazardous Tank," *New York Times*, April 16, 1989, p. 23.
12. "Sulfides," Method 9030 in Chapter 5 of *Test Methods for Evaluating Solid Wastes*, 3rd ed., SW-846, USEPA Office of Solid Waste and Emergency Response, Washington, DC, 1986.
13. "Characteristic of Corrosivity," *Code of Federal Regulations: Protection of Environment*, **40**, Part 261, Office of the Federal Register National Archives and Records Administration, Washington, D.C., 1986, pp. 373-374.
14. Eisenbroich, Christoph, and Albrecht Salzer, *Organometallics*, VCH Publishers, Inc. New York, 1989.
15. Thayer, John S., *Organometallic Chemistry*, VCH Publishers, Inc., New York, 1988.
16. Clark, Elizabeth M., Robert M. Sterritt, and John N. Lester, "The Fate of Tributyltin in the Aquatic Environment," *Environmental Science and Technology*, **22**, 600–604 (1988).
17. Seligman, Peter F., et al., "Distribution and Fate of Tributyltin in the Marine Environment," *American Chemical Society Division of Environmental Chemistry Preprint Extended Abstracts*, **28**, 573–579 (1988).

7

Toxicology of Inorganic Hazardous Wastes

7.1. INTRODUCTION

This chapter discusses the toxicities of inorganic substances. These include a variety of xenobiotic substances and some that occur naturally. The first to be discussed are elements that frequently exist as toxic compounds or (for example, ozone and white phosphorus) are toxic in their elemental forms. Heavy metals constitute an important class of toxic elements. Some specific inorganic compounds, such as cyanides, carbon monoxide, and hydrogen sulfide are notably toxic. Other significant classes of toxic inorganic species include some halogen compounds, asbestos minerals, and phosphorus compounds. Organometallic compounds and metal carbonyls constitute a final class of toxic compounds discussed in this chapter.

7.2. TOXIC ELEMENTS AND ELEMENTAL FORMS

This section discusses toxicological aspects of elements (particularly heavy metals) whose presence in a compound frequently means that the compound is toxic, as well as the toxicities of some commonly used elemental forms, such as the chemically uncombined elemental halogens. The chemistry of these substances is discussed in Chapter 6, Section 6.3.

Ozone

Ozone (O_3) is a reactive and toxic form of elemental oxygen. It is produced by ultraviolet light or electrical discharges passing through air. Its odor can be detected around inadequately vented instruments, such as spectrofluorometers, that have intense ultraviolet sources and around electrical discharges. Air pollutant ozone is largely responsible for the irritating qualities of photochemical smog.

Ozone has several toxic effects.[1] Air containing 1 ppm by volume ozone has a distinct odor. Inhalation of ozone at this level causes severe irritation and headache. Ozone irritates the eyes, upper respiratory system, and lungs. Inhalation of ozone can cause sometimes fatal pulmonary edema. Chromosomal damage has been observed in subjects exposed to ozone.

Ozone generates free radicals in tissue. These reactive species can cause lipid peroxidation, oxidation of sulfhydryl (–SH) groups, and other destructive oxidation processes. Compounds that protect organisms from the effects of ozone include radical-scavengers, anti-oxidants, and compounds containing sulfhydryl groups.

White Phosphorus

Elemental white phosphorus can enter the body by inhalation, by skin contact, or orally. It is a systemic poison that causes anemia, gastrointestinal system dysfunction, bone brittleness, and eye damage. Exposure causes **phossy jaw**, a condition in which the jawbone deteriorates and becomes fractured.

Elemental Halogens

Elemental **fluorine** (F_2) is a pale yellow highly reactive gas that is a strong oxidant. It is a toxic irritant and attacks skin and the mucous membranes of the nose and eyes.

Chlorine (Cl_2) is a greenish-yellow strongly oxidizing gas that is a strong oxidant. In water chlorine reacts to produce a strongly oxidizing solution. This reaction is responsible for some of the damage caused to the moist tissue lining the respiratory tract when the tissue is exposed to chlorine. The respiratory tract is rapidly irritated by exposure to 10-20 ppm of chlorine gas in air, causing acute discomfort that warns of the presence of the toxicant. Even brief exposure to 1,000 ppm of Cl_2 can be fatal.

Bromine (Br_2) is a volatile dark red liquid that is toxic when inhaled or ingested. Like chlorine and fluorine, it is strongly irritating to the mucous tissue of the respiratory tract and eyes and may cause pulmonary edema. The toxicological hazard of bromine is limited somewhat because its irritating odor elicits a withdrawal response.

Elemental **iodine** (I_2), a solid consisting of lustrous violet-black rhombic crystals, is irritating to the lungs much like bromine or chlorine and its general effects are similar to these elements. However, the relatively low vapor pressure of iodine limits exposure to I_2 vapor.

Heavy Metals

Some metals, commonly known as **heavy metals**, are particularly toxic in their chemically combined forms and — notably in the case of mercury — some are toxic in the elemental form. The term, heavy metal, is used loosely to refer to almost any metal with an atomic number higher than that of calcium (20); some metalloids, such as arsenic and antimony, are classified as heavy metals for discussion of their toxicities. The toxic properties of some of the most hazardous heavy metals are discussed here.

Although not truly a <u>heavy</u> metal, **beryllium** (atomic mass 9.01) is one of the more hazardous toxic elements. Skin exposed to beryllium compounds may become ulcerated and develop granulomas. The body can become hypersensitive to beryllium, resulting in skin dermatitis, acute conjunctivitis, and corneal laceration. The most serious toxic effect of beryllium is berylliosis, a condition manifested by lung fibrosis and pneumonitis, which may develop after a latency period of 5-20 years.

Cadmium adversely affects several important enzymes and it can cause painful osteomalacia (bone disease) and kidney damage. Inhalation of cadmium oxide dusts and fumes results in cadmium pneumonitis characterized by edema and pulmonary epithelium necrosis.

The danger of exposure to **lead** is higher than that for many other toxicants because it is widely distributed as metallic lead, inorganic compounds, and organometallic compounds. Lead has a number of toxic effects, including inhibition of the synthesis of hemoglobin. It also adversely affects the central and peripheral nervous systems and the kidneys.

Arsenic is a metalloid which forms a number of toxic compounds. The toxic +3 oxide, As_2O_3, is absorbed through the lungs and intestines. Biochemically, arsenic acts to coagulate proteins, forms complexes with coenzymes, and inhibits the production of adenosine triphosphate (ATP) in essential metabolic processes.

Elemental **mercury** vapor can enter the body through inhalation and be carried by the bloodstream to the brain where it penetrates the blood-brain barrier. It disrupts metabolic processes in the brain causing tremor and psychopathological symptoms such as shyness, insomnia, depression, and irritability. Divalent ionic mercury, Hg^{2+}, damages the kidney. Organometallic mercury compounds such as dimethylmercury, $Hg(CH_3)_2$, are also very toxic.

7.3. TOXIC INORGANIC COMPOUNDS

Cyanide

Both **hydrogen cyanide** (HCN) and **cyanide salts** (which contain CN^- ion) are rapidly acting poisons[2]; a dose of only 60–90 mg is sufficient to kill a human. Uses of hydrogen cyanide as a pesticidal fumigant and cyanide salt solutions in chemical synthesis and metal processing pose risks of exposure to humans. Some fire fatalities are caused by HCN evolved from the burning of nitrogen-containing polymers.

Metabolically, cyanide bonds to iron(III) in iron-containing ferricytochrome oxidase enzyme (see enzymes, Section 4.4), preventing its reduction to iron(II) in the oxidative phosphorylation process by which the body utilizes O_2. The crucial enzyme is inhibited because ferrouscytochrome oxidase, which is required to react with O_2, is not formed and utilization of oxygen in cells is prevented so that metabolic processes cease.

Carbon Monoxide

Carbon monoxide, CO, is a common cause of accidental poisonings. At CO levels in air of 10 parts per million (ppm) impairment of judgement and visual perception occur; exposure to

100 ppm causes dizziness, headache, and weariness; loss of consciousness occurs at 250 ppm; and inhalation of 1,000 ppm results in rapid death. Chronic long-term exposures to low-levels of carbon monoxide are suspected of causing disorders of the respiratory system and the heart.[3]

After entering the blood stream through the lungs, carbon monoxide reacts with hemoglobin (Hb) to convert oxyhemoglobin (O_2Hb) to carboxyhemoglobin (COHb):

$$O_2Hb + CO \rightarrow COHB + O_2 \tag{7.3.1}$$

Carboxyhemoglobin is much more stable than oxyhemoglobin so that its formation prevents hemoglobin from carrying oxygen to body tissues.

Nitrogen Oxides

The two most common toxic oxides of nitrogen are **nitric oxide** (NO) and **nitrogen dioxide** (NO_2), designated collectively as NO_x. Nitric oxide is released in large quantities from the exhausts of internal combustion and turbine engines and both oxides are common air pollutants. Regarded as the more toxic of the two gases, NO_2 causes severe irritation of the innermost parts of the lungs resulting in pulmonary edema. In cases of severe exposures, fatal bronchiolitis fibrosa obliterans may develop approximately three weeks after exposure to NO_2. Fatalities may result from even brief periods of inhalation of air containing 200–700 ppm of NO_2. Biochemically, NO_2 disrupts lactic dehydrogenase and some other enzyme systems, possibly acting much like ozone, a stronger oxidant discussed later in this chapter. Free radicals, particularly the hydroxyl radical, $HO\cdot$, (see Section 3.11) are likely formed in the body by the action of nitrogen dioxide and the compound probably causes **lipid peroxidation** in the body. This is a process that is also initiated by ozone in which the C=C double bonds in unsaturated lipids are attacked by free radicals and undergo chain reactions in the presence of O_2, resulting in their oxidative destruction.

Nitrous oxide, N_2O is used as an oxidant gas and in dental surgery as a general anesthetic. This gas was once known as "laughing gas," and was used in the late 1800s as a "recreational gas" at parties held by some of our not-so-staid Victorian ancestors. Nitrous oxide is a central nervous system depressant and can act as an asphyxiant.

Hydrogen Halides

Hydrogen halides (general formula HX, where X is F, Cl, Br, or I) are relatively toxic gases. The most widely used of these gases are HF and HCl; their toxicities are discussed here.

Hydrogen Fluoride

Hydrogen fluoride, (HF, mp -83.1°C, bp 19.5°C) is used as a clear, colorless liquid or gas or as a 30–60% aqueous solution of **hydrofluoric acid**. Hydrofluoric acid is so reactive that it is used to etch glass and clean stone. It must be kept in plastic containers because it vigorously attacks glass and other materials containing silica (SiO_2), producing gaseous silicon tetrafluoride, SiF_4.

Here reference is made to both hydrogen fluoride and hydrofluoric as HF. Both are extreme irritants to any part of the body that they contact, causing ulcers in affected areas of the upper respiratory tract. Lesions caused by contact with HF heal poorly, and tend to develop gangrene.

Fluoride ion, F^-, is toxic in soluble fluoride salts, such as NaF, causing **fluorosis**, a condition characterized by bone abnormalities and mottled, soft teeth. Livestock is especially susceptible to poisoning from fluoride fallout on grazing land; severely afflicted animals become lame and even die. Industrial pollution has been a common source of toxic levels of fluoride. Low levels of fluoride can have beneficial effects, however. About 1 ppm of fluoride used in some drinking water supplies prevents tooth decay.

Hydrogen Chloride

Colorless **hydrogen chloride** (HCl) is widely produced as a gas, pressurized liquid, or as a saturated aqueous solution containing 36% HCl called **hydrochloric acid** and commonly denoted simply as HCl. Hydrochloric acid is a major industrial chemical with U.S. production of about 2.3 million tons per year. It is much less toxic than HF and hydrochloric acid is a natural physiological fluid present as a dilute solution in the stomachs of humans and other animals. However, inhalation of HCl vapor can cause spasms of the larynx as well as pulmonary edema and even death at high levels. Hydrogen chloride vapor has such a high affinity for water that it tends to dehydrate eye and respiratory tract tissue.

Interhalogen Compounds and Halogen Oxides

Compounds formed from different halogens are called **interhalogen compounds** (Table 7.1) and compounds of halogens and oxygen are **halogen oxides**. These compounds have some chemical and toxicological similarities. They are discussed briefly here.

Table 7.1. The Major Interhalogen Compounds

Compound name	Formula	Physical properties
Chlorine monofluoride	ClF	Colorless gas, mp -154°C, bp -101°C
Chlorine trifluoride	ClF_3	Colorless gas, mp -83°C, bp 12°C
Bromine monofluoride	BrF	Pale brown gas, bp 20°C
Bromine trifluoride	BrF_3	Colorless liquid, mp 8.8°C, bp 127°C
Bromine pentafluoride	BrF_3	Colorless liquid, mp -61.3°C, bp 40°C
Bromine monochloride	$BrCl$	Red/yellow highly unstable liquid and gas
Iodine trifluoride	IF_3	Yellow solid decomposing at 28°C
Iodine pentafluoride	IF_5	Colorless liquid, mp 9.4°C, bp 100 °C
Iodine heptafluoride	IF_7	Colorless sublimable solid, mp 5.5°C
Iodine monobromide	IBr	Gray sublimable solid, mp 42°C
Iodine monochloride	ICl	Red–brown solid alpha form (mp 27°C), bp 9°C
Iodine pentabromide	IBr_5	Crystalline solid
Iodine tribromide	IBr_3	Dark brown liquid
Iodine trichloride	ICl_3	Orange–yellow solid subliming at 64°C
Iodine pentachloride	ICl_5	- - -

Interhalogen Compounds

The major interhalogen compounds listed in Table 7.1 are usually described as "fuming liquids." Most interhalogen compounds exhibit extreme reactivity and are potent oxidizing agents for organic matter and oxidizable inorganic compounds. Interhalogen compounds react with water or steam to produce hydrohalic acid solutions (HF, HCl) and nascent oxygen {O}.

Too reactive to enter biological systems in their original chemical state, interhalogen compounds tend to be powerful corrosive irritants that acidify, oxidize, and dehydrate tissue. Because of these effects skin is readily damaged by interhalogen compounds; the eyes and mucous membranes of the mouth, throat, and pulmonary systems are especially susceptible to attack. The corrosive effects of the interhalogen compounds are much like those of the elemental forms of the elements from which they are composed. Chemical reactions of interhalogen compounds tend to produce toxic products, such as HF from fluorine compounds.

Halogen Oxides

Major halogen oxides are listed in Table 7.2. These compounds tend to be unstable, highly reactive, and toxic. For the most part, the halogen oxides pose hazards similar to those of the interhalogen compounds discussed previously in this section.

Chlorine dioxide, which is usually manufactured on site because of its extreme chemical reactivity, is the most commercially important of the halogen oxides. It is employed for odor control and bleaching wood pulp. As a substitute for chlorine in water disinfection it produces fewer undesirable chemical byproducts, particularly trihalomethanes.

Oxyacids of the Halogens and their Salts

The most important of the oxyacids and their salts formed by halogens are hypochlorous acid, HOCl, and hypochlorites, such as NaOCl, used for bleaching and disinfection. The hypochlorites irritate eye, skin, and mucous membrane tissue because they react to produce active (nascent) oxygen ({O}) and acid as shown by the reaction below:

$$HClO \rightarrow H^+ + Cl^- + \{O\} \qquad\qquad (7.3.2)$$

Table 7.2. Major Oxides of the Halogens

Compound name	Formula	Physical properties
Fluorine monoxide (oxygen difluoride)	OF_2	Colorless gas, mp -224°C, bp -145°C
Chlorine monoxide	Cl_2O	Orange gas, mp -20°C, bp 2.2°C
Chlorine dioxide	ClO_2	Orange gas, mp -59°C, bp 9.9°C
Chlorine heptoxide	Cl_2O_7	Colorless oil, mp -91.5°C, bp 82°C
Bromine monoxide	Br_2O	Brown solid, Decomp. -18°C
Bromine dioxide	BrO_2	Yellow solid, Decomp. 0°C
Iodine dioxide	IO_2	Yellow solid
Iodine pentoxide	I_2O_5	Colorless oil, Decomp. 325°C

Although highly reactive oxidants when heated, perchlorates, such as ammonium perchlorate, NH_4ClO_4, are not particularly toxic. Perchloric acid, $HClO_4$, is a strong acid. The toxicities of perchlorate salts are those of the cation in the compound and tend to act as skin irritants.

Halogen Azides and Nitrogen Halides

The vapors of extremely reactive, spontaneously explosive halogen azides (XN_3, X is a halogen) are irritating. These compounds react with water to produce toxic fumes of the elemental halogen, HX, and NO_x.

Colorless, gaseous nitrogen trifluoride, NF_3, and nitrogen trichloride, NCl_3, a volatile yellow oil, are examples of nitrogen halides having the general formula N_nX_x, where X is F, Cl, Br or I. The nitrogen halides are toxic because they are irritants to eyes, skin, and mucous membranes although because of their high reactivity they are often destroyed before exposure can occur.

Monochloramine and Dichloramine

Nitrogen trichloride, monochloramine, and dichloramine are formed by the substitution of chloride for hydrogen on ammonia or ammonium ion as illustrated by the following reactions:

$$NH_4^+ + HOCl \rightarrow H^+ + H_2O + NH_2Cl \qquad (7.3.3)$$
$$\text{Monochloramine}$$

$$NH_2Cl + HOCl \rightarrow H^+ + H_2O + NHCl_2 \qquad (7.3.4)$$
$$\text{Dichloramine}$$

The chloramines are formed deliberately in the purification of drinking water to provide **combined available chlorine**. These forms of chlorine last longer in the water distribution system than Cl_2, $HOCl$, and OCl^- and act to retain disinfection throughout the water distribution system.

Inorganic Compounds of Silicon

Fortunately, many of the silicon compounds that are used in the semiconductor and related industries have relatively low toxicities. Toxicities of some of the more common inorganic silicon compounds are summarized here.

Silica

Silica, SiO_2, is a hard mineral substance known as quartz in the pure form and occurs in a variety of minerals such as sand, sandstone, and diatomaceous earth. **Silicosis** resulting from human exposure to silica dust from construction materials, sand blasting, and other sources has been a common occupational disease. A type of pulmonary fibrosis that causes lung nodules and makes victims more susceptible to pneumonia and other lung diseases, silicosis is one of the most common disabling conditions resulting from industrial exposure to hazardous substances. It can cause death from insufficient oxygen or from heart failure in severe cases.

Asbestos

Asbestos is the name given to a group of fibrous silicate minerals, typically those of the serpentine group, for which the approximate formula is $Mg_3P(Si_2O_5)(OH)_4$. Asbestos has been widely used in structural materials, brake linings, insulation, and pipe manufacture.[4] Inhalation of asbestos may cause asbestosis (a pneumonia condition), mesothelioma (tumor of the mesothelial tissue lining the chest cavity adjacent to the lungs), and bronchogenic carcinoma (cancer originating with the air passages in the lungs)[5] so that uses of asbestos have been severely curtailed and widespread programs have been undertaken to remove the material from buildings.

In 1979 560,000 metric tons of asbestos were used in the U.S. By 1988 annual consumption had dropped to 85,000 metric tons, most of it used for brake linings and pads, roofing products, cement/asbestos pipe, gaskets, heat-resistant packing, and specialty papers. In July, 1989, the U.S. Environmental Protection Agency announced regulations that would phase out most uses of asbestos by 1996.[6] However, global production of asbestos increased in 1988, largely because of increased demand from India, [7] A very large fraction of this production is in the Thetford Mines area of Quebec, which produced 705,000 metric tons of asbestos in 1988.

Alternatives to asbestos continue to be developed. Three major considerations involved in the development of alternatives are feasibility of substitution, raw materials used to manufacture substitutes, and health aspects.[8]

Silanes

Silane, SiH_4, and disilane, H_3SiSiH_3, are examples of inorganic **silanes**, which have H-Si bonds. Numerous organic ("organometallic") silanes exist in which alkyl moieties are substituted for H. Little information is available regarding the toxicities of silanes.

Silicon Halides and Halohydrides

Silicon tetrachloride, $SiCl_4$, is the only industrially significant of the **silicon tetrahalides**, a group of compounds with the general formula SiX_4, where X is a halogen. The two commercially produced **silicon halohydrides**, general formula $H_{4-x}SiX_x$, are dichlorosilane (SiH_2Cl_2) and trichlorosilane, ($SiHCl_3$). These compounds are used as intermediates in the synthesis of organosilicon compounds and in the production of high-purity silicon for semiconductors. Silicon tetrachloride and trichlorosilane, fuming liquids which react with water to give off HCl vapor, have suffocating odors and are irritants to eye, nasal, and lung tissue.

Inorganic Phosphorus Compounds

Phosphine

Phosphine (PH_3), a colorless gas that undergoes autoignition at 100°C, is used for the synthesis of organophosphorus compounds and is sometimes inadvertently produced in chemical syntheses involving other phosphorus compounds. It is a potential hazard in industrial processes and in the laboratory. Symptoms of poisoning from potentially fatal phosphine gas include pulmonary tract irritation, central nervous system depression, fatigue, vomiting, and difficult, painful breathing.

Phosphorus Pentoxide

Because of its dehydrating action and formation of acid from the reaction

$$P_4O_{10} + 6H_2O \rightarrow 4H_3PO_4 \qquad (7.3.5)$$

phosphorus pentoxide, P_4O_{10}, is a corrosive irritant to skin, eyes and mucous membranes. Phosphorus pentoxide is produced as a fluffy white powder from the combustion of elemental phosphorus and reacts with water from air to form syrupy orthophosphoric acid, H_3PO_4.

Phosphorus Halides

Phosphorus halides have the general formulas PX_3 and PX_5, where X is a halogen. The most commercially important phosphorus halide is phosphorus pentachloride used as a catalyst in organic synthesis, as a chlorinating agent and as a raw material to make phosphorus oxychloride ($POCl_3$). Because they react violently with water to produce the corresponding hydrogen halides and oxo phosphorus acids,

$$PCl_5 + 4H_2O \rightarrow H_3PO_4 + 5HCl \qquad (7.3.6)$$

the phosphorus halides are strong irritants to eyes, skin, and mucous membranes.

Phosphorus Oxyhalides

The major phosphorus oxyhalide in commercial use is phosphorus oxychloride ($POCl_3$) a faintly yellow fuming liquid. Reacting with water to form toxic vapors of hydrochloric acid and phosphonic acid (H_3PO_3), phosphorus oxyhalide is a strong irritant to the eyes, skin, and mucous membranes.

Inorganic Compounds of Sulfur

Sulfur forms several widely encountered toxic inorganic compounds. The toxicity of yellow crystalline or powdered elemental sulfur, S_8, is low, although chronic inhalation of it can irritate mucous membranes.

Hydrogen Sulfide

A colorless gas with a foul rotten-egg odor, **hydrogen sulfide** (H_2S, hazardous waste No. U135) is encountered in large quantities as a byproduct of coal coking and petroleum refining and as a constituent of sour natural gas. In some cases inhalation of hydrogen sulfide kills faster than even hydrogen cyanide; rapid death ensues from exposure to air containing more than about 1000 ppm H_2S due to asphyxiation from respiratory system paralysis. Lower doses cause symptoms that include headache, dizziness, and excitement because of damage to the central nervous system. General debility is one of the numerous effects of chronic H_2S poisoning.

Sulfur Dioxide and Sulfites

Sulfur dioxide, SO_2, dissolves in water, to produce sulfurous acid, H_2SO_3; hydrogen sulfite ion, HSO_3^-; and sulfite ion, SO_3^{2-}. Because of its water solubility, sulfur dioxide is largely removed in the upper respiratory tract. It is an irritant to the eyes, skin, mucous membranes and respiratory tract. Some individuals are hypersensitive to sodium sulfite (Na_2SO_3), which has been used as a chemical food preservative. These uses were further severely restricted in the U.S. in early 1990.

Sulfuric Acid

Number one in synthetic chemical production, **sulfuric acid** (H_2SO_4) is a severely corrosive poison and dehydrating agent in the concentrated liquid form; it readily penetrates skin to reach subcutaneous tissue causing tissue necrosis with effects resembling those of severe thermal burns. Sulfuric acid fumes and mists irritate eye and respiratory tract tissue and industrial exposure has caused tooth erosion in workers.

Sulfur Halides, Oxides, and Oxyhalides

The more important halides, oxides and oxyhalides of sulfur are listed in Table 7.3.

7.4. ORGANOMETALLIC COMPOUNDS

The toxicological properties of some organometallic compounds —pharmaceutical organoarsenicals, organomercury fungicides, and tetraethyllead antiknock gasoline additives—that have been used for many years are well known. However, toxicological experience is lacking for many relatively new organometallic compounds that are now being used in semiconductors, as catalysis, and for chemical synthesis, so they should be treated with great caution until proven safe.

Organometallic compounds often behave in the body in ways totally unlike the inorganic forms of the metals that they contain. This is due in large part to the fact that, compared to inorganic forms, organometallic compounds have an organic nature and higher lipid solubility.

Organolead Compounds

Perhaps the most notable hazardous organometallic compound is tetraethyllead (hazardous waste No. P110), the structural formula of which is shown in Figure 6.3. Tetraethyllead is a colorless, oily liquid. Although it is being phased out now, this compound was used for several decades as an octane-boosting gasoline additive, so

Table 7.3. Inorganic Sulfur Compounds

Compound name	Formula	Properties
Sulfur		
Monofluoride	S_2F_2	Colorless gas, mp -104°C, bp 99°C, toxicity similar to HF
Tetrafluoride	SF_4	Gas, bp -40°C, mp -124°C, powerful irritant
Hexafluoride	SF_6	Colorless gas, mp -51°C, surprisingly nontoxic when pure, but often contaminated with toxic lower fluorides
Monochloride	S_2Cl_2	Oily, fuming orange liquid, mp -80°C, bp 138°C, strong irritant to eyes, skin, and lungs
Tetrachloride	SCl_4	Brownish/yellow liquid/gas, mp -30°C, Decom. below 0°C, irritant
Trioxide	SO_3	Solid anhydride of sulfuric acid reacts with moisture or steam to produce sulfuric acid
Sulfuryl chloride	SO_2Cl_2	Colorless liquid, mp -54°C, bp 69°C, used for organic synthesis, corrosive toxic irritant
Thionyl chloride	$SOCl_2$	Colorless-to-orange fuming liquid, mp -105°C, bp 79°C, toxic corrosive irritant
Carbon oxysulfide	COS	Volatile liquid byproduct of natural gas or petroleum refining, toxic narcotic
Carbon disulfide	CS_2	Colorless liquid, industrial chemical, narcotic and central nervous system anesthetic

there were many opportunities for exposure in its manufacture and blending and from leaded fuels. Tetraethyllead has a strong affinity for lipids and can enter the body by inhalation, ingestion, and absorption through the skin. Tetraethyllead is highly toxic and does not act like inorganic lead compounds in the body. It affects the central nervous system with symptoms such as fatigue, weakness, restlessness, ataxia, psychosis, and convulsions. Recovery from severe lead poisoning tends to be slow. In cases of fatal tetraethyllead poisoning, death has occurred as soon as one or two days after exposure.

Organotin Compounds

The greatest number of organometallic compounds in commercial use are those of tin. Organotin compounds are used as pesticides, antifouling paints, and catalysts and are of particular environmental significance because of their increasing applications as industrial biocides.[9,10] Tributyltin chloride and related tributyltin (TBT) compounds have bactericidal, fungicidal, and insecticidal properties. Organotin compounds are readily absorbed through the skin, sometimes causing a skin rash. They probably bind with sulfur groups on proteins and appear to interfere with mitochondrial function.

Carbonyls

Metal carbonyls regarded as extremely hazardous because of their toxicities include nickel carbonyl ($Ni(CO)_4$), cobalt carbonyl, and iron pentacarbonyl. Some of the hazardous carbonyls are volatile and readily taken into the body through the respiratory tract or through the skin. The carbonyls affect tissue directly and they break down to toxic carbon monoxide and products of the metal, which have additional toxic effects.

Reaction Products of Organometallic Compounds

An example of the production of a toxic substance from the burning of an organometallic compound is provided by the oxidation of diethylzinc:

$$Zn(C_2H_5)_2 + 7O_2 \rightarrow ZnO(s) + 5H_2O(g) + 4CO_2(g) \qquad (7.4.1)$$

Zinc oxide is used as a healing agent and food additive. However, inhalation of zinc oxide fume particles produced by the combustion of zinc organometallic compounds causes zinc **metal fume fever.**

This is an uncomfortable condition characterized by elevated temperature and "chills."

A serious hazard from the combustion of an organometallic (organometalloid) occurred as the result of a chemical train derailment at Freeport, Michigan, on July 22, 1989. Flammable liquid trimethylchlorosilane ($(CH_3)_3SiCl$, (which is used in the production of high-purity silicon) spilled and burned, producing hydrogen chloride (hydrochloric acid) and silicon dioxide:

$$(CH_3)_3SiCl + 6O_2 \rightarrow 3CO_2 + 4H_2O + SiO_2 + HCl \qquad (7.4.2)$$

These products formed a dense, choking fog that obscured visibility. According to a report of the accident[11], U. S. railroads transport 4 billion tons of hazardous materials each year with 250,000 shipments each day.

LITERATURE CITED

1. Lee, S.D., M. G. Mustafa, and M. A. Mehlman, Eds., *Internatonal Symposium on the Biochemical Effects of Ozone and Related Photochemical Oxidants*, Vol. V in *Advances in Modern Environmental Toxicology*, Princeton University Press, Princeton, NJ, 1983.
2. "Cyanide," in *Clinical Toxicology of Commercial Products,* 5th ed., Robert E. Gosselin, Roger P. Smith, and Harold C. Hodge, Williams and Wilkins, Baltimore/London, 1984, pp. III-123–III-130.
3. Hodgson, Ernest, and Patricia E. Levi, *Modern Toxicology*, Elsevier, New York, 1987.
4. Steinway, Daniel M., "Scope and Numbers of Regulations for Asbestos-Containing Materials, Abatement Continue to Grow," *Hazmat World*, April, 1990, pp. 32-58.
5. Fisher, Gerald L., and Michael A. Gallo, Eds., *Asbestos Toxicity*, Marcel Dekker, New York, 1988.
6. Shabecoff, Philip, "E.P.A. to Ban Virtually All Asbestos Products by '96," *New York Times*, July 7, 1989, p. 8.
7. Swift, Allan, "Asbestos Producer Encouraged by Rise in Consumption," *Toronto Globe and Mail*, June 10, 1989, p. B3.

8. Hodgson, A. A., *Alternatives to Asbestos, the Pros and Cons*, John Wiley and Sons, Inc., Somerset, NJ, 1989.

9. Seligman, Peter F., et al., "Distribution and Fate of Tributyltin in the Marine Environment," *American Chemical Society Division of Environmental Chemistry Preprint Extended Abstracts*, **28**, 573–579 (1988).

10. Clark, Elizabeth M., Robert M. Sterritt, and John N. Lester, "The Fate of Tributyltin in the Aquatic Environment," *Environmental Science and Technology*, **22**, 600–604 (1988).

11. Cushman, John H., "Chemicals on Rails: A Growing Peril," *New York Times*, August 2, 1989, p.8.

8

Organic Hazardous Wastes

8.1. ORGANIC HAZARDOUS WASTE FORMS

Organic materials are encountered in hazardous wastes as organic solvents, solutions in organic solvents, solutions of organic compounds in water, emulsions, sludges, still bottoms, residuals, oils, greases, paint, solids, and organic pesticide wastes. There are millions of known organic compounds, most of which can be hazardous in some way and to some degree. Some of these substances are discussed in this chapter. A three-digit number following a letter (D, F, K, P, U) in parentheses after the name of a compound or waste denotes an EPA Hazardous Waste Number.

Organic/Inorganic Interactions

Organic wastes often interact strongly with inorganic constituents of hazardous wastes. Because of these interactions, organic compounds codisposed with inorganic substances often significantly affect the properties of the inorganic wastes. For example, organic surfactants, complexing agents, and chelating agents (nitrilotriacetate (NTA) ion, ethylenedinitrilotetraacetate (EDTA) anion, quadrol, citrates, gluconates) contained in metal electroplating and finishing wastes may increase the mobility and solubility of heavy metals and make the metals more difficult to remove in waste treatment processes.

165

Chemical Classification of Organic Wastes

Chemically, most organic compounds can be divided among hydrocarbons, oxygen-containing compounds, nitrogen-containing compounds, organohalides, sulfur-containing compounds, phosphorus-containing compounds, or combinations thereof. Each of these classes of organic compounds is discussed briefly here.

8.2.   HYDROCARBONS

Hydrocarbons are discussed as organic compounds in Chapter 2, Section 2.6. The major types of hydrocarbons are alkanes, alkenes, alkynes, and aryl compounds. Structural formulas of examples of each are shown in Figure 8.1

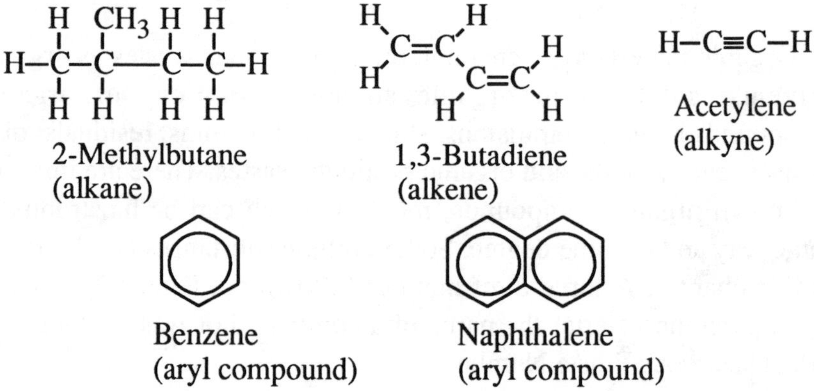

2-Methylbutane
(alkane)

1,3-Butadiene
(alkene)

Acetylene
(alkyne)

Benzene
(aryl compound)

Naphthalene
(aryl compound)

Figure 8.1. Examples of major types of hydrocarbons.

Alkanes

Alkanes, also called **paraffins** or **aliphatic hydrocarbons**, are hydrocarbons in which the C atoms are joined by single covalent bonds (sigma bonds) consisting of two shared electrons (see Section 2.6). The carbon atoms in hydrocarbons may form straight chains or branched chains. As shown in Figure 8.1, a typical branched chain alkane is 2-methylbutane, a volatile, highly flammable liquid. It is a component of gasoline, which may explain why it was the most abundant hydrocarbon (other than methane) found in a detailed study of ambient air hydrocarbons in 39 U.S. cities.[1] Alkanes may also have cyclic structures, as in cyclohexane (C_6H_{12}). Each of the 6 carbon atoms in a cyclohexane molecule has 2 H atoms bonded to it. The general molecular formula for straight- and branched-chain alkanes is C_nH_{2n+2}, and that of cyclic alkanes is C_nH_{2n}.

Reactions of Alkanes

One of the more significant chemical reactions of alkanes is **oxidation** with molecular oxygen in air as shown for the following combustion reaction of propane:

$$C_3H_8 + 5O_2 \rightarrow 3CO_2 + 4H_2O + \text{heat} \qquad (8.2.1)$$

Common alkanes are highly flammable and the more volatile lower molecular mass alkanes form explosive mixtures with air. Furthermore, combustion of alkanes in an oxygen-deficient atmosphere or in an automobile engine produces significant quantities of carbon monoxide, CO, the toxic properties of which are discussed in Section 7.3.

Alkanes also undergo **substitution reactions** in which one or more H atoms on an alkane are replaced by atoms of another element. The most common such reaction is the replacement of H by chlorine, to yield **organohalide** compounds. For example, methane reacts with chlorine to give carbon tetrachloride, an **organochlorine** compound:

$$
\underset{\displaystyle \overset{\textstyle H}{\underset{\textstyle H}{|}}{\overset{\textstyle |}{H-C-H}}}{}
\; + \; 4Cl_2 \;\longrightarrow\;
\underset{\displaystyle \overset{\textstyle Cl}{\underset{\textstyle Cl}{|}}{\overset{\textstyle |}{Cl-C-Cl}}}{}
\; + \; 4HCl \qquad (8.2.2)
$$

Alkenes and Alkynes

Alkenes or **olefins** are hydrocarbons that have double bonds consisting of 4 shared electrons. The simplest and most widely manufactured alkene is ethylene,

$$
\overset{\textstyle H}{} \underset{\textstyle H}{} C=C \overset{\textstyle H}{} \underset{\textstyle H}{} \quad \text{Ethylene (ethene)}
$$

used for the production of polyethylene polymer. Another example of an important alkene is 1,3-butadiene (Figure 8.1), widely used in the manufacture of polymers, particularly synthetic rubber. The lighter alkenes, including ethylene and 1,3-butadiene, are highly flammable and form explosive mixtures with air. This was illustrated tragically by a massive explosion and fire involving leaking ethylene that destroyed a 20 billion pound per year Phillips Petroleum Company plastics plant in Pasadena, Texas, on October 23, 1989, killing more than 20 workers.[2]

Acetylene (Figure 8.1) is an **alkyne,** a class of hydrocarbons characterized by carbon-carbon triple bonds consisting of 6 shared electrons. Highly flammable acetylene is used in large quantities as a chemical raw material and fuel for oxyacetylene torches. It forms dangerously explosive mixtures with air.

Addition Reactions

Alkenes and alkynes both undergo **addition reactions** in which pairs of atoms are added across unsaturated bonds as shown in the reaction of ethylene with hydrogen to give ethane,

$$
\begin{array}{ccc}
\text{H}\diagdown & \diagup\text{H} \\
\quad\text{C=C} & + \;\; \text{H–H} \;\; \rightarrow \;\;
\begin{array}{c}
\text{H H} \\
| \; | \\
\text{H–C–C–H} \\
| \; | \\
\text{H H}
\end{array} \\
\text{H}\diagup & \diagdown\text{H}
\end{array}
\qquad (8.2.3)
$$

or that of HCl gas with acetylene to give vinyl chloride:

$$
\begin{array}{ccc}
\text{H}\diagdown & \diagup\text{H} \\
\quad\text{C=C} & + \;\; \text{H–H} \;\; \rightarrow \;\;
\begin{array}{c}
\text{H H} \\
| \; | \\
\text{H–C–C–H} \\
| \; | \\
\text{H H}
\end{array} \\
\text{H}\diagup & \diagdown\text{H}
\end{array}
\qquad (8.2.4)
$$

This kind of reaction, which is not possible with alkanes, adds to the chemical and metabolic versatility of compounds containing unsaturated bonds and is a factor contributing to their generally higher toxicities.

Aromatic Hydrocarbons

Benzene and naphthalene shown in Figure 8.1 are **aromatic** or **aryl** hydrocarbons, a class of compounds discussed in Section 2.8. Aryl compounds have special characteristics of **aromaticity**, which include a low hydrogen:carbon atomic ratio; C–C bonds that are quite strong and of intermediate length between such bonds in alkanes and those in alkenes; tendency to undergo substitution reactions rather than the addition reactions characteristic of alkenes; and delocalization of π electrons over several carbon atoms resulting in resonance stabilization of the molecule.

Benzene (U019) is a volatile, colorless, highly flammable liquid that is consumed as a raw material for the manufacture of phenolic and polyester resins, polystyrene plastics, alkylbenzene surfactants, chlorobenzenes, insecticides, and dyes. It is hazardous both for its ignitability and toxicity (exposure to benzene causes blood abnormalities that may develop into leukemia). **Naphthalene** (U165) is the simplest member of a large number of multicyclic aromatic hydrocarbons having two or more fused rings. It is a volatile white crystalline solid with a characteristic odor and has been used to make mothballs. The most important of the many chemical derivatives made from naphthalene is phthalic anhydride, from which phthalate ester plasticizers are synthesized.

Polycyclic Aromatic Hydrocarbons

Benzo(a)pyrene,

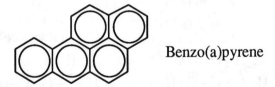

Benzo(a)pyrene

is the most studied of the polycyclic aromatic hydrocarbons (PAHs), which are characterized by condensed ring systems ("chicken wire" structures). These compounds are formed by the incomplete combustion of other hydrocarbons, a process that consumes hydrogen in preference to carbon. The carbon residue is left in the thermodynamically favored condensed aromatic ring system of the PAH compounds.

Because there are so many partial combustion and pyrolysis processes that favor production of PAHs, these compounds are encountered abundantly in the atmosphere, soil, and elsewhere in the environment from sources that include engine exhausts, wood stove smoke, cigarette smoke, and char-broiled food. Coal tars and petroleum residues such as road and roofing asphalt have high levels of PAHs. Some PAH compounds, including benzo(a)pyrene, are of toxicological concern because they are precursors to cancer-causing metabolites.

8.3. ORGANOOXYGEN COMPOUNDS

As shown in Figure 8.2, numerous hazardous organic compounds contain oxygen in various fuctional groups. The major types of these

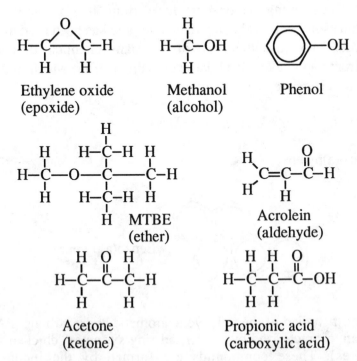

Figure 8.2. Examples of oxygen-containing hazardous waste compounds.

compounds are epoxides, alcohols, phenols, ethers, aldehydes, ketones, and carboxylic acids. The functional groups characteristic of these compounds are illustrated by the examples of oxygen-containing compounds shown in Figure 8.2.

Ethylene oxide (U115) is a moderately to highly toxic sweet-smelling, colorless, flammable, explosive gas used as a chemical intermediate, sterilant, and fumigant. It is a mutagen and a carcinogen to experimental animals. It is classified as hazardous for both its toxicity and ignitability. **Methanol** (U0154) is a clear, volatile, flammable liquid alcohol used for chemical synthesis, as a solvent, and as a fuel. It is being advocated strongly in some quarters as an alternative to gasoline that would result in significantly less photochemical smog formation than currently used gasoline formulations.[3] Ingestion of methanol can be fatal and blindness can result from sublethal doses. **Phenol** is a dangerously toxic aryl alcohol widely used for chemical synthesis and polymer manufacture. **Methyltertiarybutyl ether**, MTBE, is an ether that has become the octane booster of choice to replace tetraethyllead in gasoline. (Petro-Canada is building a huge MTBE plant in Edmonton, Alberta, to begin operation in 1991 using feedstock from natural gas liquids.[4] In April, 1990, Shell Oil Co. announced that it would be selling a new low-polluting, lower-vapor-pressure premium gasoline containing MTBE in place of volatile butane in Hartford, New York City, Philadelphia, Baltimore, Chicago, Milwaukee, Houston, Los Angeles, and San Diego.[5]) **Acrolein**, (P003) is an alkenic aldehyde and a volatile, flammable, highly reactive chemical. It forms explosive peroxides upon prolonged contact with O_2. An extreme lachrimator and strong irritant, acrolein is quite toxic by all routes of exposure. **Acetone** is the lightest of the ketones. **Propionic acid** is a typical organic carboxylic acid.

8.4. ORGANONITROGEN COMPOUNDS

Figure 8.3 shows examples of three major classes of the many kinds of compounds that contain N (amines, nitrosamines, and nitro compounds). Nitrogen occurs in many functional groups in organic compounds, some of which contain nitrogen in ring structures, or along with oxygen.

Methylamine is a colorless, highly flammable gas with a strong odor. It is a severe irritant affecting eyes, skin, and mucous mem-

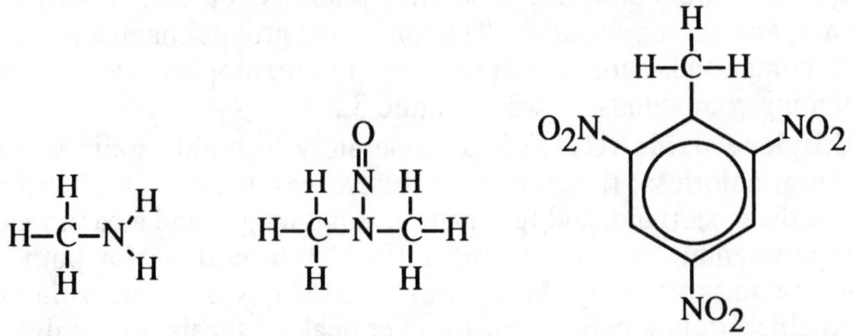

Methylamine   Dimethylnitrosamine   Trinitrotoluene (TNT)
                (N-nitrosodimethylamine)

Figure 8.3.   Examples of hazardous waste organic compounds containing nitrogen.

branes. Methylamine is the simplest of the **amine** compounds, which have the general formula,

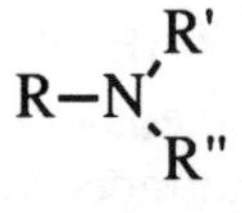

$$R-N\begin{matrix} R' \\ R'' \end{matrix}$$

where at least one of the R's is a hydrocarbon group.

Dimethylnitrosamine is an N-nitroso compound, all of which contain the N-N=O functional group. It was once widely used as an industrial solvent, but was observed to cause liver damage and jaundice in exposed workers. Subsequently numerous other N-nitroso compounds, many produced as byproducts of industrial operations and food and alcoholic beverage processing, were found to be carcinogenic.[5]

Solid **trinitrotoluene** (TNT) has been widely used as a military explosive. TNT is moderately to very toxic and has caused toxic hepatitis or aplastic anemia in exposed individuals, a few of whom have died from its toxic effects. It belongs to the general class of nitro compounds characterized by the presence of $-NO_2$ groups bonded to a hydrocarbon structure.

Some organonitrogen compounds are chelating agents that bind strongly to metal ions and play a role in the solubilization and transport of heavy metal wastes. Prominent among these are salts of the aminocarboxylic acids which, in the acid form, have $-CH_2CO_2H$

groups bonded to nitrogen atoms. A prominent example of such a compound is the monohydrate of trisodium nitrilotriacetate (NTA):

$$
\begin{array}{c}
\text{H O} \\
\text{C–C–O}^-\text{Na}^+ \\
\end{array}
$$

Na$^+$-O–C–C——N •H$_2$O

This compound is widely used in Canada as a substitute for detergent phosphates to bind to calcium ion and make the detergent solution basic. NTA is used in metal plating formulations. It is highly water soluble and quickly eliminated with urine when ingested. It has a low acute toxicity and no chronic effects have been shown for plausible doses. However, concern does exist over its interaction with heavy metals in waste treatment processes and in the environment.

8.5. ORGANOHALIDE COMPOUNDS

Organohalides exhibit a wide range of physical and chemical properties. These compounds consist of halogen-substituted hydrocarbon molecules, each of which contains at least one atom of F, Cl, Br, or I. They may be saturated (**alkyl halides**), unsaturated (**alkenyl halides**), or aromatic (**aryl halides**). The most widely manufactured organohalide compounds are chlorinated hydrocarbons, many of which are listed as hazardous substances and hazardous wastes.[6]

Alkyl Halides

Substitution of halogen atoms for one or more hydrogen atoms on alkanes gives **alkyl halides**, example structural formulas of which are given in Figure 8.4. Most of the commercially important alkyl halides are derivatives of alkanes of low molecular mass. A brief discussion of the uses of the compounds listed in Figure 8.4 is given here to provide an idea of the versatility of the alkyl halides.

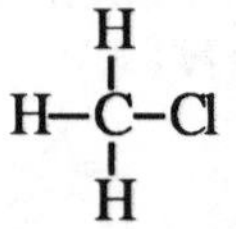

H
|
H–C–Cl
|
H

Chloromethane
(fp -98°C, bp -24°C)

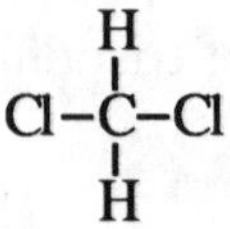

H
|
Cl–C–Cl
|
H

Dichloromethane
(methylene chloride,
fp -97°C, bp 40°C)

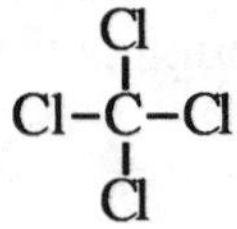

Cl
|
Cl–C–Cl
|
Cl

Carbon tetrachloride
(fp -23°C, bp 77°C)

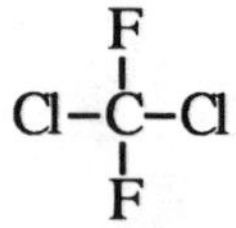

F
|
Cl–C–Cl
|
F

Dichlorodifluoro-
methane ("Freon-12,"
fp -158°C, bp -29°C)

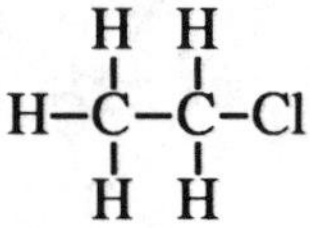

H H
| |
H–C–C–Cl
| |
H H

Chloroethane (ethyl-
ene chloride, fp
-139°C, bp 12°C)

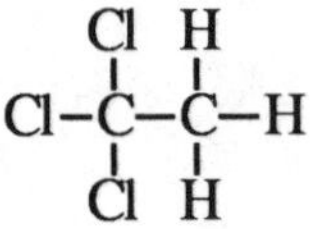

Cl H
| |
Cl–C–C–H
| |
Cl H

1,1,1–Trichloroethane
(methyl chloroform,
fp -33°C, bp 74°C)

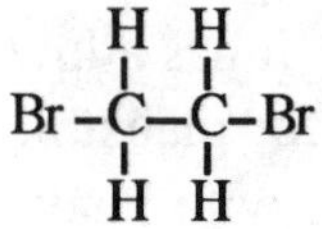

H H
| |
Br–C–C–Br
| |
H H

1,2–Dibromoethane (ethylene dibromide,
fp 9.3°C, bp 131°C)

Figure 8.4. Some typical low-molecular-mass alkyl halides.

Volatile **chloromethane** (methyl chloride) is consumed in the manufacture of silicones. **Dichloromethane** is a volatile liquid with excellent solvent properties for nonpolar organic solutes. It has been used as a solvent for the decaffeination of coffee, in paint strippers, as a blowing agent in urethane polymer manufacture, and to depress vapor pressure in aerosol formulations. Once commonly sold as a solvent and stain remover, highly toxic **carbon tetrachloride** is now largely restricted to uses as a chemical intermediate under controlled conditions, primarily to manufacture chlorofluorocarbon refrigerant fluid compounds, which are also discussed in this section. **Chloro-ethane** is an intermediate in the manufacture of tetraethyllead and is an ethylating agent in chemical synthesis. One of the more common industrial chlorinated solvents is **1,1,1-trichloroethane**. Insecticidal **1,2-dibromoethane** has been consumed in large quantities as a lead scavenger in leaded gasoline and to fumigate soil, grain, and fruit (Fumigation with this compound has been discontinued because of toxicological concerns). An effective solvent for resins, gums, and waxes, it serves as a chemical intermediate in the syntheses of some pharmaceutical compounds and dyes.

Alkenyl Halides

Viewed as hydrocarbon-substituted derivatives of alkenes, the **alkenyl** or **olefinic organohalides** contain at least one halogen atom and at least one carbon–carbon double bond. The most significant of these are the lighter chlorinated compounds, such as those illustrated in Figure 8.5.

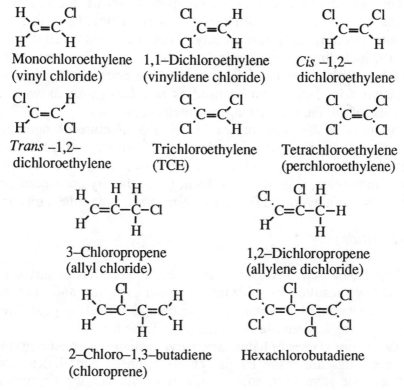

Monochloroethylene (vinyl chloride)

1,1–Dichloroethylene (vinylidene chloride)

Cis –1,2–dichloroethylene

Trans –1,2–dichloroethylene

Trichloroethylene (TCE)

Tetrachloroethylene (perchloroethylene)

3–Chloropropene (allyl chloride)

1,2–Dichloropropene (allylene dichloride)

2–Chloro–1,3–butadiene (chloroprene)

Hexachlorobutadiene

Figure 8.5. The more common low-molecular-mass alkenyl chlorides.

Vinyl chloride is consumed in large quantities as a raw material to manufacture pipe, hose, wrapping, and other products fabricated from polyvinylchloride plastic. This highly flammable, volatile, sweet-smelling gas is a known human carcinogen.

As shown in Figure 8.5, there are three possible dichloroethylene compounds, all clear, colorless liquids. Vinylidene chloride forms a copolymer with vinyl chloride used in some kinds of coating materials. The geometrically isomeric 1,2-dichloroethylenes are used as organic synthesis intermediates and as solvents.

Trichloroethylene is a clear, colorless, nonflammable, volatile liquid. It is an excellent degreasing and drycleaning solvent and has been used as a household solvent and for food extraction (for example, in decaffeination of coffee). Colorless, non-flammable liquid **tetrachloroethylene** has properties and uses similar to those of trichloroethylene.

The two chlorinated propene compounds shown are colorless liquids with pungent, irritating odors. **Allyl chloride** is an intermediate in the manufacture of allyl alcohol and other allyl compounds, including pharmaceuticals, insecticides, and thermosetting varnish and plastic resins. **Dichloropropene** compounds, of which one isomer is shown, can be used as soil fumigants, as well as solvents for oil, fat, drycleaning, and metal degreasing.

Produced in large quantities for the manufacture of neoprene rubber, **chloroprene** is a colorless liquid with an ethereal odor. **Hexachlorobutadiene**, a colorless liquid with an odor somewhat like that of turpentine, is used as a solvent for higher hydrocarbons and elastomers, as a hydraulic fluid, in transformers, and for heat transfer.

Aryl Halides

Aryl halide derivatives of benzene and toluene have many uses, which have resulted in substantial human exposure and environmental contamination. Some of these compounds, their properties, and major applications are summarized in Table 8.1.

Two major classes of halogenated aryl compounds containing two benzene rings are made by the chlorination of naphthalene and biphenyl and have been sold as mixtures with varying degrees of chlorine content. Examples of chlorinated naphthalenes, and polychlorinated biphenyls (PCBs discussed further in Section 8.8), are shown in Figure 8.6. The less highly chlorinated of these compounds are liquids and those with higher chlorine contents are solids. Because of their physical and chemical stabilities and other desirable qualities, these compounds have had many uses, including heat transfer fluids, hydraulic fluids, and dielectrics. Polybrominated biphenyls (PBBs) have served as flame retardants. However, because chlorinated naphthalenes, PCBs, and PBBs are environmentally extremely persistent, their uses have been severely curtailed.

Table 8.1. Examples of Single-Ring Aryl Halides

Structural formula	Name	Properties
[benzene ring]—Cl	Monochlor-obenzene	Flammable liquid (fp -45°C, bp 132°C), solvent, heat transfer fluid, synthetic reagent
[benzene ring]—Cl with Cl	1,2-Dichlor-obenzene	Solvent for degreasing hides and wool, synthetic reagent for dye manufacture
Cl—[benzene ring]—Cl	1,4-Dichlor-obenzene	White sublimable solid, dye manufacture, germicide, moth repellant
Cl—[benzene ring]—Cl with Cl	1,2,4-Trichlor-obenzene	Liquid (fp 17°C, bp 213°C), solvent, lubricant, dielectric fluid, formerly used as a termiticide
Cl—[benzene ring]—Cl with Cl, Cl, Cl, Cl	Hexachlor-obenzene	High-melting-point solid, seed fungicide, wood preservative, intermediate for organic synthesis
[benzene ring]—Br	Bromobenzene	Liquid (fp -31°C, bp 156°C), solvent, motor oil additive, intermediate for organic synthesis
[benzene ring]—CH$_3$ with Cl	1-Chloro-2-methylbenzene	Intermediate for the synthesis of 1-chlorobenzo-trifluoride

Chemical Reactivities of Organohalides

Although the alkyl halides have generally low reactivities, they may pyrolyze in flames to produce HCl gas and other hazardous products. Alkenyl halides are more reactive. As shown by the example below, they may burn to produce highly toxic phosgene.

$$\begin{array}{c} Cl \quad\quad Cl \\ \diagdown C = C \diagup \\ H \diagup \quad\quad \diagdown Cl \end{array} + O_2 \longrightarrow HCl + \overset{\displaystyle O}{\underset{\displaystyle \|}{Cl-C-Cl}} + CO \quad\quad (8.5.1)$$

2-Chloronaphthalene

Polychlorinated
naphthalenes

Polychlorinated
biphenyls (PCBs)

Polybrominated
(PBBs)

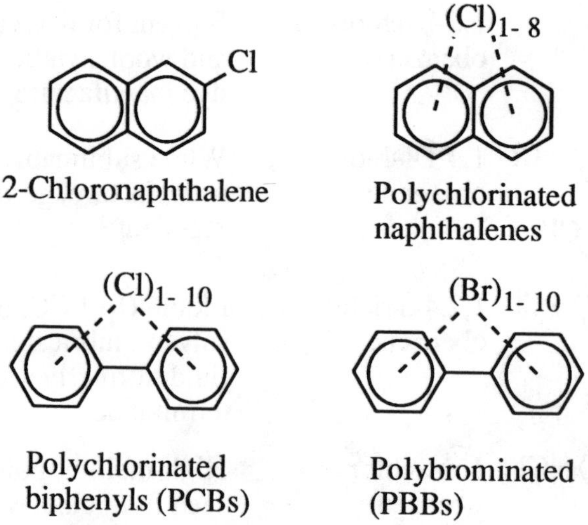

Figure 8.6. Halogenated naphthalenes and biphenyls.

Chlorofluorocarbons, Halons, and Hydrogen-Containing Chlorofluorocarbons

Chlorofluorocarbons (CFCs) are volatile 1- and 2-carbon compounds that contain Cl and F bonded to carbon. These compounds are notably stable and non-toxic. They have been widely used in recent decades in the fabrication of flexible and rigid foams and as fluids for refrigeration and air conditioning. The most widely manufactured of these compounds are CCl_3F (CFC-11), CCl_2F_2 (CFC-12), $C_2Cl_3F_3$ (CFC-113), $C_2Cl_2F_4$ (CFC-114), and C_2ClF_5 (CFC-115).

Halons are related compounds that contain bromine and are used in fire extinguisher systems. The major commercial halons are $CBrClF_2$ (Halon-1211), $CBrF_3$ (Halon-1301), and $C_2Br_2F_4$ (Halon-2402), where the sequence of numbers denotes the number of carbon, fluorine, chlorine, and bromine atoms, respectively, per molecule.

Halons are particularly effective fire extinguishing agents because of the way in which they stop combustion. Some fire supressants, such as carbon dioxide, act by depriving the flame of oxygen by a smothering effect, whereas water cools a burning substance to a temperature below which combustion is supported. Halons act by chain reactions (see Section 3.11) that destroy hydrogen atoms which sustain combustion. The basic sequence of reactions involved is outlined below:

$$CBrClF_2 + H \cdot \longrightarrow CClF_2 \cdot + HBr \qquad (8.5.2)$$

$$HBr + H \cdot \longrightarrow Br \cdot + H_2 \qquad (8.5.3)$$

Chain reaction

$$Br \cdot + H \cdot \longrightarrow HBr \qquad (8.5.4)$$

Halons are used in automatic fire extinguishing systems, such as those located in flammable solvent storage areas, and in specialty fire extinguishers, such as those on aircraft. As of 1989, there were no substitutes available for halons that had their same excellent performance characteristics.[7]

All of the chlorofluorocarbons and halons discussed above have been implicated in the halogen-atom-catalyzed destruction of atmospheric ozone. As a result of U. S. Environmental Protection Agency regulations imposed in accordance with the 1986 Montreal Protocol on Substances that Deplete the Ozone Layer, production of CFCs and halocarbons in the U. S. was curtailed starting in 1989.[8] The most likely substitutes for these halocarbons are hydrogen-containing chlorofluorocarbons (HCFCs) and hydrogen-containing fluorocarbons (HFCs). The substitute compounds most likely to be produced commercially first are CH_2FCF_3 (HFC-134a, a substitute for CFC-12 in automobile air conditioners and refrigeration equipment), $CHCl_2CF_3$ (HCFC-123, substitute for CFC-11 in plastic foam-blowing), CH_3CCl_2F (HCFC-141b, substitute for CFC-11 in plastic foam-blowing), $CHClF_2$ (HCFC-22, air conditioners and manufacture of plastic foam food containers). Because of the more readily broken H-C bonds that they contain, these compounds are more easily destroyed by atmospheric chemical reactions (particularly with hydroxyl radical, see Section 3.11) before they reach the stratosphere. Relative to a value of 1.0 for CFC-11, the ozone-depletion potentials of these substitutres are HFC-134a, 0; HCFC-123, 0.016; HCFC-141b, 0.081; and HCFC-22, 0.053. Concern has been expressed over toxicities and fire hazards of HCFCs, which are more reactive both chemically and biochemically than the CFCs and halons that they are designed to replace. The four leading substitutes, like the CFCs that they are designed to replace, have shown no evidence of causing skin or eye irritation, birth defects, or other short term toxic effects.[9]

Industrial concerns have undertaken a program to replace CFCs in manufacturing processes.[10] In August, 1989, The American Telephone and Telegraph Company, which consumes 3 million pounds of CFCs per year to manufacture circuit boards and electronic chips, stated that it will cut consumption in half by 1991 and totally by 1994. Other major companies that have announced similar intentions include Japan's Seiko Epson Company and Canada's Northern Telecom.

The Du Pont Company, which introduced chlorofluorocarbons in the 1930s and is the largest manufacturer of them, has announced that it intends to cease production shortly after the year 2000. The company has plans to manufacture HFC and HCFC substitutes during the early 1990s, including HFC-134a in a plant in Corpus Christi, Texas, HCFC-123 in a plant in Maitland, Ontario, and HCFC-141b in a plant in Montague, Michigan. ICI intends to produce HFC-134a in St. Gabriel, Louisiana, in the U.S. and in Runcorn, England.

Chlorinated Phenols

The chlorinated phenols, particularly **pentachlorophenol**,

Pentachlorophenol

and the trichlorophenol isomers are significant hazardous wastes. These compounds are biocides that are used to treat wood to prevent rot by fungi and to prevent termite infestation. They are toxic, causing liver malfunction and dermatitis; contaminant polychlorinated dibenzodioxins may be responsible for some of the observed effects.

Wood preservative chemicals such as pentachlorophenol may be encountered at hazardous waste sites in wastewaters and sludges. (Wood preservatives from wood treatment sites can be very troublesome organic hazardous wastes. One of the materials widely used for wood treatment is cresote. A fluid derived from coal coking, creosote is a mixture of almost 300 organic compounds, including a number of polycyclic aromatic hydrocarbons).

8.6. ORGANOSULFUR COMPOUNDS

Sulfur is chemically similar to, but more diverse than oxygen. Whereas, with the exception of peroxides, most chemically combined organic oxygen is in the -2 oxidation state, sulfur occurs in the -2, +4, and +6 oxidation states. Many organosulfur compounds are noted for their foul, "rotten egg" or garlic odors.

Thiols and Thioethers

Substitution of alkyl or aryl hydrocarbon groups such as phenyl and methyl for H on hydrogen sulfide, H_2S, leads to a number of different organosulfur **thiols** (mercaptans, R–SH) and **sulfides**, also called thioethers (R–S–R). Structural formulas of examples of these compounds are shown in Figure 8.7.

Methanethiol and other lighter alkyl thiols are fairly common air pollutants that have "ultragarlic" odors; both 1- and 2-butanethiol are associated with skunk odor. Gaseous methanethiol and volatile liquid ethanethiol are used as odorant leak-detecting additives for natural gas, propane, and butane; they are also employed as intermediates in pesticide synthesis. Although information about their toxicities to humans is lacking, methanethiol, ethanethiol, and 1-propanethiol should be considered dangerously toxic, especially by inhalation.

A toxic, irritating volatile liquid with a strong garlic odor, 2-propene-1-thiol (allyl mercaptan) is a typical alkenyl mercaptan. Alpha-toluenethiol (benzyl mercaptan, bp 195°C) is very toxic and is an experimental carcinogen. Benzenethiol (phenyl mercaptan), is the simplest of the aryl thiols. It is a toxic liquid with a severely "repulsive" odor.

Alkyl sulfides or thioethers contain the C-S-C functional group. The lightest of these compounds is dimethyl sulfide, a volatile liquid (bp 38°C) that is moderately toxic by ingestion. Cyclic sulfides contain the C-S-C group in a ring structure. The most common of these compounds is thiophene, a heat-stable liquid (bp 84°C) with a solvent action much like that of benzene, that is used in the manufacture of pharmaceuticals, dyes, and resins. Its saturated analog is tetrahydrothiophene, or thiophane.

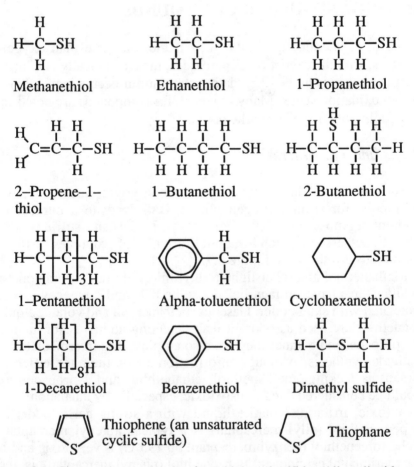

Figure 8.7. Common low-molecular-mass thiols and sulfides. All are liquids at room temperature, except for methanethiol, which boils at 5.9°C.

Nitrogen-Containing Organosulfur Compounds

Many important organosulfur compounds also contain nitrogen. One such compound is **thiourea**, the sulfur analog of urea. Its structural formula is shown in Figure 8.8 along with other thiourea compounds. Thiourea and **phenylthiourea** have been used as rodenticide. Commonly called ANTU, **1-naphthylthiourea** is an excellent rodenticide that is virtually tasteless and has a very high rodent:human toxicity ratio.

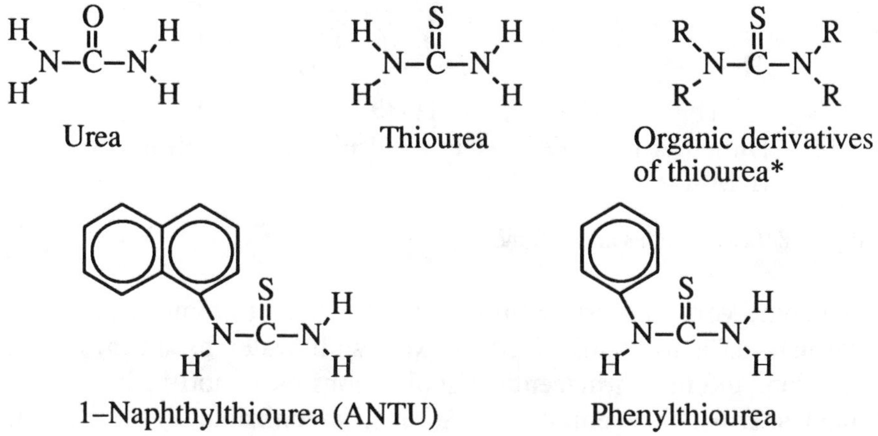

Urea   Thiourea   Organic derivatives
                                                   of thiourea*

1–Naphthylthiourea (ANTU)   Phenylthiourea

Figure 8.8. Structural formulas of urea, thiourea, and organic derivatives of thiourea.

* At least one R group is an alkyl, alkenyl, or aryl substituent.

Substitution of hydrocarbon groups such as the methyl group for H on thiocyanic acid (HSCN) yields organic **thiocyanates**. First used for insect control during the 1930s, some thiocyanates are regarded as the first synthetic organic insecticides. Volatile methyl, ethyl, and isopropyl thiocyanates kill insects upon contact and are effective fumigants for insect control.

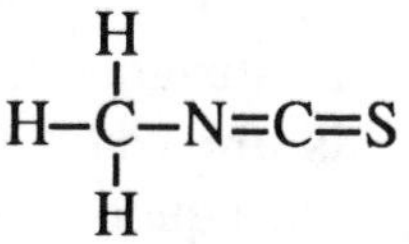

Methylisothiocyanate (structure below),

also known as methyl mustard oil, and its ethyl analog have been developed as military poisons. Both are powerful irritants to eyes, skin, and respiratory tract. When decomposed by heat, these compounds emit sulfur oxides and hydrogen cyanide. Other common compounds in this class are allyl and phenyl isothiocyanates.

Sulfoxides and Sulfones

Sulfoxides and **sulfones** (Figure 8.9) contain both sulfur and oxygen. **Dimethylsulfoxide** (DMSO) is a liquid with numerous uses

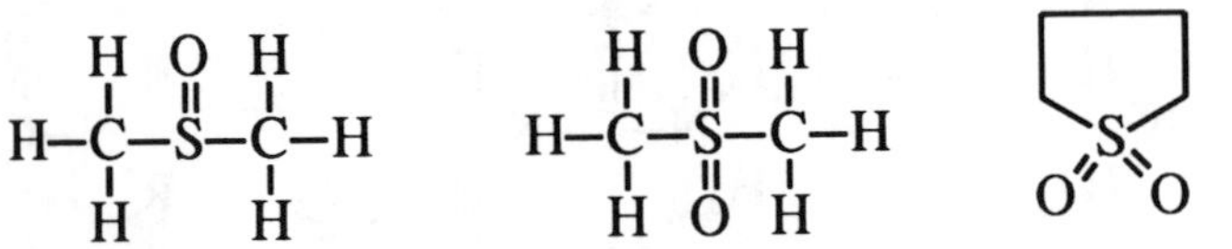

Dimethylsulfoxide (DMSO)   Dimethylsulfone   Sulfolane

Figure 8.9. Sulfoxides and sulfones.

and some very interesting properties. It is used to remove paint and varnish, as a hydraulic fluid, mixed with water as an antifreeze solution, and in pharmaceutical applications as an anti-inflammatory and bacteriostatic agent. A polar aprotic (no ionizable H) solvent with a relatively high dielectric constant, **sulfolane** dissolves both organic and inorganic solutes. It is the most widely produced sulfone because of its use in an industrial process called BTX processing in which it selectively extracts benzene, toluene, and xylene from aliphatic hydrocarbons; as the solvent in the Sulfinol process by which thiols and acidic compounds are removed from natural gas; as a solvent for polymerization reactions; and as a polymer plasticizer.

Sulfonic Acids, Salts, and Esters

Sulfonic acids and sulfonate salts contain the $-SO_3H$ and $-SO_3^-$ groups, respectively, attached to a hydrocarbon moiety. The structural formula of two sulfonic acids and of sodium 1-(p-sulfophenyl)decane, a biodegradable detergent surfactant, are shown in Figure 8.10. The common sulfonic acids are water-soluble strong acids that lose virtually all ionizable H^+ in aqueous solution. They are used commercially to hydrolyze fat and oil esters to fatty acids and glycerol.

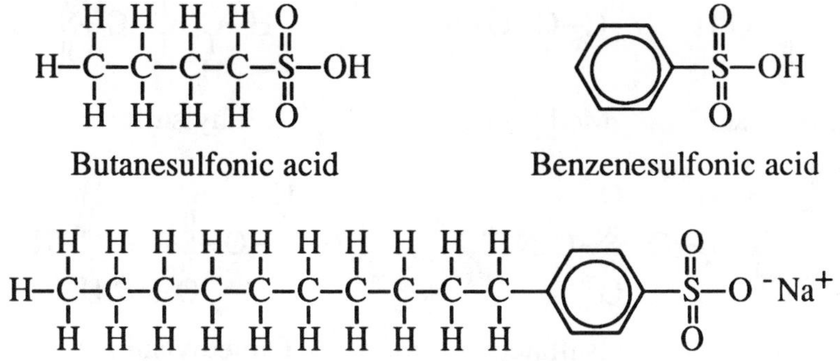

Butanesulfonic acid   Benzenesulfonic acid

Sodium 1–(p–sulfophenyl)decane

Figure 8.10. Sulfonic acids and a sulfonate salt.

Sulfonic acids form esters, such as methylmethane sulfonate:

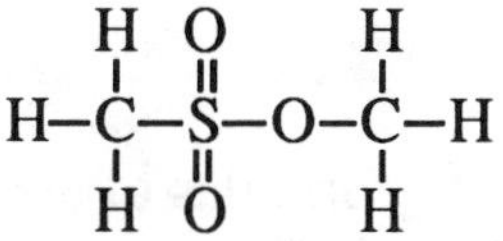

H–C–S–O–C–H   Methylmethane sulfonate

This compound is especially dangerous because it is a primary or direct-acting carcinogen that does not require metabolic conversion to cause cancer.[11]

Organic Esters of Sulfuric Acid

Replacement of 1 H on sulfuric acid, H_2SO_4, yields an acid ester and replacement of both yields an ester. Examples of these esters are shown in Figure 8.11.

Sulfuric acid esters are used as alkylating agents, which act to attach alkyl groups (such as methyl) to organic molecules, in the manufacture of agricultural chemicals, dyes, and drugs. **Methylsulfuric acid** and **ethylsulfuric acid** are oily water-soluble liquids that are strong irritants to skin, eyes, and mucous tissue. Liquid **dimethylsulfate** is colorless, odorless, and—like methylmethane sulfonate—a primary carcinogen.

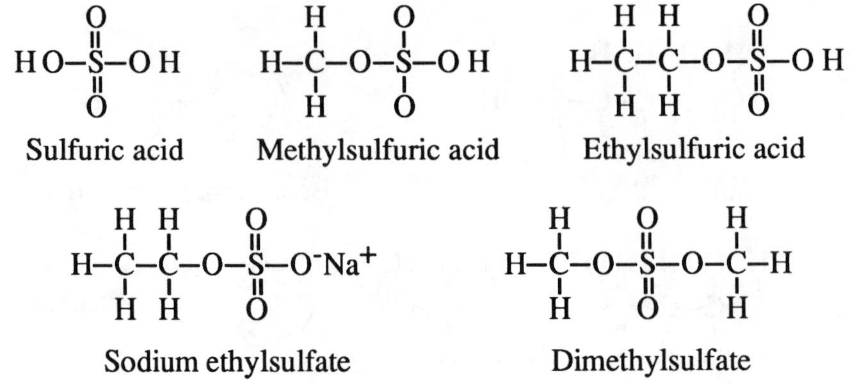

$$O$$
$$HO-\overset{\overset{\displaystyle O}{\|}}{\underset{\underset{\displaystyle O}{\|}}{S}}-OH$$

$$H-\overset{\overset{\displaystyle H}{|}}{\underset{\underset{\displaystyle H}{|}}{C}}-O-\overset{\overset{\displaystyle O}{\|}}{\underset{\underset{\displaystyle O}{\|}}{S}}-OH$$

$$H-\overset{\overset{\displaystyle H}{|}}{\underset{\underset{\displaystyle H}{|}}{C}}-\overset{\overset{\displaystyle H}{|}}{\underset{\underset{\displaystyle H}{|}}{C}}-O-\overset{\overset{\displaystyle O}{\|}}{\underset{\underset{\displaystyle O}{\|}}{S}}-OH$$

Sulfuric acid   Methylsulfuric acid   Ethylsulfuric acid

$$H-\overset{\overset{\displaystyle H}{|}}{\underset{\underset{\displaystyle H}{|}}{C}}-\overset{\overset{\displaystyle H}{|}}{\underset{\underset{\displaystyle H}{|}}{C}}-O-\overset{\overset{\displaystyle O}{\|}}{\underset{\underset{\displaystyle O}{\|}}{S}}-O^{-}Na^{+}$$

$$H-\overset{\overset{\displaystyle H}{|}}{\underset{\underset{\displaystyle H}{|}}{C}}-O-\overset{\overset{\displaystyle O}{\|}}{\underset{\underset{\displaystyle O}{\|}}{S}}-O-\overset{\overset{\displaystyle H}{|}}{\underset{\underset{\displaystyle H}{|}}{C}}-H$$

Sodium ethylsulfate   Dimethylsulfate

Figure 8.11. Sulfuric acid and organosulfate esters.

8.7. ORGANOPHOSPHORUS COMPOUNDS

Alkyl and Aryl Phosphines

As shown in Figure 8.12, the structural formulas of alkyl and aryl phosphine compounds may be derived by substituting organic groups for the H atoms in phosphine (PH_3), the hydride of phosphorus discussed as a toxic inorganic compound in Section 6.4. **Methylphosphine** is a colorless, reactive gas and **dimethylphosphine** is a colorless, reactive, volatile liquid (bp 25°C). **Trimethylphosphine** is a colorless, volatile, reactive, spontaneously ignitable liquid (bp 42°C). **Phenylphosphine** (phosphaniline) is a reactive, moderately flammable liquid (bp 16°C). All of these compounds should be regarded as having high toxicities. Crystalline, solid **triphenylphosphine** has a low reactivity and moderate toxicity when inhaled or ingested.

As shown by the reaction,

$$4C_3H_9P + 26O_2 \rightarrow 12O_2 + 18H_2O + P_4O_{10} \qquad (8.7.1)$$

combustion of aryl and alkyl phosphines produces P_4O_{10}, a corrosive irritant toxic substance, or droplets of corrosive orthophosphoric acid, H_3PO_4.

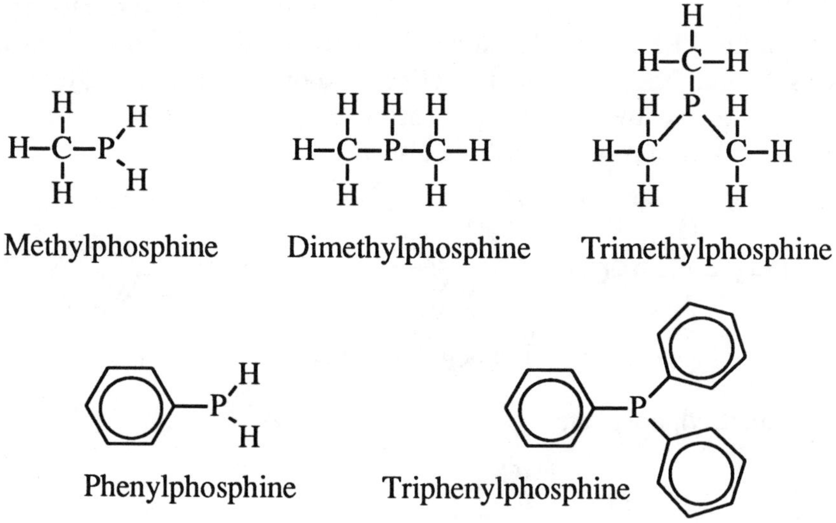

Methylphosphine   Dimethylphosphine   Trimethylphosphine

Phenylphosphine   Triphenylphosphine

Figure 8.12. Some of the more significant alkyl and aryl phosphines.

Phosphine Oxides and Sulfides

Structural formulas of a typical common phosphine oxide and a typical phosphine sulfide are the following:

$$C_2H_5-\overset{\overset{\displaystyle O}{\|}}{\underset{\underset{\displaystyle C_2H_5}{|}}{P}}-C_2H_5 \qquad C_4H_9-\overset{\overset{\displaystyle S}{\|}}{\underset{\underset{\displaystyle C_4H_9}{|}}{P}}-C_4H_9$$

Triethylphosphine oxide   Tributylphosphine oxide

The phosphine oxides and sulfides tend to be toxic. When burned, they give off dangerous phosphorus oxide fumes and, in the case of phosphine sulfides, sulfur oxides.

Organophosphate Esters

The structural formulas of three esters of orthophosphoric acid (H_3PO_4) and an ester of pyrophosphoric acid ($H_4P_2O_6$) are shown in Figure 8.13. Although **trimethylphosphate** and **triphenylphosphate** are considered to be only moderately toxic, **tri-*o*-cresyl-phosphate,**

TOCP, has a notorious record of poisonings. **Tetraethylpyrophosphate, TEPP,** was developed in Germany during World War II as a substitute for insecticidal nicotine. Although it is a very effective insecticide, its use in that application was of very short duration because it kills almost everything else, too.

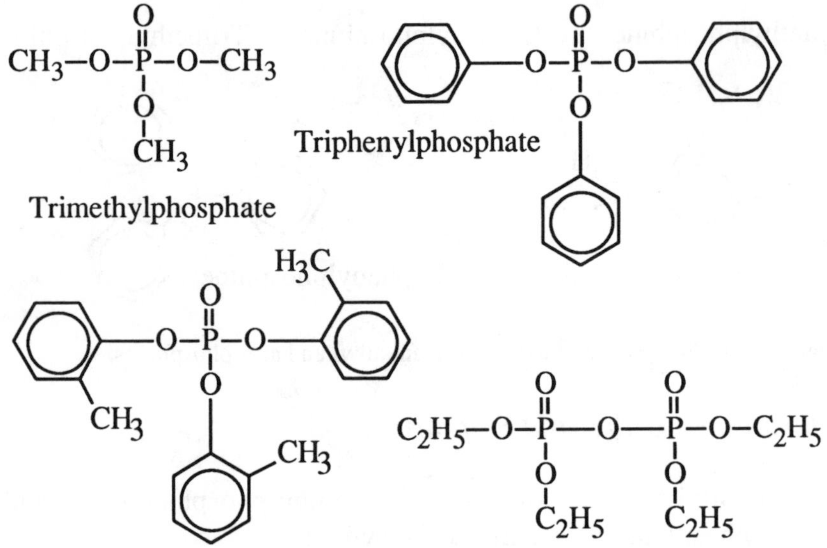

Tri-*o*-cresylphosphate, TOCP   Tetraethylpyrophosphate

Figure 8.13. Phosphate esters.

Phosphorothionate and Phosphorodithioate Esters

Thiophosphate esters are used as insecticidal acetylcholinesterase inhibitors. The general formulas of insecticidal **phosphorothionate** and **phosphorodithioate** esters are shown in Figure 8.14, where R is usually a methyl ($-CH_3$) or ethyl ($-C_2H_5$) group and Ar is a moiety of more complex structure, frequently aromatic. Phosphorothionate and phosphorodithioate esters contain the P=S (thiono) group, which increases their insect:mammal toxicity ratios and decreases their tendency to undergo non-enzymatic hydrolysis compared to their analogous compounds that contain the P=O functional group. The metabolic oxidative desulfuration conversion of P=S to P=O in organisms converts the phosphorothionate and phosphorodithioate esters to species that have insecticidal activity.

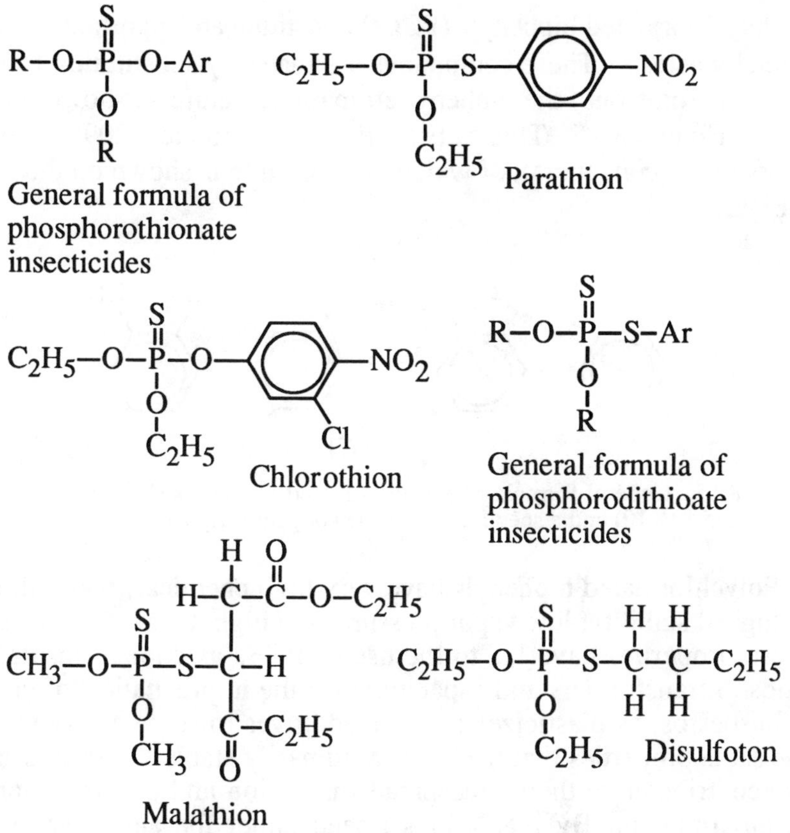

Figure 8.14. General formulas and specific examples of phosphorothionate and phosphorodithioate organophosphate insecticides.

Since the first organophosphate insecticides were developed in Germany during the 1930s and 1940s, many insecticidal organophosphate compounds have been synthesized. One of the earliest and most successful of these is **parathion**, *O,O*-diethyl-*O*-*p*-nitrophenyl-phosphorothionate. From a long-term environmental standpoint organophosphate insecticides are superior to the organohalide insecticides that they largely displaced because the organophosphates readily undergo biodegradation and do not bioaccumulate.

8.8. POLYCHLORINATED BIPHENYLS

Polychlorinated biphenyls (PCBs) constitute an important class of special wastes.[12] These compounds are made by substituting from 1 to 10 Cl atoms onto the biphenyl aromatic structure as shown on the left in Figure 8.15. This substitution can produce 209 different compounds (congeners), of which one example is shown on the right in Figure 8.15.

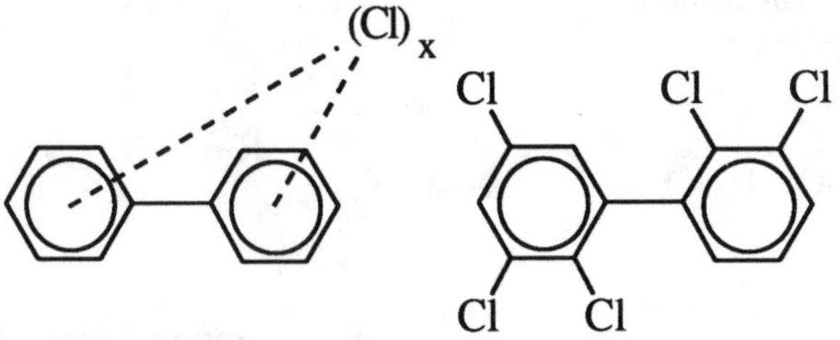

Figure 8.15. General formula of polychlorinated biphenyls (left, where X may range from 1 to 10) and a specific 5-chlorine congener (right).

Polychlorinated biphenyls have very high chemical, thermal, and biological stability; low vapor pressure; and high dielectric constants. These properties have led to the use of PCBs as coolant-insulation fluids in transformers and capacitors; for the impregnation of cotton and asbestos; as plasticizers; and as additives to some epoxy paints. The same properties that made extraordinarily stable PCBs so useful also contributed to their widespread dispersion and accumulation in the environment. By regulations issued under the authority of the Toxic Substances Control Act passed in 1976, the manufacture of PCBs was discontinued in the U.S. and their uses and disposal were strictly controlled.

Askarel

Askarel is the generic name of PCB-containing dielectric fluids in transformers. These fluids are 50-70 percent PCBs and may contain 30-50 percent trichlorobenzenes (TCBs). As of 1989 an estimated 100,000 askarel-containing transformers were still in use in the U.S.[13] Although the dielectric fluid may be replaced in these transformers enabling their continued use, PCBs tend to leach into the replacement fluid from the transformer core and other parts of the transformer over a several month period.

8.9. DIOXINS IN HAZARDOUS WASTES

Whereas PCBs discussed in the preceding section are derived from the two-ring hydrocarbon biphenyl, chlorinated dibenzo-*p*-dioxins are derived from dibenzo-*p*-dioxin shown in Figure 8.16. The chlorinated derivatives are commonly referred to as "dioxins." They have a high environmental and toxicological significance.[14]

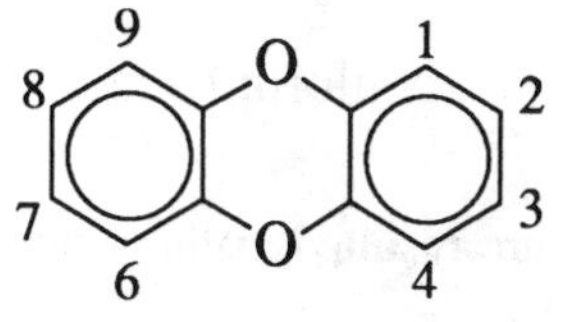

Dibenzo-*p*-dioxin 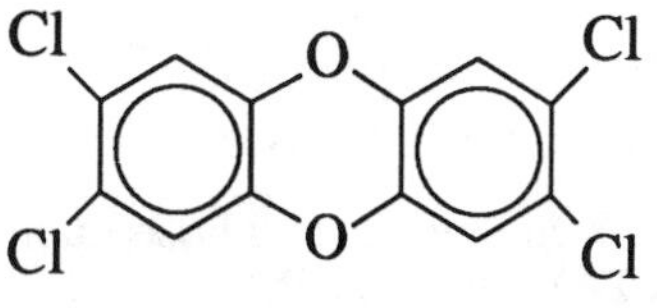   2,3,7,8-Tetrachlorodibenzo-
                                        p-dioxin

Figure 8.16.   Dibenzo-*p*-dioxin and 2,3,7,8-tetrachlorodibenzo-*p*-dioxin (TCDD), often called simply "dioxin." In the structure of dibenzo-*p*-dioxin each number refers to a numbered carbon atom to which an H atom is bound and the names of derivatives are based upon the carbon atoms where another group has been substituted for the H atoms, as is seen by the structure and name of 2,3,7,8-tetrachlorodibenzo-*p*-dioxin.

From 1 to 8 Cl atoms may be substituted for H atoms on dibenzo-*p*-dioxin, giving a total of 75 possible chlorinated derivatives. Of these, the most notable hazardous waste compound is 2,3,7,8-tetrachlorodibenzo-*p*-dioxin (TCDD), often referred to simply as "dioxin." This compound, which is one of the most toxic of all synthetic substances to some animals, was produced as a low-level contaminant in the manufacture of some aromatic, oxygen-containing organohalide compounds such as chlorophenoxy herbicides and hexachlorophene (Figure 8.17) manufactured by processes used until the 1960s.

The chlorophenoxy herbicides, such as 2,4,5-trichlorophenoxy-acetic acid (2,4,5-T) shown in Figure 8.17, were manufactured on a large scale for weed and brush control and as military defoliants. Fungicide and bactericidal hexachlorophene was once widely applied to crops in the production of vegetables and cotton and was used as an antibacterial agent in personal care products, an application that has been discontinued because of toxic effects and possible TCDD contamination.

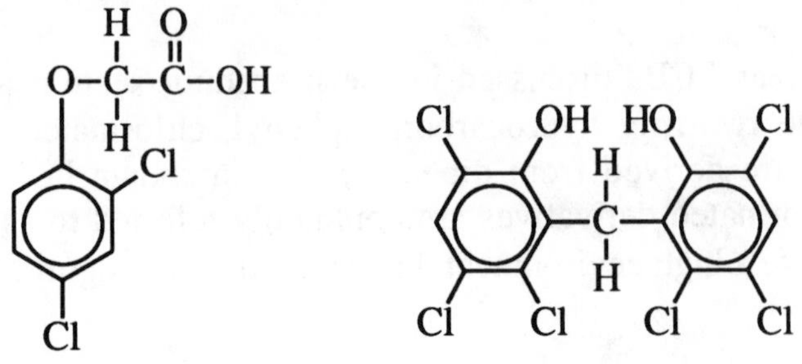

2,4–Dichlorophenoxy-
acetic acid (and esters)

Hexachlorophene

Figure 8.17. Two chemicals whose manufacture resulted in the production of byproduct TCDD contaminant.

TCDD has a very low vapor pressure of only 1.7×10^{-6} mm Hg at 25°C, a high melting point of 305°C, and a water solubility of only 0.2 μg/L. It is stable thermally up to about 700°C, has a high degree of chemical stability, and is poorly biodegradable. It is very toxic to some animals, with an LD_{50} of only about 0.6 μg/kg body mass in male guinea pigs. (The type and degree of its toxicity to humans is largely unknown; it is known to cause a severe skin condition called chloracne). Because of its properties, TCDD is a stable, persistent environmental pollutant and hazardous waste constituent of considerable concern. It has been the subject of several widely publicized environmental incidents including the contamination of the city of Times Beach, Missouri, by TCDD-containing waste oil in the early 1970s and release from a massive industrial accident at the Givaudan-La Roche Icmesa manufacturing plant near Seveso, Italy, in 1976.

LITERATURE CITED

1. Seinfeld, John H., "Urban Air Pollution: State of the Science," *Science* **243**, 745-752 (1989).
2. Suro, Roberto, "Many are Missing as Plastics Plant Explodes in Texas, *New York Times*, October 24, 1989, p. 8.
3. Stevenson, Richard W., "Facing Up to a Clean Air Pact," *New York Times*, March 3, 1989, p. 25.
4. "Petro-Canada in Octane Plan," *New York Times*, June 9, 1989, p. 37.

5. "Shell Offers New Premium Gasoline," *St. Louis Post-Dispatch*, April 12, 1990, p. 7B.
6. "EPA Expands Rules in Battle to Control Water Contamination," *Wall Street Journal*, March 7, 1990, p. 1.
7. Bradsher, Keith, "The Danger of a Firefighting Wonder," *New York Times*, Aug. 9, 1989, p. 25.
8. Zurer, Pamela S., "Producers, Users Grapple with Realities of CFC Phaseout," *Chemical and Engineering News*, July 24, 1989, pp. 7-13.
9. Zurer, Pamela S., "CFC Substitutes: Candidates Pass Early Toxicity Tests," *Chemical and Engineering News*, October 9, 1989, p. 4.
10. Shabecoff, Philip, "A.T. & T. Barring Chemicals Depleting Ozone Layer," *New York Times*, Aug. 2, 1989, p. 8.
11. Levi, Patricia E., "Toxic Action," Chapter 6 in *Modern Toxicology*, Ernest Hodgson and Patricia E. Levi, Eds., Elsevier, New York, 1987, pp. 133–184.
12. McCoy, Drew E., "PCB Wastes," Section 4.2 in *Standard Handbook of Hazardous Waste Treatment and Disposal*, Harry M. Freeman, Ed., McGraw Hill, New York, 1989, pp. 4.13-4.23.
13. Bishop, Jim, "Cleaning up Askarel," *Hazmat World*, June, 1989, p. 29.
14. Espositio, M. Pat, "Dioxin Wastes," Section 4.3 in *Standard Handbook of Hazardous Waste Treatment and Disposal*, Harry M. Freeman, Ed., McGraw Hill, New York, 1989, pp. 4.25-4.34.

9

Toxicology of Organic Hazardous Wastes

9.1. INTRODUCTION

Many organic compounds are encountered in hazardous wastes. The toxicities of these compounds to exposed humans and biota are of utmost concern and the subject of intense study. This chapter summarizes the toxicological aspects of organic compounds in hazardous wastes.

As with all substances, estimating probable human health effects from exposures to low levels of organic compounds can be an uncertain undertaking involving extrapolation from effects of relatively high doses on animals to probable human health effects of low doses. Some of the uncertainties of this approach may be seen from a detailed comparison of the carcinogenic risks from formaldehyde and benzene.[1] Formaldehyde is known to cause a minimal occurrence of nasopharynx cancer in rats exposed to 6-15 ppm of the compound in air, but not at 2 ppm. However, convincing evidence does not exist for formaldehyde-induced cancer in humans. In contrast, although animal studies of benzene carcinogenesis have been inconclusive, there is reasonable documentation of myelocytic or acute nonlymphocytic leukemia in humans resulting from the toxic effects of benzene upon bone marrow.

195

9.2.   HYDROCARBONS

Alkanes

Methane, ethane, n-butane, and isobutane (both C_4H_{10}) are regarded as **simple asphyxiants**; air containing high levels of simple asphyxiants does not contain sufficient oxygen to support respiration. Although simple asphyxiant gases are not known to have major systemic toxicological effects, they may have subtle effects that are hard to detect. A high concentration of asphyxiant propane (C_3H_8) gas also affects the central nervous system.

Inhalation of volatile liquid 5–8 carbon n-alkanes and branched-chain alkanes may cause central nervous system depression manifested by dizziness and loss of coordination. Workplace exposure to vapors of n-hexane, a widely used solvent and reaction medium, has caused multiple disorders of the nervous system, a condition called **polyneuropathy**.[2] Exposure to n-hexane results in loss of myelin (a fatty substance constituting a sheath around certain nerve fibers) and degeneration of axons (part of a nerve cell through which nerve impulses are transferred out of the cell). This has resulted in symptoms including muscle weakness and impaired sensory function of the hands and feet. The cyclic alkane cyclohexane (C_6H_{12}) is a weak anesthetic with effects similar to, but more pronounced than those of n-hexane. The most common toxicological occupational problem associated with the use of hydrocarbon liquids in the workplace is dermatitis caused by dissolution of the fat portions of the skin and characterized by inflamed, dry, scaly skin.

Although alkanes higher than C_8 (kerosene, jet fuel, diesel fuel, mineral oil, fuel oil) are not regarded as very toxic, their inhalation can cause dizziness, headache, and stupor. Coma and death have resulted from extreme exposures. A lung condition called aspiration pneumonia has resulted from inhalation of mists or aspiration of vomitus containing higher alkane liquids.

Alkenes and Alkynes

Ethylene,

$$\underset{H}{\overset{H}{\diagdown}}C=C\underset{H}{\overset{H}{\diagup}}$$

a colorless gas with a somewhat sweet odor, the most widely used organic chemical, consumed in large quantities as a chemical feedstock for the manufacture of polyethylene and other organic chemicals, acts as a simple asphyxiant and anesthetic to animals. Ethylene is phytotoxic (toxic to plants). The toxicological properties of propylene (C_3H_6) are very similar to those of ethylene. Colorless, odorless gaseous 1,3-butadiene is an irritant to eyes and respiratory system mucous membranes; at higher levels it can cause unconsciousness and even death. Liquid produced by release of pressurized gaseous 1,3-butadiene can cause frostbite-like burns on exposed flesh.

Acetylene, H-C≡C-H, is a colorless gas with an odor resembling garlic. It acts as an asphyxiant and narcotic, causing headache, dizziness, and gastric disturbances. Some of these effects may be due to the presence of impurities in the commercial product.

Benzene and Aromatic Hydrocarbons

Benzene, toluene (Figure 2.10), and the xylenes (benzene rings with two -CH_3 groups attached) are volatile liquids widely used in chemical synthesis, as solvents, and in unleaded gasoline formulations. Their toxicological characteristics are discussed here.

Benzene

Inhaled benzene is readily absorbed by blood, from which it is strongly taken up by fatty tissues. For the non-metabolized compound, the process is reversible and benzene is excreted through the lungs. As shown in Figure 9.1, benzene undergoes a Phase I oxidation reaction (see Section 4.8) in the liver, producing phenol. The phenol is converted by a Phase II conjugation reaction to water-soluble glucuronide or sulfate, both of which are readily eliminated through the kidneys. The benzene epoxide intermediate in the oxidative metabolism of benzene is probably responsible for the unique toxicity of benzene, which involves damage to bone marrow.

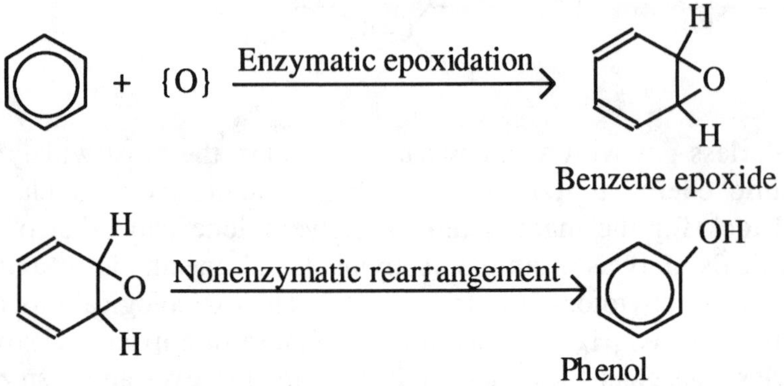

Figure 9.1. Conversion of benzene to phenol in the body.

Benzene is a skin irritant, and progressively higher local exposures can cause skin redness (erythema), burning sensations, fluid accumulation (edema) and blistering. Inhalation of air containing about 7 g/m^3 of benzene causes acute poisoning within an hour, because of a narcotic effect upon the central nervous system manifested progressively by excitation, depression, respiratory system failure, and death. Inhalation of air containing more than about 60 g/m^3 of benzene can be fatal within a few minutes.

Long-term exposures to lower levels of benzene cause nonspecific symptoms, including fatigue, headache, and appetite loss. Chronic benzene poisoning causes blood abnormalities, including a lowered white cell count, an abnormal increase in blood lymphocytes (colorless corpuscles introduced to the blood from the lymph glands), anemia, a decrease in the number of blood platelets required for clotting (thrombocytopenia), and damage to bone marrow. It is thought that preleukemia, leukemia, or cancer may result.

Toluene

Toluene, a colorless liquid boiling at 101.4°C, is classified as moderately toxic through inhalation or ingestion; it has a low toxicity by dermal exposure. Toluene can be tolerated without noticeable ill effects in ambient air up to 200 ppm. Exposure to 500 ppm may cause headache, nausea, lassitude, and impaired coordination without detectable physiological effects. Massive exposure to toluene has a narcotic effect, which can lead to coma. Because it possesses an aliphatic side-chain that can be oxidized enzymatically leading to products that are readily excreted from the body (see the metabolic reaction scheme in Figure 9.2), toluene is much less toxic than benzene.

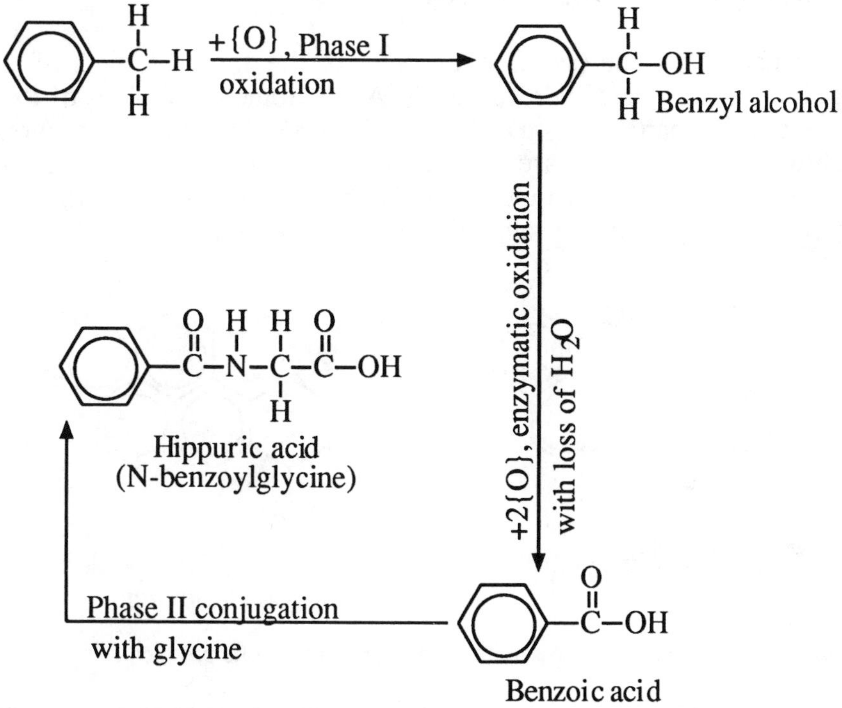

Figure 9.2. Metabolic oxidation of toluene with conjugation to hippuric acid, which is excreted with urine.

Naphthalene

Naphthalene — a volatile white crystalline solid with a characteristic odor once used to make mothballs — and its alkyl derivatives are important industrial chemicals. As is the case with benzene, naphthalene undergoes a Phase I oxidation reaction that places an epoxide group on the aromatic ring. This process is followed by Phase II conjugation reactions to yield products that can be eliminated from the body.

Exposure to naphthalene can cause anemia and marked reductions in red cell count, hemoglobin, and hematocrit in genetically susceptible individuals. Naphthalene causes skin irritation or severe dermatitis in sensitized individuals. Headaches, confusion, and vomiting may result from inhalation or ingestion of naphthalene. Death from kidney failure occurs in severe instances of poisoning.

Polycyclic Aromatic Hydrocarbons

Benzo(a)pyrene (see Section 8.2) is the most studied of the polycyclic aromatic hydrocarbons (PAHs). Some metabolites of PAH compounds, particularly the 7,8-diol-9,10 epoxide of benzo(a)pyrene shown in Figure 9.3 are known to cause cancer. There are two stereoisomers of this metabolite, both of which are known to be potent mutagens and presumably can cause cancer.

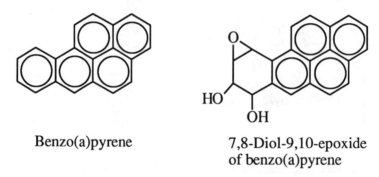

Benzo(a)pyrene

7,8-Diol-9,10-epoxide
of benzo(a)pyrene

Figure 9.3. Benzo(a)pyrene and its carcinogenic metabolic product.

9.3. OXYGEN-CONTAINING ORGANIC COMPOUNDS

Oxides

Hydrocarbon **oxides**, which are characterized by an **epoxide** functional group consisting of an oxygen atom bridging between two adjacent C atoms as shown for ethylene oxide in Figure 8.2, are significant for both their uses and their toxic effects. Ethylene oxide, a gaseous colorless, sweet-smelling, flammable, explosive gas used as a chemical intermediate, sterilant, and fumigant, has a moderate to high toxicity, is a mutagen, and is carcinogenic to experimental animals. Inhalation of relatively low levels of ethylene oxide results in respiratory tract irritation, headache, drowsiness, and dyspnea, whereas exposure to higher levels causes cyanosis, pulmonary edema, kidney damage, peripheral nerve damage, and even death.

Propylene oxide,

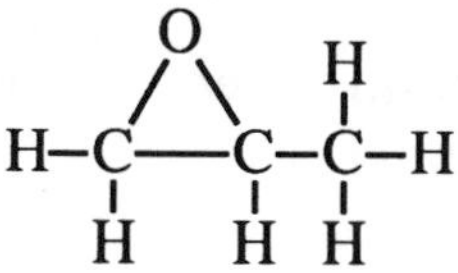

$$\begin{array}{ccc} & O & H \\ & /\backslash & | \\ H-C & C-C-H \\ | & | & | \\ H & H & H \end{array}\quad \text{Propylene oxide}$$

is a colorless, reactive, volatile liquid (bp 34°C) with uses similar to those of ethylene oxide and similar, though less severe, toxic effects. The toxicity of 1,2,3,4-butadiene epoxide, the oxidation product of 1,3-butadiene, is notable in that it is a direct–acting (primary) carcinogen.

Alcohols

As shown by the structural formulas in Figure 9.4, **alcohols** are oxygenated compounds in which the hydroxyl functional group is attached to an aliphatic or olefinic hydrocarbon skeleton. Human exposure to the three light alcohols shown in Figure 9.4 is common because they are widely used industrially and in consumer products.

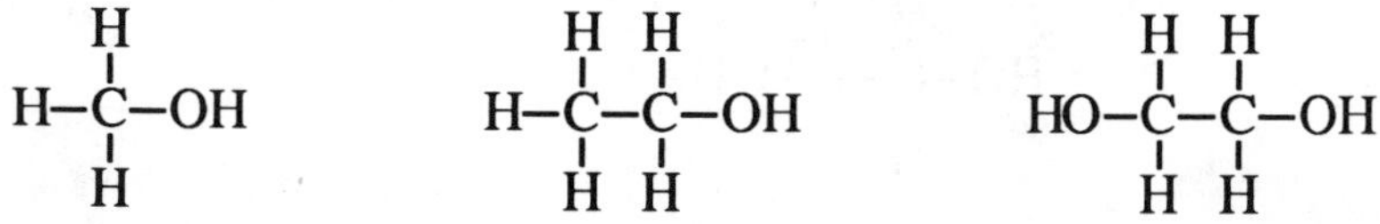

$$\begin{array}{c} H \\ | \\ H-C-OH \\ | \\ H \end{array}\qquad \begin{array}{c} H\ \ H \\ |\ \ | \\ H-C-C-OH \\ |\ \ | \\ H\ \ H \end{array}\qquad \begin{array}{c} H\ \ H \\ |\ \ | \\ HO-C-C-OH \\ |\ \ | \\ H\ \ H \end{array}$$

Methanol   Ethanol   Ethylene glycol

Figure 9.4. Three lighter alcohols with particular toxicological significance.

Methanol has caused many fatalities when ingested accidentally or consumed as a substitute for beverage ethanol. Metabolically, methanol is oxidized to formaldehyde and formic acid, toxicants discussed later in this section. In addition to causing acidosis, these products affect the central nervous system and the optic nerve. Acute exposure to lethal doses causes an initially mild inebriation, followed in about 10–20 hours by unconsciousness, cardiac depression, and death. Sublethal exposures can cause blindness from deterioration of the optic nerve and retinal ganglion cells. Inhalation of methanol fumes may result in chronic, low level exposure.

Ethanol has a range of acute effects resulting from central nervous system depression. These effects are the following:

- At 0.05% blood ethanol–mild effects such as decreased inhibitions and slowed reaction times
- At 0.15–0.3% blood ethanol–intoxication
- At 0.3–0.5% blood ethanol–stupor
- More than 0.5% blood ethanol–coma, death

Ethanol is usually ingested through the gastrointestinal tract. However, it can be absorbed by the alveoli of the lungs and symptoms of intoxication can be observed from inhalation of air containing more than 1000 ppm ethanol. Ethanol is oxidized metabolically more rapidly than methanol, first to acetaldehyde (discussed later in this section), then to CO_2.

Despite its widespread use in automobile cooling systems, exposure to ethylene glycol is limited by its low vapor pressure. However, inhalation of droplets of ethylene glycol can be very dangerous. In the body, ethylene glycol initially stimulates the central nervous system, then depresses it. Glycolic acid,

$$\underset{\underset{H}{|}}{\overset{\overset{H}{|}}{HO-C}}-\overset{\overset{O}{\|}}{C}-OH \quad \text{Glycolic acid}$$

formed as an intermediate metabolite in the metabolism of ethylene glycol, may cause acedemia. Kidney damage occurs in later stages of ethylene glycol poisoning because of the deposition of insoluble solid calcium oxalate, CaC_2O_4, precipitated by the reaction of calcium ion with oxalic acid produced from the metabolic oxidation of ethylene glycol; deposits of solid calcium oxalate may also be formed in liver and brain tissue of victims of ethylene glycol poisoning.

Higher alcohols containing three or more carbon atoms are widely used for a variety of purposes, including solvents and chemical intermediates, which has led to significant exposure to these compounds. For example, 1-butanol (butyl alcohol or *n*–butanol, formula, $CH_3(CH_2)_2CH_2OH$) is an irritant, but its toxicity is limited by its low vapor pressure. Unsaturated (olefinic) allyl alcohol, $CH_2=CHCH_2OH$, has a pungent odor and is strongly irritating to eyes, mouth, and lungs.

The 8-carbon alcohol, ***n*-octanol**, $CH_3(CH_2)_6CH_2OH$, is used in environmental toxicology studies for the measurement of the **octanol-water partition coefficient**. This parameter is taken as an approximate estimate of the tendency of organic toxicants to partition from water to lipids, an indication of of the toxicant's abilities to cross cell membranes and cause toxic effects.

Phenols

Figure 9.5 shows some of the more important phenolic compounds, aryl analogs of alcohols which have properties much different from those of the aliphatic and olefinic alcohols. Nitro groups ($-NO_2$) and halogen atoms (particularly Cl) bonded to the aromatic rings strongly affect the chemical and toxicological behavior of phenolic compounds.

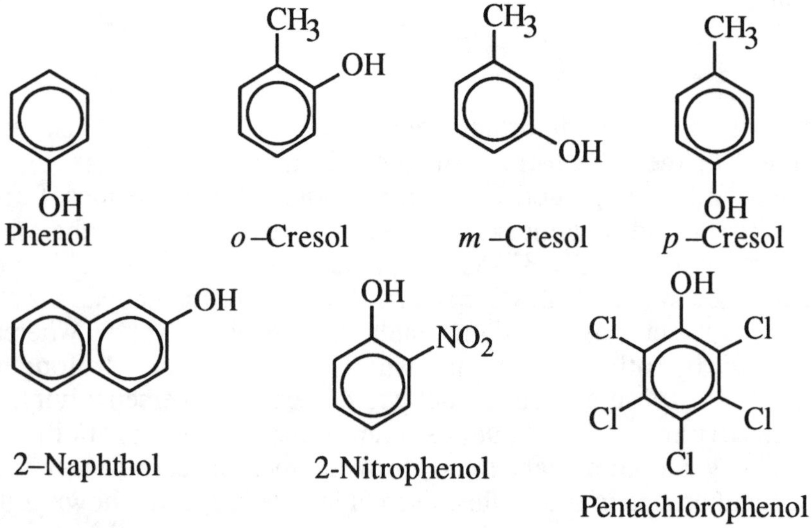

Figure 9.5. Some phenols and phenolic compounds.

The phenols have generally similar toxicological effects. Although it was the original antiseptic used on wounds and in surgery, starting with the work of Lord Lister in 1885, phenol is a protoplasmic poison that damages all kinds of cells and is alleged to have caused "an astonishing number of poisonings" since it came into general use.[3] The acute toxicological effects of phenol are predominantly upon the central nervous system and death can occur as soon as one-half hour after exposure. Acute poisoning by phenol can cause severe gastrointestinal disturbances, kidney malfunction, circulatory system failure, lung edema, and convulsions. Fatal doses of phenol may be absorbed through the skin. Key organs damaged by chronic exposure to phenol include the spleen, pancreas, and kidneys.

Aldehydes and Ketones

Formaldehyde

Aldehydes and ketones are compounds that contain the carbonyl (C=O) group. The simplest of the carbonyl compounds is **formaldehyde** (U122),

$$\underset{H \quad \quad H}{\overset{\overset{\displaystyle O}{\|}}{C}} \qquad \text{Formaldehyde}$$

which is uniquely important because of its widespread use and toxicity. In the pure form formaldehyde is a colorless gas with a pungent, suffocating odor. **Formalin**, employed in antiseptics, fumigants, tissue and biological specimen preservatives, and embalming fluid, is marketed as a 37–50% aqueous solution of formaldehyde containing some methanol. Exposure to inhaled formaldehyde via the respiratory tract is usually to molecular formaldehyde vapor, whereas exposure by other routes is usually to formalin. Prolonged, continuous exposure to formaldehyde can cause hypersensitivity. A severe irritant to the mucous membrane linings of both the respiratory and alimentary tracts, formaldehyde reacts strongly with functional groups in molecules. Formaldehyde has been shown to be a lung carcinogen in experimental animals. The toxicity of formaldehyde is largely due to its metabolic oxidation product, formic acid (see below).

Humans may be exposed to formaldehyde in the manufacture and use of phenol, urea, and melamine resin plastics and from formaldehyde-containing adhesives in pressed wood products, such as particle board, used in especially large quantities in mobile home construction.[4] However, significantly improved manufacturing processes have greatly reduced formaldehyde emissions from these synthetic building materials.

Some Significant Aldehydes and Ketones

The structures of some important aldehydes and ketones are shown in Figure 9.6.

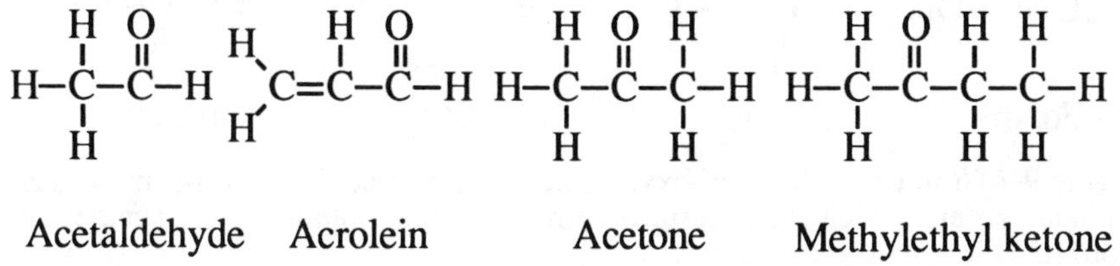

Acetaldehyde   Acrolein   Acetone   Methylethyl ketone

Figure 9.6. Commercially and toxicologically significant aldehydes and ketones.

The lower aldehydes are relatively water soluble and intensely irritating. These compounds attack exposed moist tissue, particularly the eyes and mucous membranes of the upper respiratory tract. (Some of the irritating properties of photochemical smog, Section 3.11, are due to the presence of aldehydes.) However, aldehydes that are relatively less soluble can penetrate further into the respiratory tract and affect the lungs. Of the two aldehydes shown in Figure 9.6, colorless, liquid acetaldehyde is relatively less toxic and acts as an irritant and systemically as a narcotic to the central nervous system. Toxic exposure to acrolein, a colorless to light yellow liquid alkenic aldehyde, may occur by all routes of contact and ingestion. Extremely irritating, lachrimating acrolein vapor has a choking odor and inhalation of it can cause severe damage to respiratory tract membranes. Tissue exposed to acrolein may undergo severe necrosis, and direct contact with the eye can be especially hazardous.

The ketones shown in Figure 9.6 are relatively less toxic than the aldehydes. Pleasant smelling acetone can act as a narcotic and causes dermatitis by dissolving fats from skin. Not many toxic effects have been attributed to methylethyl ketone exposure. It is suspected of having caused neuropathic disorders in shoe factory workers.

Carboxylic Acids

Carboxylic acids, oxidation products of aldehydes, contain the –C(O)OH functional group bound to an aliphatic, olefinic, or aromatic hydrocarbon moiety as shown by the examples in Figure 9.7. Simple carboxylic acids are common natural products. Some of the higher carboxylic acids are constituents of oil, fat, and wax esters and can be prepared by hydrolysis of these esters.

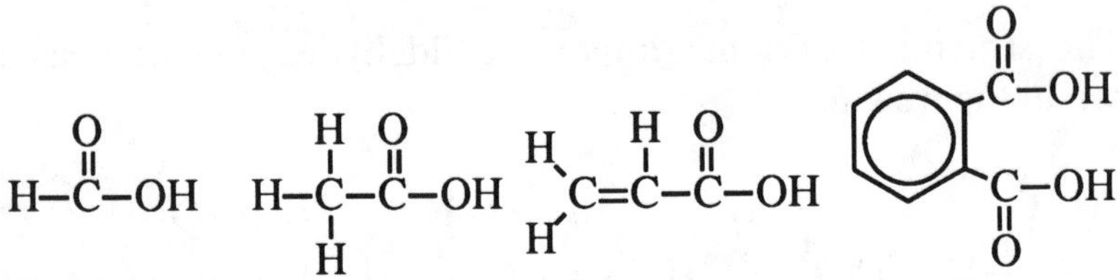

Figure 9.7. Some common carboxylic acids. Phthalic acid is significant because phthalate esters, which are widespread environmental pollutants, are synthesized from it.

Formic acid is a relatively strong acid; it is corrosive to tissue, much like strong mineral acids. In Europe decalcifier formulations for removing mineral scale that contain about 75% formic acid are sold and children ingesting these solutions have suffered corrosive lesions to mouth and esophageal tissue. Although acetic acid as a 4–6% solution in vinegar is an ingredient of many foods, pure acetic acid (glacial acetic acid) is extremely corrosive to tissue that it contacts. Ingestion of, or skin contact with acrylic acid can cause severe damage to tissues.

Ethers

The common ethers (Figure 9.8) have relatively low toxicities because of the low reactivity of the C–O–C functional group which has very strong carbon-oxygen bonds. Exposure to volatile diethyl ether is usually by inhalation and about 80% of this compound that gets into the body is eliminated unmetabolized as the vapor through the lungs. Diethyl ether depresses the central nervous system and is a depressant widely used as an anesthetic for surgery. Low doses of diethyl ether causes drowsiness, intoxication, and stupor, whereas higher exposures cause unconsciousness and even death. Several other volatile ethers affect the central nervous system.

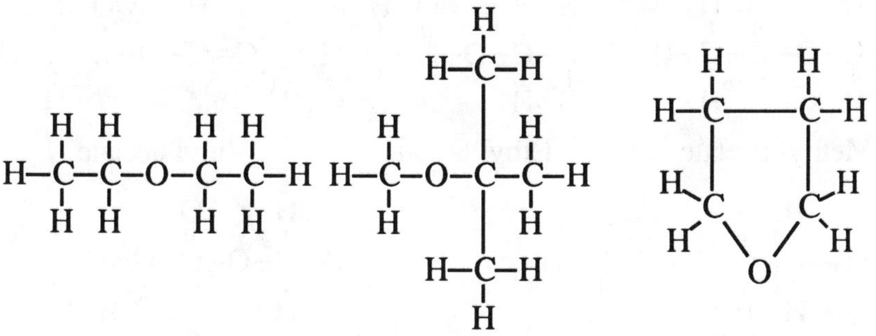

Diethyl ether   Methyl*tert*- butyl ether   Tetrahydrofuran (cyclic ether)

Figure 9.8. Examples of ethers.

Acid Anhydrides

The most important carboxylic **acid anhydride**, acetic anhydride,

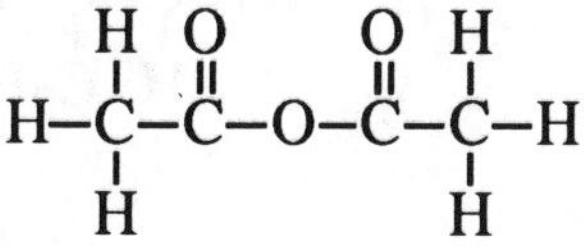

H–C–C–O–C–C–H Acetic anhydride

is a systemic poison and especially corrosive to the skin, eyes, and upper respiratory tract, causing blisters and burns that heal only slowly. Levels in the air should not exceed 0.04 mg/m^3 and adverse effects to the eyes have been observed at about 0.4 mg/m^3. A powerful lachrimator, acetic anhydride has a very strong acetic acid odor. Inhalation of acetic anhydride causes an intense burning sensation in the nose and throat that is accompanied by coughing, which causes exposed individuals to withdraw from the source of exposure.

Esters

There are many naturally occurring esters in fats, oils, and waxes and in compounds responsible for odors and flavors of fruits, flowers, and other natural products. Synthetic esters are used as ingredients of solvents, plasticizers, lacquers, soaps, and surfactants. Figure 9.9 shows some representative esters.

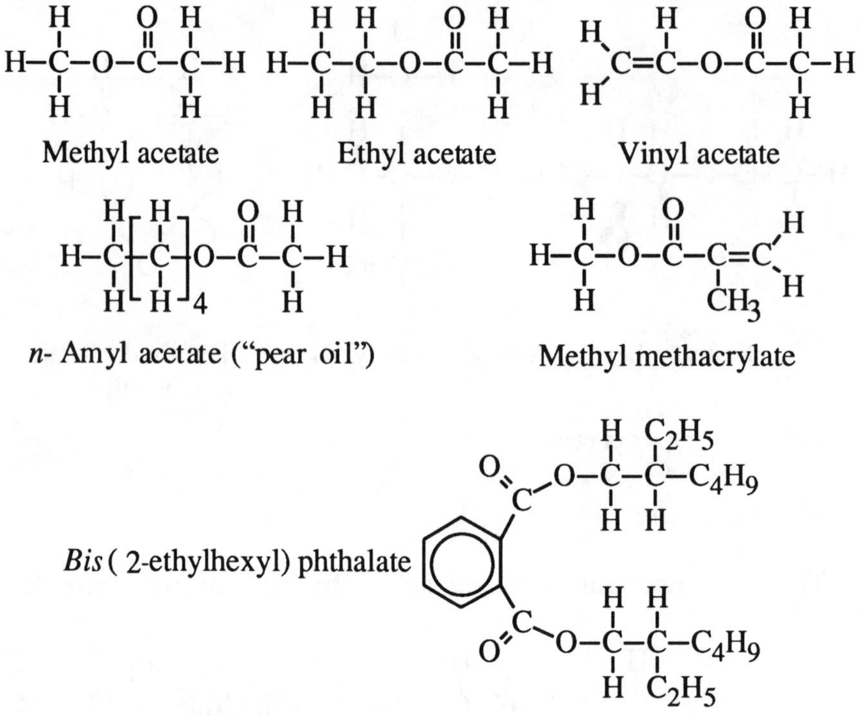

Methyl acetate   Ethyl acetate   Vinyl acetate

n- Amyl acetate ("pear oil")   Methyl methacrylate

Bis (2-ethylhexyl) phthalate

Figure 9.9. Examples of esters.

Many esters have relatively high volatilities so that the pulmonary system is a major route of exposure. Because of their generally good solvent properties, esters penetrate tissues and tend to dissolve body lipids. For example, vinyl acetate acts as a skin defatting agent. Because esters hydrolyze (split apart) in water, their toxicities tend to be the same as the toxicities of the acids and alcohols from which they were formed. Many volatile esters exhibit asphyxiant and narcotic action. Whereas many of the naturally occurring esters have insignificant toxicities at low doses, allyl acetate and some of the other synthetic esters are relatively toxic.

Bis(2-ethylhexyl) phthalate (diethylhexyl phthalate, DEHP) is a widely used and persistent ester of considerable environmental concern. Other significant phthalate esters are dimethyl, diethyl, di-n-butyl, di-n-octyl, and butylbenzyl phthalates. Because of their widespread use as plasticizers to improve the qualities of plastics and their poor degradation properties, these compounds have become persistent pollutants found in many parts of the environment. Fortunately, most of the commonly used phthalates have low toxicity ratings of 2 or 3.

9.4.   ORGANONITROGEN COMPOUNDS

Aliphatic Amines

Figure 9.10 shows structures of some toxicologically significant amines. All of the aliphatic amines have strong odors. The lower

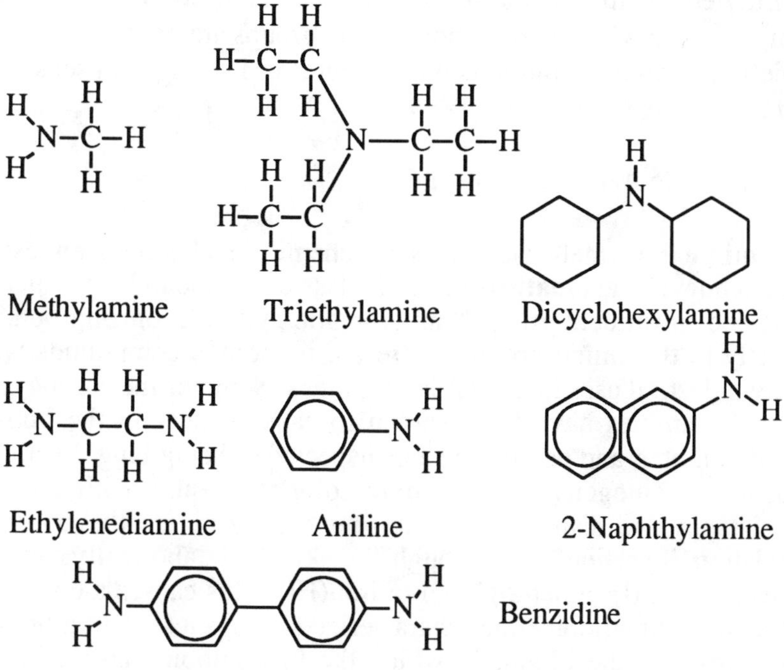

Figure 9.10. Structural formulas of some toxicologically significant amines.

amines, such as the methylamines, are rapidly and easily taken into the body by all common exposure routes. They are basic and react with water in tissue,

$$R_3N + H_2O \longrightarrow R_3N^+ + OH^- \tag{9.4.1}$$

raising the pH of the tissue to harmful levels, acting as corrosive poisons (especially to sensitive eye tissue), and causing tissue necrosis at the point of contact. Among the systemic effects of amines are necrosis of the liver and kidneys, lung hemorrhage and edema, and sensitization of the immune system. The lower amines are among the more toxic substances in routine, large-scale use.

Dicyclohexylamine is a dangerous compound that is readily absorbed through the skin because of its relatively less polar, more lipid-soluble nature. It is caustic to eyes, mucous membranes, and skin. As a systemic poison, it causes nausea (vomiting), anxiety, restlessness, and drowsiness and it can damage the female reproductive system.

Ethylenediamine is the most common of the **alkyl polyamines**, compounds in which two or more amino groups are bonded to alkane moieties. Its toxicity rating is only 3, but it is a strong skin sensitizer and can damage eye tissue.

Carbocyclic Aromatic Amines

Aniline is a widely used industrial chemical and is the simplest of the **carbocyclic aromatic amines**, a class of compounds in which at least one substituent group is an aromatic hydrocarbon ring bonded directly to the amino group. There are numerous compounds with many industrial uses in this class of amines. Some of the carbocyclic aromatic amines have been shown to cause cancer in the human bladder, ureter, and pelvis, and are suspected of being lung, liver, and prostate carcinogens. A very toxic colorless liquid with an oily consistency and distinct odor, aniline readily enters the body by inhalation, ingestion, and through the skin.[5] Metabolically, aniline converts iron(II) in hemoglobin to iron(III). This causes a condition called methemoglobinemia, characterized by cyanosis and a brown-black color of the blood, in which the hemoglobin can no longer transport oxygen in the body. This condition is not reversed by oxygen therapy.

Both **1-naphthylamine** (alpha-naphthylamine) and **2-naphthyl-amine** (beta-naphthylamine) are proven human bladder carcinogens. In addition to being a proven human carcinogen, **benzidine**, *p*-aminodiphenyl, is highly toxic and has systemic effects that include blood hemolysis, bone marrow depression, and kidney and liver damage. It can be taken into the body orally, by inhalation, and by skin sorption.

Pyridine

Pyridine, a colorless liquid with a sharp, penetrating, "terrible"odor, is an aromatic amine in which an N atom is part of a

Pyridine

6-membered ring. This widely used industrial chemical is only moderately toxic with a toxicity rating of 3. Symptoms of pyridine poisoning include anorexia, nausea, fatigue, and, in cases of chronic poisoning, mental depression. In a few rare cases pyridine poisoning has been fatal.

Nitriles

Nitriles, such as acetonitrile and acrylonitrile,

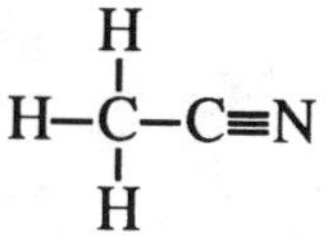

contain the $-C\equiv N$ functional group. Colorless, liquid **acetonitrile** (U003) is a good solvent for many organic and inorganic compounds that has many industrial uses. Because of its low boiling point, it is used as a reaction medium that can be recovered. With a toxicity rating of 3–4, acetonitrile is considered relatively safe, although it has caused human deaths, perhaps by metabolic release of cyanide.

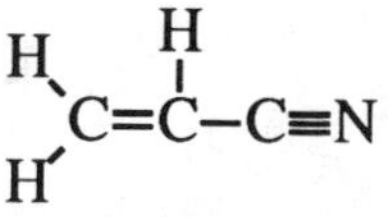

Acrylonitrile (U009) is used in large quantities in the manufacture of acrylic fibers, dyes, and pharmaceutical chemicals. A colorless liquid with a peach-seed (cyanide) odor, highly reactive acrylonitrile contains both nitrile and C=C groups. It is a highly reactive compound with a strong tendency to polymerize. Ingested, absorbed through the skin, or inhaled as vapor, acrylonitrile metabolizes to release HCN, which it resembles toxicologically. Acrylonitrile has a toxicity rating of 5.

Acetone cyanohydrin (structure below)

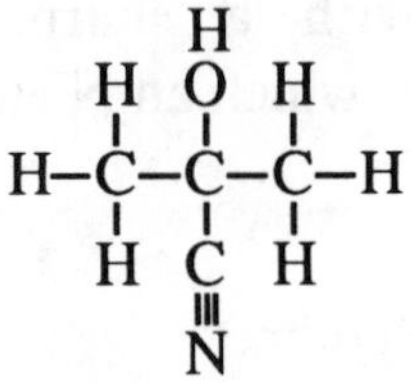

$$
\begin{array}{ccc}
 & \overset{\displaystyle H}{\underset{\displaystyle |}{}} & \\
\overset{\displaystyle H}{\underset{\displaystyle |}{}} & \overset{\displaystyle O}{\underset{\displaystyle |}{}} & \overset{\displaystyle H}{\underset{\displaystyle |}{}} \\
H-\overset{\displaystyle |}{\underset{\displaystyle |}{C}}-\overset{\displaystyle |}{\underset{\displaystyle |}{C}}-\overset{\displaystyle |}{\underset{\displaystyle |}{C}}-H \\
 & \overset{\displaystyle H}{} \; \overset{\displaystyle C}{} \; \overset{\displaystyle H}{} & \\
\end{array}
\quad \text{Acetone cyanohydrin}
$$

is an oxygen-containing nitrile with several important industrial applications. This colorless liquid is readily absorbed through the skin. In the body it releases deadly hydrogen cyanide, to which it should be considered toxicologically equivalent on a molecule-per-molecule basis.

Nitro Compounds

Nitro compounds, such as those shown in Figure 9.11, contain the $-NO_2$ functional group. **Nitromethane** is an oily liquid that has a

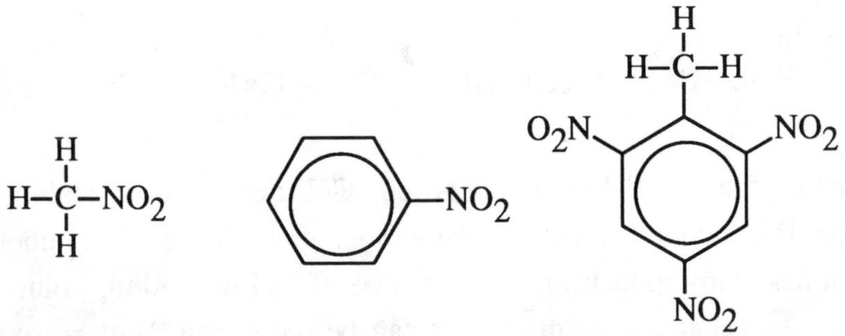

Figure 9.11. Some major nitro compounds.

toxicity rating of 3; it causes anorexia, diarrhea, nausea, and vomiting and damages the kidneys and liver. **Nitrobenzene** (U169), a pale yellow oily liquid with an odor of bitter almonds or shoe polish, has a toxicity rating of 5. It can enter the body through all routes and has a toxic action much like that of aniline. In the body nitrobenzene converts hemoglobin to methemoglobin, which cannot carry oxygen to body tissue. Nitrobenzene poisoning is manifested by cyanosis. **Trinitrotoluene** (TNT), a widely used military explosive, has a toxicity rating of 3–4. It can damage bone marrow, kidney, and liver cells. Major toxic effects of trinitrotoluene poisoning are hepatitis and aplastic anemia.

Nitrosamines

N-nitroso compounds (nitrosamines) have been used as solvents and as intermediates in chemical synthesis. As shown by the examples below, N-nitroso compounds contain the -N–N=O functional group. N-nitroso compounds have been found in a variety of

Dimethylnitrosamine
(N–nitrosodimethylamine)

N–nitrosopiperidine

materials to which humans may be exposed, including beer, whiskey, and cutting oils used in machining.

Cancer may result from exposure to a single large dose or from chronic exposure to relatively small doses of some nitrosamines. The carcinogenicity of these compounds was first revealed for dimethylnitrosamine. Once widely used as an industrial solvent and known to cause liver damage and jaundice in exposed workers,[6] dimethylnitrosamine was shown to be carcinogenic from studies starting in the 1950s. Different nitrosamines cause cancer in different organs.

Isocyanates and Methyl Isocyanate

Compounds with the general formula R–N=C=O, isocyanates are noted for the high chemical and metabolic reactivity of their characteristic functional group. They have numerous uses in chemical synthesis, particularly in the manufacture of specialty polymers with carefully tuned properties. Methyl isocyanate (Figure 9.12) was the toxic agent involved in the catastrophic industrial poisoning in Bhopal, India on December 2, 1984, the worst industrial

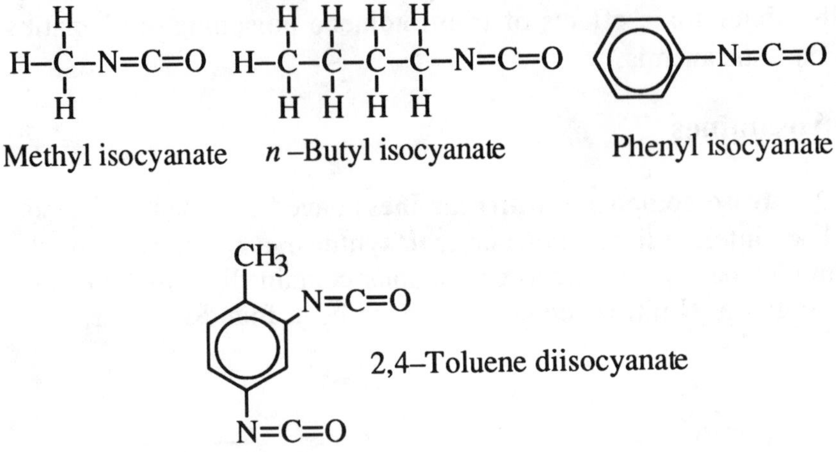

Methyl isocyanate   *n* –Butyl isocyanate   Phenyl isocyanate

2,4–Toluene diisocyanate

Figure 9.12. Example isocyanate compounds.

accident in history. In this incident several tons of methyl isocyanate were released, killing 2,000 people and affecting about 100,000. The lungs of victims were attacked; survivors suffered long-term shortness of breath and weakness from lung damage as well as numerous other toxic effects including nausea and bodily pain.[7]

Organonitrogen Pesticides

Six of the numerous organonitrogen compounds used as pesticides are shown in Figure 9.13. They are discussed below.

Carbamates

Carbamates are characterized by the structural skeleton of carbamic acid outlined by the dashed box in the structural formula of carbaryl in Figure 9.13. Widely used on lawns and gardens, insecticidal **carbaryl** has a low toxicity to mammals. Highly water soluble **carbofuran** is taken up by the roots and leaves of plants. It is a systemic insecticide in that insects are poisoned by the carbamate compound in the plant material on which they feed. The toxic effects to animals of carbamates are due to the fact that they inhibit acetylcholinesterase directly without the need to first undergo biotransformation. This effect is relatively reversible because of metabolic hydrolysis of the carbamate ester.

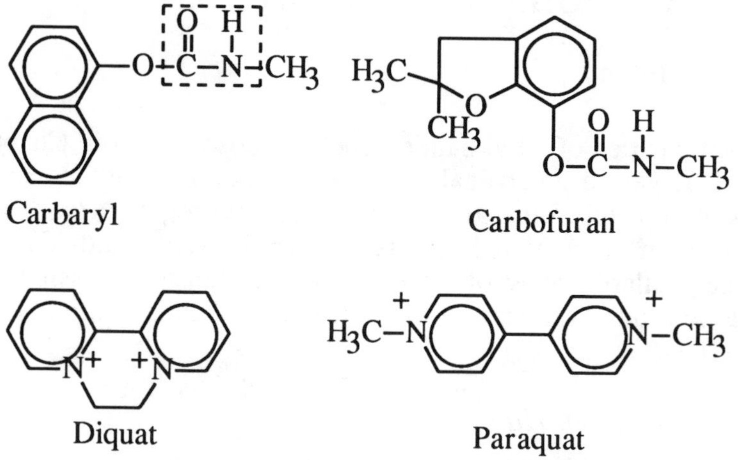

Figure 9.13. Carbamic acid and examples of organonitrogen pesticides.

Bipyridilium Herbicides

Herbicidal bipyridilium **diquat** and **paraquat** (Figure 9.13), which contain 2 pyridine rings per molecule, are applied directly to plant tissue. These herbicides rapidly destroy plant cells, giving the plant a frost-bitten appearance. Reputed to have "been responsible for hundreds of human deaths,"[8] paraquat has a toxicity rating of 5. Although the chronic effects of exposure to low levels of paraquat over extended periods of time are not well known, dangerous or even fatal acute exposures can occur by all pathways, including inhalation of spray, skin contact, ingestion, and even suicidal hypodermic injections. Although paraquat can be corrosive at the point of contact, it is a systemic poison that affects enzyme activity and is devastating to a number of organs. Pulmonary fibrosis results in animals that have inhaled paraquat aerosols, and the lungs are also adversely affected by non-pulmonary exposure. Acute exposure may cause variations in the levels of catecholamine, glucose, and insulin. The most prominent initial symptom of poisoning is vomiting, followed within a few days by dyspnea, cyanosis, and evidence of impairment of the kidneys, liver, and heart. Pulmonary fibrosis, often accompanied by pulmonary edema and hemorrhaging, is observed in fatal cases.

9.5.   ORGANOHALIDE COMPOUNDS

Alkyl Halides

The toxicities of alkyl halides, such as those shown in Chapter 8, Table 8.1, vary a great deal with the compound. Once considered almost completely safe, alkyl halides are now regarded with much more caution as additional health and animal toxicity study data have become available. Most of these compounds cause depression of the central nervous system, and individual compounds exhibit specific toxic effects.

Carbon Tetrachloride

Used for many years as a degreasing solvent, in home fire extinguishers, and for other industrial and consumer product applications, carbon tetrachloride (U211) compiled a grim record of toxic effects which led the U. S. Food and Drug Administration (FDA) to prohibit its household use in 1970. Carbon tetrachloride is a systemic poison that affects the nervous system when inhaled and the gastrointestinal tract, liver, and kidneys when ingested. The biochemical mechanism of carbon tetrachloride toxicity involves reactive radical species including,

$$
\begin{array}{ccc}
 & \mathrm{Cl} & \mathrm{Cl} \\
 & | & | \\
\mathrm{Cl-C\cdot} & & \cdot\mathrm{OO-C-Cl} \\
 & | & | \\
 & \mathrm{Cl} & \mathrm{Cl}
\end{array}
$$

Unpaired
electrons

that react with biomolecules, such as proteins and DNA. The most damaging such reaction occurs in the liver as **lipid peroxidation**, consisting of the attack of free radicals on unsaturated lipid molecules, followed by oxidation of the lipids through a free radical mechanism.

Alkenyl Halides

The most significant **alkenyl** or **olefinic organohalides** are the lighter chlorinated compounds, such as those illustrated in Chapter 8,

Figure 8.5. Because of their widespread use and disposal in the environment, the numerous acute and chronic toxic effects of the alkenyl halides are of considerable concern.

The central nervous system, respiratory system, liver, and blood and lymph systems are all affected by vinyl chloride (U043) exposure, which has been widespread because of this compound's use in polyvinylchloride manufacture. Most notably, vinyl chloride is carcinogenic, causing a rare angiosarcoma of the liver. This deadly form of cancer has been observed in workers chronically exposed to vinyl chloride while cleaning autoclaves in the polyvinylchloride fabrication industry. The alkenyl organohalide, 1,1-dichloroethylene, is a suspect human carcinogen based upon animal studies and its structural similarity to vinyl chloride The toxicities of both 1,2-dichloroethylene isomers are relatively low. These compounds act in different ways in that the *cis* isomer is an irritant and narcotic, whereas the *trans* isomer affects both the central nervous system and the gastrointestinal tract, causing weakness, tremors, cramps, and nausea. A suspect human carcinogen, trichloroethylene (U228) has caused liver carcinoma in experimental animals and is known to affect numerous body organs. Like other organohalide solvents, trichloroethylene causes skin dermatitis from dissolution of skin lipids and it can affect the central nervous and respiratory systems, liver, kidneys, and heart. Symptoms of exposure include disturbed vision, headaches, nausea, cardiac arrhythmias, and burning/tingling sensations in the nerves (paresthesia). Tetrachloroethylene (U210) damages the liver, kidneys, and central nervous system. It is a suspect human carcinogen based upon studies with mice. The chlorinated propenes irritate eyes, skin, and respiratory tract and cause skin rashes, blisters, and burns. Chronic exposure to allyl chloride causes pulmonary edema and aching muscles and bones, as well as damage to the liver, lungs, and kidney. Chloroprene is an eye and respiratory system irritant that causes skin dermatitis, alopecia (a condition characterized by hair loss in the affected skin area), nervousness, and irritability. Cells in the liver and kidney are adversely affected by hexachlorobutadiene in the body and the compound is a suspect human carcinogen.

Once an important industrial chemical used directly as an agricultural fumigant and as an intermediate in the manufacture of insecticides, hexachlorocyclopentadiene,

Hexachlorocyclopentadiene

is a liquid cyclic alkenyl halide with two double bonds. Large quantitities of this compound and still bottoms from its manufacture have been disposed in hazardous waste chemical sites such as the one at Love Canal. A very toxic compound with a toxicity rating of 4, hexachlorocyclopentadiene gives off fumes that are strongly lachrimating, and irritating to skin, eye, and mucuous membrane tissue. It has been found to damage most major organs of experimental animals, including the kidney, heart, brain, adrenal glands, and liver.

Aryl Halides

Figure 8.6 in Chapter 8 gives the structural formulas of some important aryl halides. Individuals exposed to irritant monochlorobenzene by inhalation or skin contact suffer symptoms to the respiratory system, liver, skin, and eyes. Ingestion of this compound causes effects similar to those of toxic aniline, including incoordination, pallor, cyanosis, and eventual collapse.

The dichlorobenzenes are irritants that affect the same organs as monochlorobenzene; the 1,4- isomer has been known to cause profuse rhinitis (running nose), nausea, jaundice, liver cirrhosis, and weight loss associated with anorexia. *Para*-dichlorobenzene (1,2-dichlorobenzene), a chemical used in air fresheners and mothballs, has become the center of a controversy regarding the evaluation of carcinogenicity. Based upon animal tests that involved subjecting rats to large amounts of the chemical, then extrapolating to humans with adjustments for differences in body size and many orders of magnitude in dose, the U. S. Department of Health and Human Service's National Toxicology Program has classified 1,2-dichlorobenzene as a potential cancer-causing substance. Some industry groups, including the Synthetic Organic Chemicals Association, have filed suit contending that the procedures used are inaccurate and out of date. The suit is viewed as a general challenge to the current use of animal studies to establish possible carcinogenicity of chemicals.[9] Specifically, the suit asks that the report acknowledge that the protein through which the chemical acts in rats does not exist in humans.

Many authorities contend that more sophisticated alternatives to animal studies should be used to predict human carcinogenicity of chemicals, especially biochemical investigations of the interactions of suspect chemicals and their metabolites in the body, pathways through the body, and chemical (structural and functional) similarities to known carcinogens.

A notorious compound in the annals of toxicology, hexachlorobenzene was involved in the tragic poisoning of 3,000 people in Turkey during the period 1955–1959;[10] some of the observed effects may have been due to the presence of manufacturing byproduct impurity polychlorinated dibenzodioxins. As a consequence of eating seed wheat that had been treated with 10% hexachlorobenzene to deter fungal growth, the victims developed **porphyria cutanea tarda**, a condition in which the skin becomes blistered, fragile, photosensitive, and subject to excessive hair growth. Other symptoms included skin damage, eye dysfunction, weight loss from anorexia, and wasting of skeletal muscles.

Little information is available regarding the human toxicity of bromobenzene, although it has been shown to damage the livers of rats used in animal tests. This compound can enter the body through the respiratory tract, gastrointestinal tract, or skin.

The widely variable toxicities reported for chlorinated naphthalenes suggest that some of the effects observed have been due to the presence of manufacturing impurities. Inhalation of the more highly chlorinated naphthalenes has caused chloracne rash and debilitating liver necrosis in humans. Feed contaminated by polychlorinated naphthalene caused the deaths of several hundred thousand cattle in the 1940s and early 1950s.

Because of their once widespread use in electrical equipment, as hydraulic fluids, and in many other applications, polychlorinated biphenyls (PCBs) became widespread, extremely persistent environmental pollutants.[11] Polybrominated biphenyl analogs (PBBs) were much less widely used and distributed. However, PBBs were involved in one major incident that resulted in catastrophic agricultural losses when livestock feed contaminated with PBB flame retardant caused massive livestock poisoning in Michigan in 1973.

Organohalide Insecticides

Structural formulas of some of the more significant organochlorine insecticides are given in Figure 9.14. The four major chem-

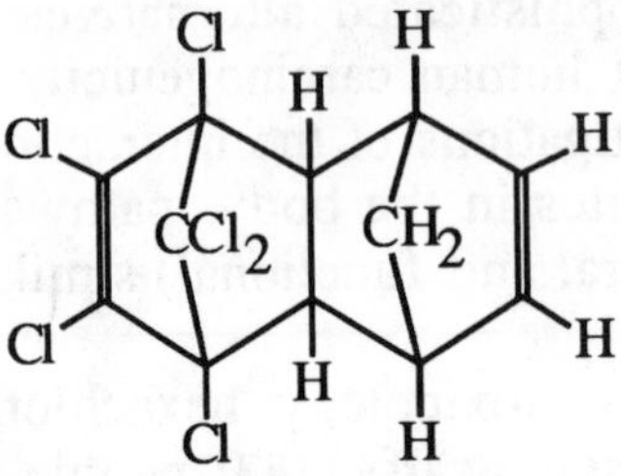

Aldrin

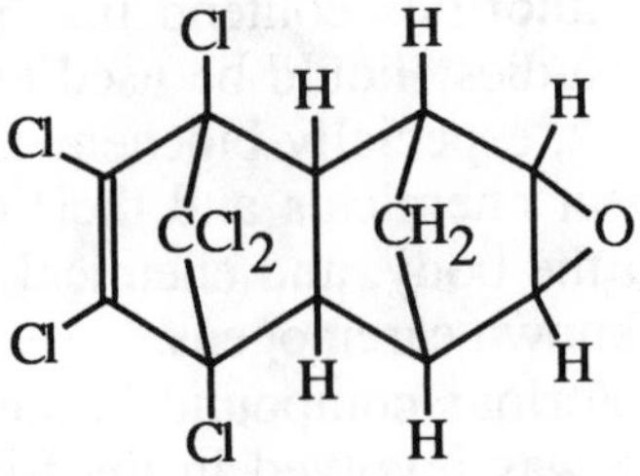

Dieldrin

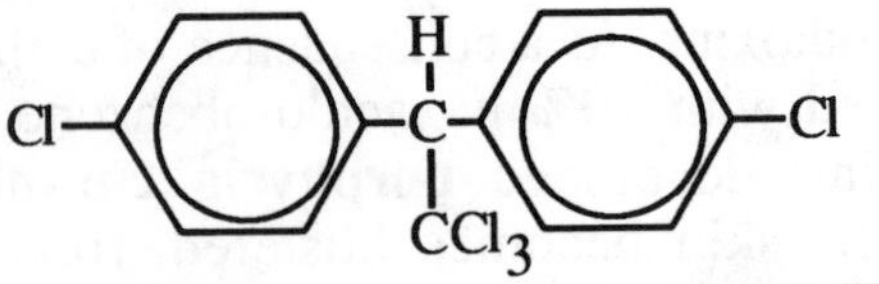

DDT

Methoxychlor

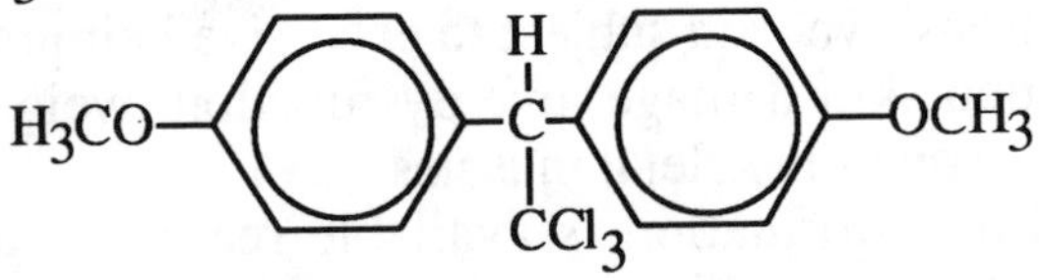

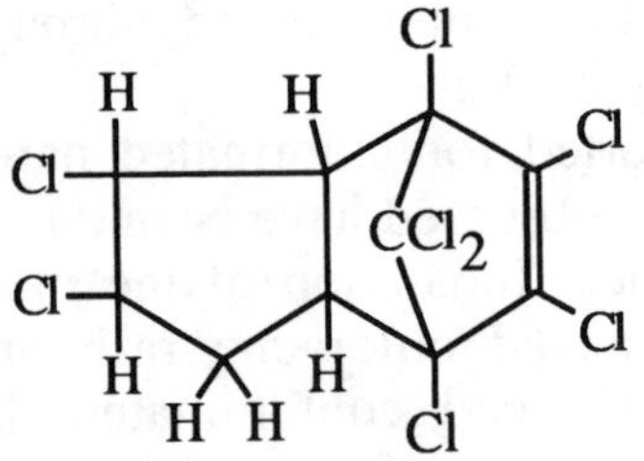

Chlordane

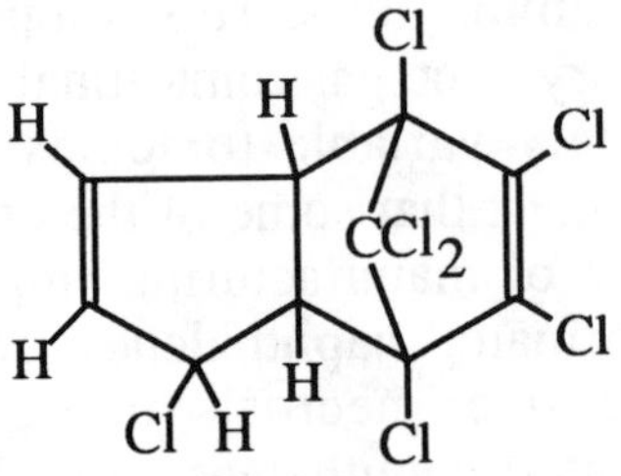

Heptachlor

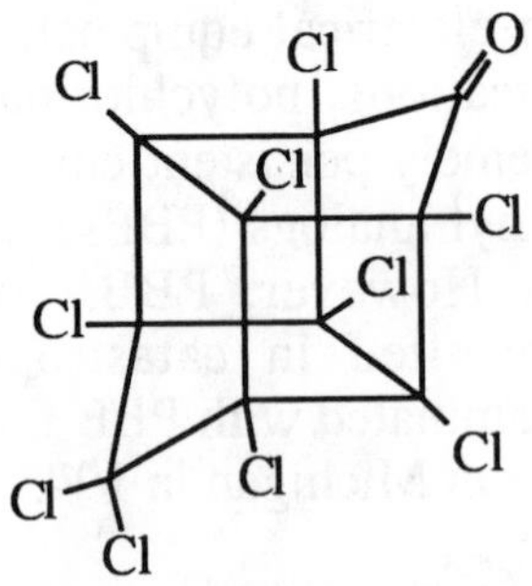

Chlordecone (Kepone); Mirex is structurally similar with 2 Cl's in place of =O.

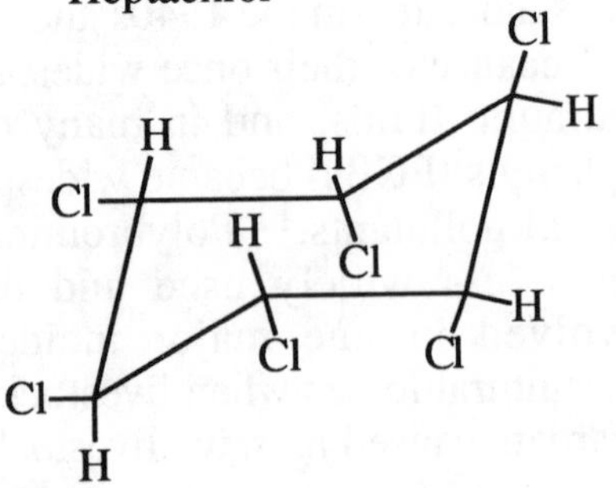

The gamma isomer of hexachlorocyclohexane (lindane)

Figure 9.14. Some organohalide insecticides

ical classes of organochlorine insecticides are the chloroethylene derivatives (DDT and methoxychlor), chlorinated cyclodiene compounds (aldrin, dieldrin, and heptachlor), the hexachlorocyclohexane stereoisomers (lindane), and a group known collectively as toxaphene.

Toxicities of Organohalide Insecticides

Exhibiting a wide range of kind and degree of toxic effects, many organohalide insecticides affect the central nervous system, causing symptoms such as tremor, irregular jerking of the eyes, changes in personality, and loss of memory. Such symptoms are characteristic of acute DDT poisoning. However, the acute toxicity of DDT to humans is very low and it was used for the control of typhus and malaria in World War II by large scale direct application to people. The chlorinated cyclodiene insecticides — aldrin, dieldrin, endrin, chlordane, heptachlor, endosulfan, and isodrin — act on the brain, releasing betaine esters and causing headaches, dizziness, nausea, vomiting, jerking muscles, and convulsions. Dieldrin, chlordane, and heptachlor have caused liver cancer in test animals and some chlorinated cyclodiene insecticides are teratogenic or fetotoxic. Because of these effects, aldrin, dieldrin, heptachlor, and — more recently — chlordane have been prohibited from use in the U. S.

Application of environmentally damaging Mirex and Kepone has been restricted in the U. S. and the uses of these insecticides are now restricted to eradication of fire ants in the southeastern states. A significant number of human exposures to these compounds have occurred. A flawed Kepone manufacturing operation in Hopewell, Virginia, resulted in the discharge of about 53,000 kg of this compound to the James River through the city sewage system during the 1970s. Kepone affects the central nervous system with symptoms of irritability, tremor, and hallucinations. It also causes adverse effects to sperm and damage to the nerves and muscles. A teratogen to test animals, this compound causes liver cancer in rodents.

The gamma isomer of **hexachlorocyclohexane** shown in Figure 9.14 is the major ingredient of insecticidal **lindane**. Individuals exposed to toxic doses of lindane have experienced degeneration of kidney tubules, liver damage associated with fatty tissue, and hystoplastic anemia.

Toxaphene is an insecticidal chlorinated camphene often represented by the empirical formula $C_{10}H_{10}Cl_8$ and consisting of a mixture of more than 170 compounds containing 10 C atoms and 6–10 Cl atoms per molecule. It was once the most widely used insecticide in the U. S. with annual consumption of about 40 million kg. Wide variations exist in the toxicities of the individual compounds contained in toxaphene; many tend to produce epileptic-type convulsions in exposed mammals.

Chlorophenoxy Compounds

The major **chlorophenoxy** herbicides (Figure 9.15) are 2,4-dichlorophenoxyacetic acid (2,4-D), 2,4,5-trichlorophenoxyacetic

2,4–Dichlorophenoxy-
acetic acid (and esters)

2,4,5–Trichlorophenoxy-
acetic acid (and esters)

Silvex

2,3,7,8–Tetrachloro–*p*–dioxin

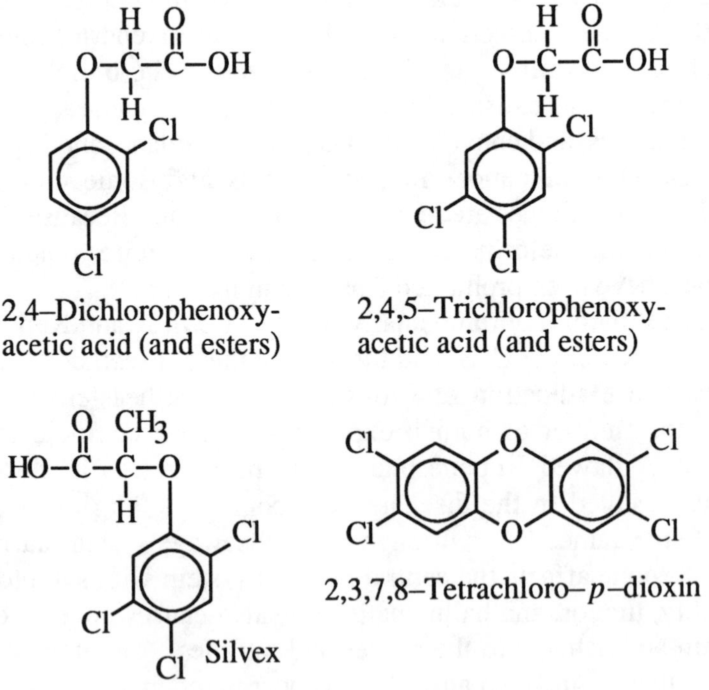

Figure 9.15. Herbicidal chlorophenoxy compounds and TCDD manufacturing by-product.

acid (2,4,5-T or Agent Orange), and Silvex. Large doses of 2,4-dichlorophenoxyacetic acid have been shown to cause nerve damage, (peripheral neuropathy), convulsions, and brain damage. According to a National Cancer Institute study,[12] Kansas farmers who had handled 2,4-D extensively have suffered 6 to 8 times the incidence of non-Hodgkins lymphoma as comparable unexposed populations. With a toxicity somewhat less than that of 2,4-D, Silvex is largely excreted unchanged in the urine. The toxic effects of 2,4,5-T (used as a herbicidal warfare chemical called "Agent Orange") have resulted from the presence of 2,3,7,8-tetrachloro-p-dioxin (TCDD, commonly known as "dioxin"), a manufacturing by-product. Autopsied carcasses of sheep poisoned by this herbicide have exhibited nephritis, hepatitis, and enteritis.

TCDD

As discussed in Section 8.9, **polychlorinated dibenzodioxins** have the same basic structure as that of TCDD (2,3,7,8-tetrachlorodibenzo-p-dioxin shown in Figure 9.15), but different numbers and arrangements of chlorine atoms on the ring structure. Polychlorinated dibenzodioxins exhibit varying degrees of toxicity. Although TCDD is unquestionably extremely toxic to some animals — its acute LD_{50} to male guinea pigs is only 0.6 µg/kg of body mass — its degree of toxicity and types of toxic effects to humans are both rather uncertain; it is known to cause a skin condition called chloracne. TCDD has been a manufacturing byproduct of some commercial products (see the discussion of 2,4,5-T, above), contaminant identified in some municipal incineration emissions, and widespread environmental pollutant from improper waste disposal (for example, the infamous "dioxin" spread from waste oil at Times Beach, Missouri). This compound has been released in a number of industrial accidents, the most massive of which exposed several tens of thousands of people to a cloud of chemical emissions spread over an approximately 3-square-mile area at the Givaudan-La Roche Icmesa manufacturing plant near Seveso, Italy, in 1976. On an encouraging note from a toxicological perspective, no abnormal occurrences of major malformations were found in a study of 15,291 children born in the area within 6 years after the release.[13]

Chlorinated Phenols

The chlorinated phenols used in largest quantities have been **pentachlorophenol** (Chapter 8, page 180) and the trichlorophenol isomers. These compounds have been used as wood preservatives that prevent wood rot through their fungicidal action and prevent termite infestation because of their insecticidal properties. Although exposure to these compounds has been correlated with liver malfunction and dermatitis, contaminant polychlorinated dibenzodioxins may have caused some of the observed effects.

9.6. ORGANOSULFUR COMPOUNDS

The structural formulas of examples of the organosulfur compounds discussed in this section are given in Figure 9.16. Despite the high toxicity of H_2S, not all organosulfur compounds are particularly toxic. Their hazards are often reduced by their strong, offensive odors that warn of their presence.

Thiols and Sulfides

Inhalation of even very low concentrations of the alkyl **thiols** (mercaptans) can cause nausea and headaches; higher levels can cause increased pulse rate, cold hands and feet, and cyanosis and, in extreme cases, unconsciousness, coma, and death. Like H_2S, the alkyl thiols are precursors to cytochrome oxidase poisons.

A typical alkenyl mercaptan, 2-propene-1-thiol (allyl mercaptan) is an odorous volatile liquid that is highly toxic and strongly irritating to mucous membranes. Alpha-toluenethiol (benzyl mercaptan) is very toxic orally and it is an experimental carcinogen. The foul-smelling aryl thiol, benzenethiol (phenyl mercaptan), causes headache and dizziness; skin exposure results in severe contact dermatitis.

Sulfides and Disulfides

Dimethyl sulfide, an alkyl sulfide or thioether, is a volatile liquid that is moderately toxic by ingestion. Organic disulfides, such as *n*-butyldisulfide and diphenyldisulfide, may act as allergens that produce dermatitis in contact with skin. Animal studies suggest that these compounds may have several toxic effects, including homolytic anemia.

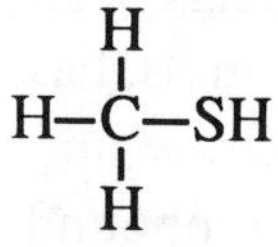

Methanethiol 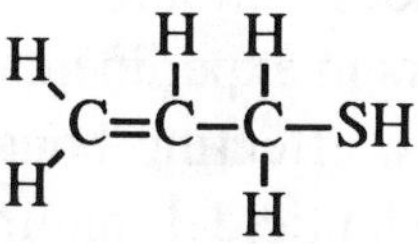   2–Propene–1–thiol 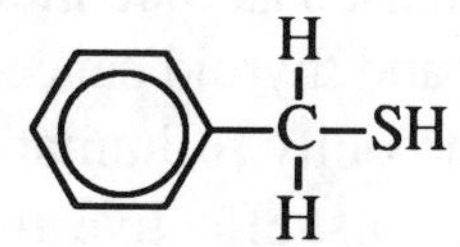   Alpha-toluenethiol 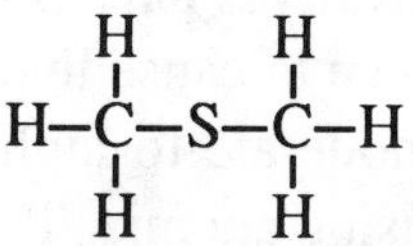   Dimethyl sulfide

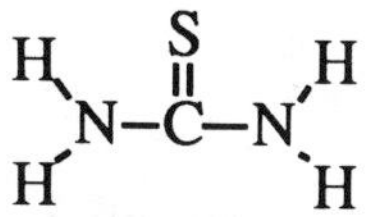

Thiourea

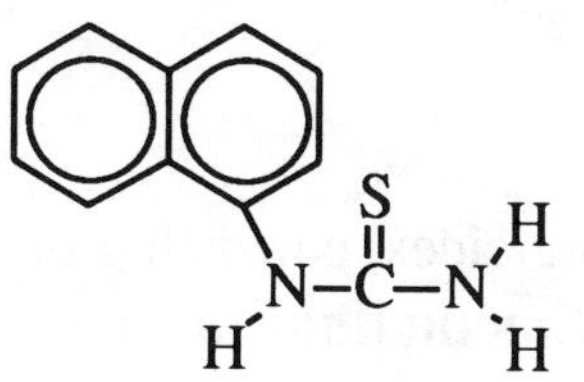

1–Naphthylthiourea (ANTU) 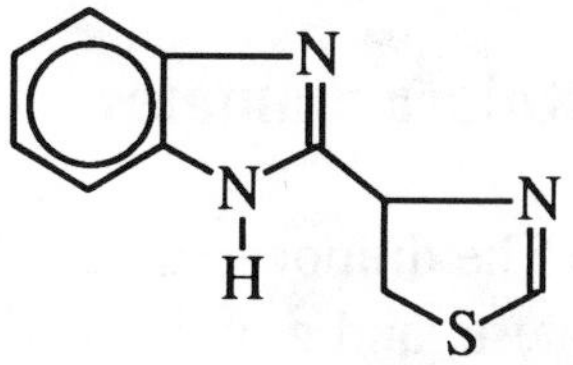   Thiabendazole

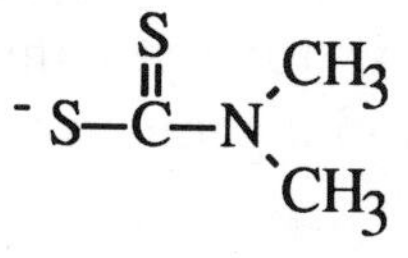

Dimethyldithiocarbamate anion 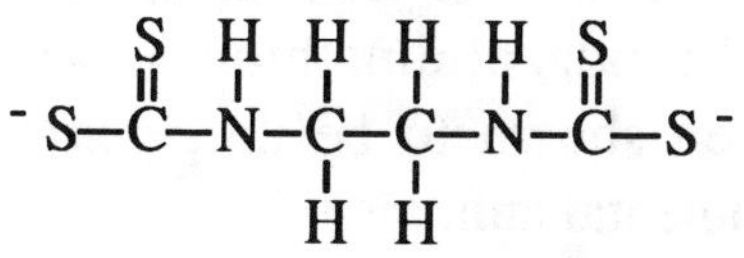   Ethylenebisdithiocarbamate anion 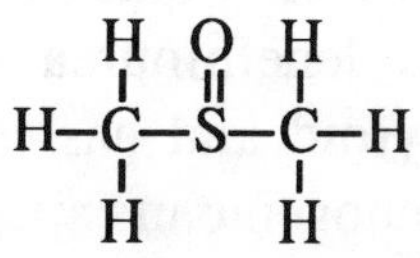   Dimethylsulfoxide (DMSO)

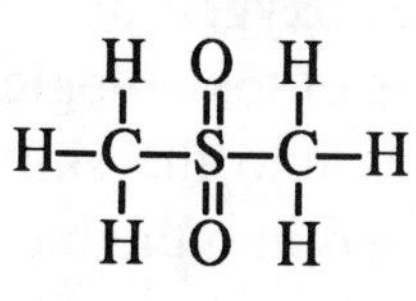

Dimethylsulfone 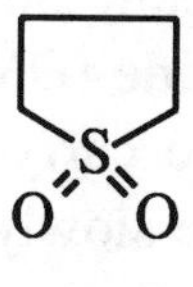   Sulfolane 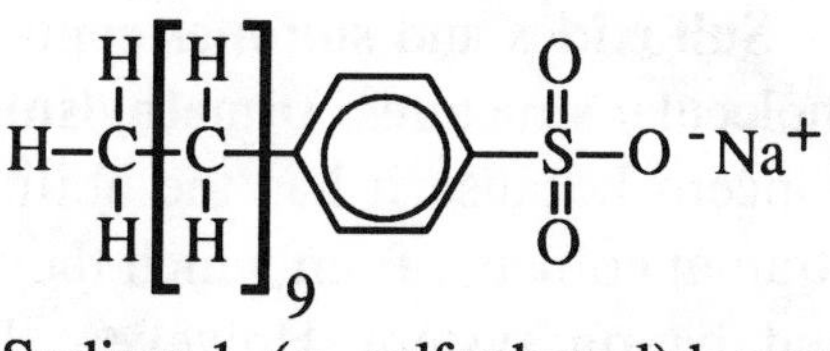   Sodium 1–(p–sulfophenyl)decane

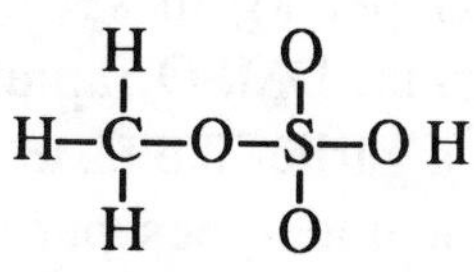

Methylsulfuric acid 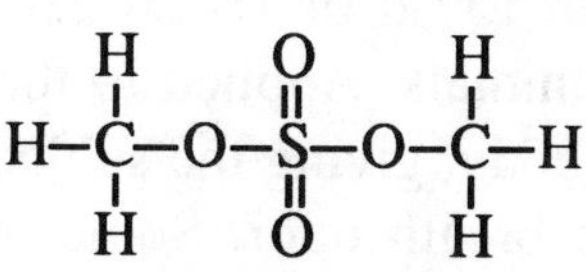   Dimethylsulfate 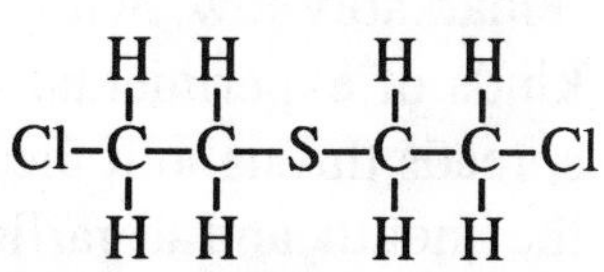   Mustard oil (H)

Figure 9.16. Structural formulas of some important toxic organosulfur compounds.

Thiourea Compounds

Thiourea (U219), the sulfur analog of urea, and some of its derivatives have been used as rodenticides. Thiourea, which has been shown to cause liver and thyroid cancers in experimental animals, has a moderate-to-high toxicity to humans, affecting bone marrow and causing anemia. The most effective rodenticidal thiourea compound is ANTU, **1-naphthylthiourea**, is a virtually tasteless compound with a very high rodent:human toxicity ratio.

Dithiocarbamates

The dithiocarbamate fungicides, consisting of sodium, iron, manganese, and zinc salts of **dimethylthiocarbamate** and **ethylenebis-dithiocarbamate** anions, have been popular for agricultural use because of their effectiveness and relatively low toxicities to animals. However, some of their environmental breakdown products may be toxic environmental pollutants. The most significant of these is ethylenethiourea (2-imidazolidinethione), which is toxic to the thyroid and has been shown to be mutagenic, carcinogenic, and teratogenic in experimental animals.

Sulfoxides and Sulfones

Sulfoxides and **sulfones** contain both sulfur and oxygen in their molecular structures. **Dimethylsulfoxide** (DMSO) is of toxicological concern because it has the ability to carry solutes into the skin's stratum corneum from which they are slowly released into the blood and lymph system. However, this compound's acute toxicity is remarkably low, with an LD$_{50}$ of 10–20 *grams* per kg in several kinds of experimental animals. Applied to the skin, DMSO rapidly spreads throughout the body, giving the subject a garlic-like taste in the mouth and a garlic breath odor. Some of it undergoes partial metabolism to dimethylsulfide and dimethylsulfone and some is excreted directly in the urine.

Sulfolane is the most widely used sulfone. Although it can cause eye and skin irritation, its overall toxicity is relatively low.

Sulfonic Acids, Salts, and Esters

Sulfonic acids and their salts, such as sodium 1-(p-sulfophenyl) decane contain the $-SO_3H$ or $-SO_3^-$ groups, respectively, attached to a hydrocarbon moiety. Strongly acidic sulfonic acids should be treated with the precautions due strong acids. Skin, eyes, and mucous membranes are subject to strong irritation when exposed to benzenesulfonic acid or p-toluenesulfonic acid. Methylmethane sulfonate ester is particularly dangerous because it is a primary or direct-acting carcinogen that does not require metabolic conversion to cause cancer.[14]

Organic Esters of Sulfuric Acid

Replacement of one of the H atoms on H_2SO_4 by a hydrocarbon substituent (for example, $-CH_3$) produces an acid ester of sulfuric acid and replacement of both H's yields an ester. As discussed in Section 4.5, water-soluble, readily eliminated acid ester products of xenobiotic compounds (such as phenol) are produced by phase II reactions in the body.

An oily water-soluble liquid, **methylsulfuric acid** is a strong irritant to skin, eyes, and mucous tissue. Colorless, odorless **dimethylsulfate** is highly toxic and, like methylmethane sulfonate, a primary carcinogen. Skin or mucous membranes exposed to dimethylsulfate develop conjunctivitis and inflammation of nasal tissue and respiratory tract mucous membranes following an initial latent period during which few symptoms are observed. Damage to the liver and kidney, pulmonary edema, cloudiness of the cornea, and death within 3–4 days can result from heavier exposures.

Sulfur Mustards

A typical example of deadly **sulfur mustards**, compounds used as military poisons, or "poison gases," is mustard oil (bis(2-chloroethyl)sulfide).[15] An experimental mutagen and primary carcinogen, mustard oil produces vapors that penetrate deep within tissue, resulting in destruction and damage at some depth from the point of contact; penetration is very rapid, so that efforts to remove the toxic agent from the exposed area are ineffective after 30 minutes. This military "blistering gas" poison, causes tissue to become severely inflamed with lesions that often become infected. These lesions in the lung can cause death.

Sulfur in Pesticides

In addition to rodenticidal thioureas, insecticidal thiocyanates, and fungicidal dithiocarbamates, there are other classes of pesticides containing sulfur. The most notable of these are the organophosphate insecticides discussed in Section 9.7.

9.7. ORGANOPHOSPHORUS COMPOUNDS

Organophosphorus compounds have varying degrees of toxicity. Some of these compounds, such as the "nerve gases" produced as industrial poisons, are deadly in minute quantities. The toxicities of major classes of organophosphate compounds are discussed in this section.

Organophosphate Esters

Some organophosphate esters are shown in Figure 9.17. **Tri-**

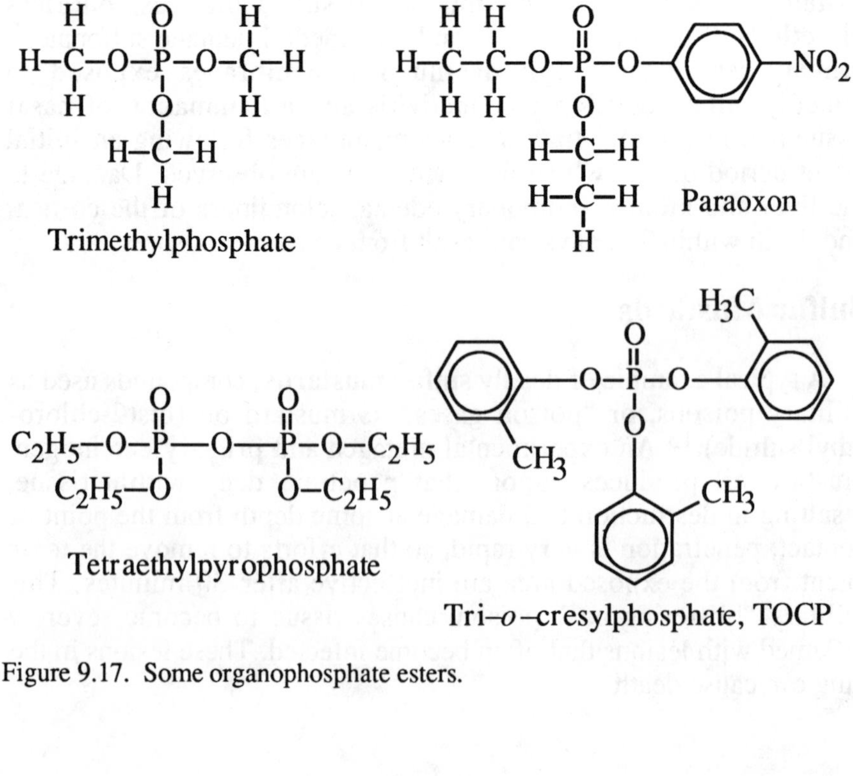

Figure 9.17. Some organophosphate esters.

methylphosphate is probably moderately toxic when ingested or absorbed through the skin, whereas moderately toxic **triethyl-phosphate**, $(C_2H_5O)_3PO$, damages nerves and inhibits acetylcholinesterase. Notoriously toxic **tri-*o*-cresylphosphate, TOCP,** apparently is metabolized to products that inhibit acetylcholinesterase. Exposure to TOCP causes degeneration of the neurons in the body's central and peripheral nervous systems with early symptoms of nausea, vomiting, and diarrhea accompanied by severe abdominal pain.[16] About 1–3 weeks after these symptoms have subsided, peripheral paralysis develops manifested by "wrist drop" and "foot drop," followed by slow recovery, which may be complete or leave a permanent partial paralysis.

Tetraethylpyrophosphate

Briefly used in Germany as a substitute for insecticidal nicotine, **tetraethylpyrophosphate, TEPP**, is a very potent acetylcholinesterase inhibitor. With a toxicity rating of 6 (supertoxic), TEPP is deadly to humans and other mammals.

Phosphorothionate and Phosphorodithioate Ester Insecticides

Because esters containing the P=S (thiono) group are resistant to non-enzymatic hydrolysis and are not as effective as P=O compounds in inhibiting acetylcholinesterase, they exhibit higher insect: mammal toxicity ratios than their non-sulfur analogs. Therefore, **phosphorothionate** and **phosphorodithioate** esters (Figure 9.18) are widely used as insecticides. The insecticidal activity of these compounds requires metabolic conversion of P=S to P=O (oxidative desulfuration). Environmentally, organophosphate insecticides are superior to many of the organochlorine insecticides because the organophosphates readily undergo biodegradation and do not bioaccumulate.

Parathion

The first commercially successful phosphorothionate and phosphorodithioate ester insecticide was **parathion**, *O,O*-diethyl-*O-p*-nitrophenylphosphorothionate, first licensed for use in 1944. This insecticide has a toxicity rating of 6 (supertoxic). Since its use began, several hundred people have been killed by parathion, including 17 of 79 people exposed to contaminated flour in Jamaica in 1976. As little as 120 mg of parathion has been known to kill an adult human and a dose of 2 mg has been fatal to a child. Most accidental poisonings have occurred by absorption through the skin. Methylparathion (a closely related compound with methyl groups instead of ethyl groups) is regarded as extremely toxic.

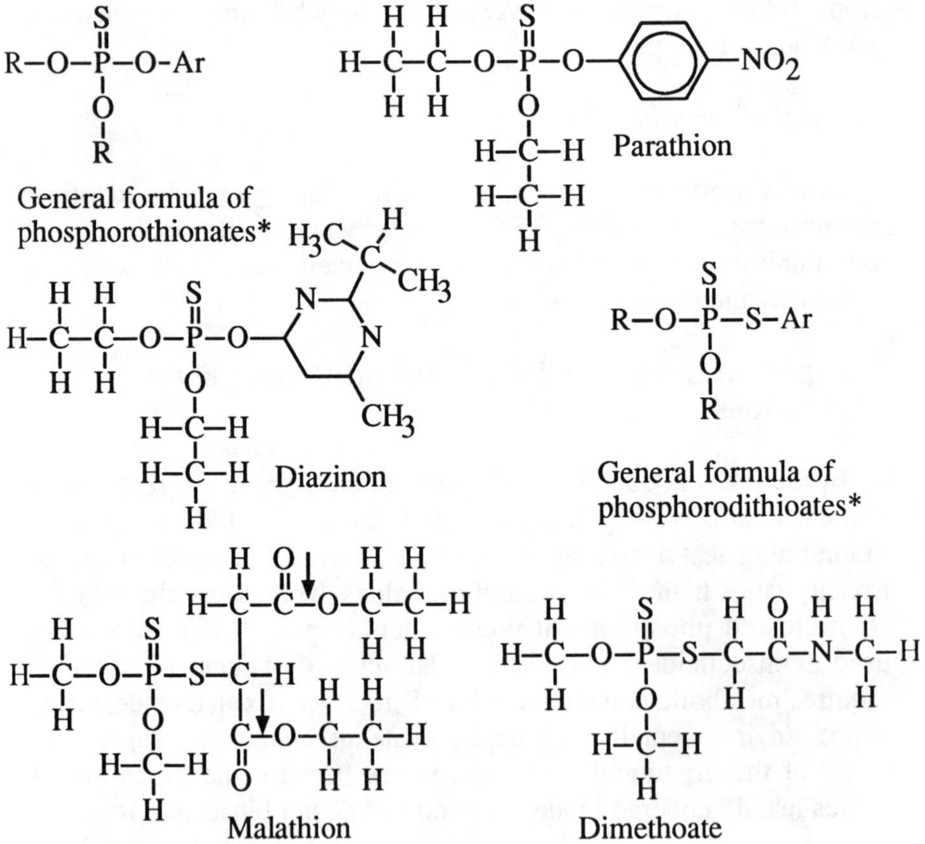

* R is a methyl ($-CH_3$) group or ethyl ($-C_2H_5$) group; Ar is a more complex, frequently aromatic, entity

Figure 9.18. Phosphorothionate and phosphorodithioate ester insecticides. Arrows point to the hydrolyzable carboxyester linkages in malathion (see text).

In order for parathion to have a toxic effect, it must be converted metabolically to paraoxon (Figure 9.17), which is a potent inhibitor of acetylcholinesterase. Because of the time required for this conversion, symptoms develop several hours after exposure, whereas the toxic effects of TEPP or paraoxon develop much more rapidly. Humans poisoned by parathion exhibit skin twitching and respiratory distress. In fatal cases, respiratory failure occurs due to central nervous system paralysis.

Phosphorodithioate Insecticides

Malathion is the best known of the phosphorodithioate insect-icides. It has a relatively high insect:mammal toxicity ratio because of its two carboxyester linkages which are hydrolyzable by carboxyl-ase enzymes (possessed by mammals, but not insects) to relatively non-toxic products. For example, although malathion is a very effective insecticide, its LD_{50} for adult male rats is about 100 times that of parathion.

Organophosphorus Military Poisons

Powerful inhibitors of acetylcholinesterase enzyme, organophos-phorus "nerve gas" military poisons such as **Sarin** and **VX** (Figure 9.19) are among the most toxic synthetic compounds ever made.[15] A

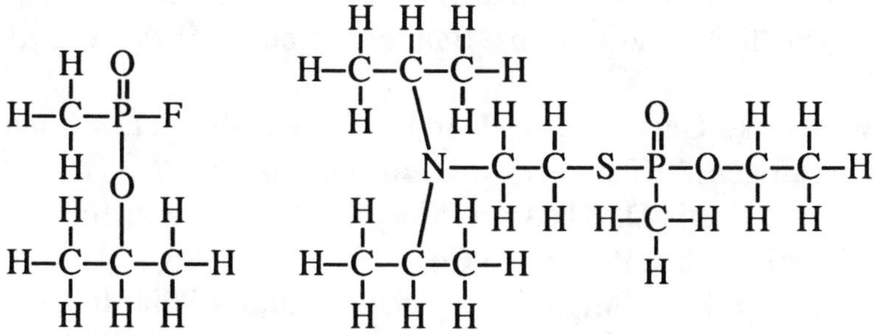

Figure 9.19. Two examples of organophosphate military poisons.

systemic poison to the central nervous system that is readily absorbed as a liquid through the skin, Sarin may be lethal at doses as low as about 0.01 mg/kg; a single drop can kill a human. Other examples of these deadly poisons are **Tabun** (O-ethyl N,N-dimethylphosphor-amidocyanidate), **Soman** (*o*-pinacolyl methylphosphonofluoridate), and "**DF**" (methylphosphonyldifluoride).

LITERATURE CITED

1. Graham, John D., Laura C. Green, and Marc J. Roberts, *In Search of Safety; Chemicals and Cancer Risk*, Harvard University Press, Cambridge, Mass., 1988.
2. Cornish, Herbert H., "Solvents and Vapors," Chapter 18 in *Casarett and Doull's Toxicology*, 2nd ed., John Doull, Curtis D. Klaassen and Mary O. Amdur, Eds., Macmillan Publishing Co., New York, 1980.
3. Gosselin, Robert E., Roger P. Smith, and Harold C. Hodge, "Phenol," in *Clinical Toxicology of Commercial Products*, 5th ed., Williams and Wilkins, Baltimore/London, 1984, pp. III-344–III-348.
4. Gammage, R. G., and C. C. Travis, "Formaldehyde Exposure and Risk in Mobile Homes," Chapter 17 in *The Risk Assessment of Environmental and Human Health Hazards: A Textbook of Case Studies*, John Wiley and Sons, New York, 1989, pp. 601-611.
5. Gosselin, Robert E., Roger P. Smith, and Harold C. Hodge, "Aniline," in *Clinical Toxicology of Commercial Products,* 5th ed., Williams and Wilkins, Baltimore/London, 1984, pp. III-32–III-36.
6. Williams, Gary M., and John H. Weisburger, "Chemical Carcinogens," Chapter 5 in *Casarett and Doull's Toxicology*, 3rd ed., Curtis D. Klaassen, Mary O. Amdur, and John Doull, Eds., Macmillan Publishing Co., New York, 1986, p. 117.
7. Lepowski, Wil, "Methyl Isocyanate: Studies Point to Systemic Effects," *Chemical and Engineering News*, June 13, 1988, p. 6.
8. Gosselin, Robert E., Roger P. Smith, and Harold C. Hodge, "Paraquat," in *Clinical Toxicology of Commercial Products,* 5th ed., Williams and Wilkins, Baltimore/London, 1984, pp. III-328–III-336.
9. Shabecoff, Philip, "Industry Fights Use of Animal Tests to Assess Cancer Risk," *New York Times*, July 25, 1989, p. 20.

10. Stopford, Woodhall, "The Toxic Responses of Pesticides," Chapter 11 in *Industrial Toxicology*, Phillip L. Williams and James L. Burson, Eds., Van Nostrand Reinhold Co., New York, 1985, pp. 211–229.
11. Safe, S., Ed., *Polychlorinated Biphenyls (PCBs): Mammalian and Environmental Toxicology*, Springer-Verlag, New York, 1987.
12. Silberner, J., "Common Herbicide Linked to Cancer," *Science News*, **130**(11), 167–174 (1986).
13. "Dioxin is Found Not to Increase Birth Defects," *New York Times*, March 18, 1988, p. 12.
14. Levi, Patricia E., "Toxic Action," Chapter 6 in *Modern Toxicology*, Ernest Hodgson and Patricia E. Levi, Eds., Elsevier, New York, 1987, pp. 133-184.
15. "Global Experts Offer Advice on Chemical Weapons Treaty," *Chemical and Engineering News*, July 27, 1987, pp. 16-27.
16. Gosselin, Robert E., Roger P. Smith, and Harold C. Hodge, "Tri-*ortho*-cresyl Phosphate," in *Clinical Toxicology of Commercial Products,* 5th ed., Williams and Wilkins, Baltimore/London, 1984, pp. III-388–III-393.

10

Biohazards

10.1.INTRODUCTION

In this book **biohazard** is used as a general term to refer to potentially harmful substances produced by organisms or by facilities that deal with organisms. For example, a toxin produced by bacteria or waste materials produced by laboratories that deal with research animals are both discussed as "biohazards."

Biomedical waste[1] is a term that has come into use to describe wastes that originate from sources involved with the treatment of, or research on, humans or animals. Facilities that generate biomedical wastes obviously include hospitals and clinics, as well as pathology laboratories, nursing homes, dialysis centers, veterinary facilities, research laboratories, and pharmaceutical, cosmetics, and food industries. Recently, concern has increased over biomedical wastes because of reports that such materials have been found on beaches, in refuse containers and other locations where exposure to the public might occur. During the summer of 1988, discarded hypodermic syringes and blood vials were washed up on U. S. East Coast beaches, adding to public concern. The U. S. Government estimates that quantities of biomedical wastes exceed 3 million tons per year in the U.S. at an estimated cost of $3.7 billion in 1988, projected to exceed $10 billion per year in 1991. The fear of AIDS infections from contaminated hypodermic needles and blood has also added to concern over biomedical wastes.

The disposal of biomedical wastes has become more of a problem in recent years with the increase of single-use disposable medical items. For example, diabetics in the U. S. may use as many as 1 billion disposable syringes each year. (Proposals have been made to charge deposits on disposable syringes to encourage their return for proper disposal.) Increased use of disposable syringes, "sharps," and other single-use materials has increased the proportion of biomedical wastes designated as infectious (see Section 10.2). Another factor leading to an increase in quantities of biomedical wastes has been the growth of facilities outside of hospitals, including nursing homes, kidney dialysis centers, blood banks, home-care sickrooms, and walk-in medical and dental clinics. The special problems of disposal by smaller clinics have been detailed in a report on biomedical wastes.[2]

Disposable plastics now account for about 1/3 of hospital wastes. Many of the plastics in biomedical wastes are organochlorine polymers, which complicates their destruction by burning because of emissions of acidic HCl and perhaps minute amounts of chlorinated dioxins. Nevertheless, incineration remains the major disposal option for biomedical wastes from hospitals.

Hospitals and clinics produce a large variety of waste materials. These include plastics, glass, paper, pads, swabs, gauze, disposable clothing, disinfectants, fluids, fecal matter, and even anatomical parts. Veterinary clinics and animal research facilities produce similar kinds of products as well as bedding material and shavings.

The amount of biomedical wastes produced by a hospital, clinic, or research facility depends upon what the facility does. For example, a hospital in which a great deal of surgery is performed will produce different quantities and a different mix of wastes compared to a nursing home. Typically a hospital will generate about 13 pounds per day per patient compared to about 3 pounds for a rest home and 0.5 pounds for a pathology laboratory.

In addition to wastes of biological origin, hospitals, clinics, and animal research laboratories produce a variety of solvent and chemical wastes. These include hydrocarbon solvents, such as cyclohexane, pentane petroleum ether, and xylene; alcohols, including methanol, ethanol, and butanol; diethyl ether; acetone; tetrahydrofuran; and methyl cellusolve. Other waste chemicals from biomedical sources include acids, alkalis, chromates, disinfectants, formalin, pharmaceutical agents, chemical reagent and staining solutions, ethylene oxide, and nitrous oxide. Radioisotopes are also produced as byproducts of nuclear medical diagnostic and treatment procedures. Radioisotope and regulated chemical wastes must usually be sent to a licensed treatment and disposal facility.

10.2. INFECTIOUS WASTES

Infectious wastes are those that are capable of causing disease in individuals that come into contact with the wastes.[3] The U.S. EPA has placed infectious wastes into several major categories.[4] Two of these consist of isolation wastes and pathological wastes. Another category is made up of cultures and stock of infectious agents and biological materials associated with these materials. Because of the danger of infection from blood-borne diseases such as AIDS and hepatitis, human blood and blood products are in a separate class. Contaminated sharps (hypodermic needles) are classified as infectious wastes. Another category consists of animal carcasses, body parts, and bedding and a final one of miscellaneous contaminated wastes.

Infectious wastes now make up 10 to 15 percent of total hospital wastes. About 20% of infectious wastes consist of plastic (compared to only about 7% of general municipal wastes).

The most obvious sources of infectious wastes are hospitals and other health care facilities. Examples of areas in these facilities that contribute to infectious wastes are surgery, pathology and histology laboratories, oncology, blood bank, dialysis, and the morgue. Veterinary care and pharmaceutical industries also contribute to infectious wastes, as do academic and non-academic research laboratories, food processing, and the cosmetics industries. For ease of identification, infectious wastes are collected in red plastic bags of a special kind, which has given rise to the name of "red-bag wastes."

Disposal of infectious waste has become a major problem for hospitals and other facilities that produce it and a subject of much concern for environmentalists, regulatory agencies and waste handlers. Many landfill operations will not accept even sterilized medical waste and virtually none of it can be recycled. As discussed in Section 10.3, incineration is about the only available option.

Treatment of Infectious Wastes

Several methods are available to treat infectious wastes to make them non-hazardous. Incineration (see Section 10.3 and Chapter 14) is very effective, and both destroys the infectious agents and gets rid of the combustible fraction of the wastes. Infectious agents can also be destroyed by chemical disinfection or irradiation. Sterilization in an autoclave (under pressurized steam) effectively destroys pathogens, although it is relatively expensive and limited in capacity.

Recent Legislation

Based on earlier New Jersey and New York state legislation, the Federal Medical Waste Tracking Act of 1988 was passed as a 2-year demonstration program to develop means of dealing with medical wastes. In addition to New Jersey and New York, this legislation applies to the District of Columbia, Connecticut, Puerto Rico, Rhode Island, and Louisiana. The act, which went into effect in June, 1989, prescribes systems for handling medical wastes and for tracking the wastes from their origin through various handlers and to their final disposal. It requires even small producers of medical wastes to keep records pertaining to waste disposal.

10.3. INCINERATION OF BIOMEDICAL WASTES

Incineration has been practiced for many years as a means of disposing of potentially hazardous biomedical wastes and is likely to remain so for at least the next decade. Medical wastes incinerated by the generating facility are not required to be included under the Federal Medical Waste Tracking Act (see above), giving even more incentive for incineration. Many hospital incinerators were built and installed before rigorous regulations regarding their design and operation were in place. As a result of deficiencies and obsolescence in design as well as improper operation, some biomedical waste incinerators have discharged soot, ash, and smoke and have been a source of unpleasant odors, causing an unfavorable image with the public. The location of most hospital incinerators in densely populated areas has aggravated the problem.

Many hospitals in the U. S. are now faced with making a decision regarding whether or not to keep existing incinerators, upgraded to meet current standards. Other options include construction of a new incinerator serving one hospital (plus associated clinics and physicians' offices), construction of a centralized facility serving several hospitals, or contracting with commercial concerns for biomedical waste disposal. In areas of relatively high population density, construction of a regional facility is often the most attractive in terms of overall cost and efficiency. However, such a facility tends to generate adverse public reaction.

Fortunately, biomedical waste incinerators need not be large. A capacity of several tons per day is adequate for many such facilities. When hazardous waste is incinerated along with nonhazardous material, all of the ash is deemed hazardous (unless it can be delisted) from the standpoint of RCRA regulation, which can limit the incineration of biomedical wastes.

Many biomedical waste incinerators are retort incinerators, often fired manually. These incinerators may have two chambers. Typically wastes are introduced into a primary combustion chamber in which most of the volatile portion of the waste is driven off and partially burned. A secondary combustion chamber fired with a supplemental fuel and excess air provides the residence time, high temperature, and oxygen needed to destroy vaporized or entrained solid materials prior to discharge through the stack. The other major kind of incinerator used to destroy biomedical wastes is the rotary kiln type described in Section 14.4.

The operation of a small biomedical waste incinerator is complicated by the variable and often unknown composition of the wastes. Usually "red bag" wastes are fed to an incinerator without opening the bags and the operator may not know how well the contents will burn. The heat content of the waste constituents may be as high as almost 20,000 Btu/Lb for polyethylene plastics or as low as 1,000 Btu/Lb for pathological wastes.

Byproducts and emissions from hospital incinerators are coming under increasing control. The major solid byproduct is ash, which is usually relatively safe, but may contain heavy metals or other hazardous substances. The incinerator temperature should be kept below 1,000°C to prevent the ash from slagging and forming harmful deposits in the incinerator. Stack emissions may contain acid gases and particulate matter. The predominant acid gas from the burning of biomedical wastes is usually hydrogen chloride, HCl, produced in the combustion of chloride-containing plastics (polyvinylchloride). Most of the particulate matter in stack gas is carbonaceous material, which can be removed by fabric filters, venturi scrubbers, electrostatic precipitators or other devices.

10.4. HAZARDOUS NATURAL PRODUCTS

Organisms of various kinds produce a variety of natural products that are hazardous because of their toxicities.[5] Natural products may even present other hazards. Some plant materials are dangerously flammable when dry, and grain dusts have caused many fatal explosions in grain elevators. Perhaps the most acutely toxic substance known is botulism toxin produced by the anaerobic bacterium *Clostridium botulinum* (see Section 10.5). Mycotoxins generated by fungi (molds) can cause a number of human maladies; some mycotoxins (aflatoxins) are carcinogenic to experimental animals. Allergy-causing pollens are much more likely than hazardous wastes to inflict misery on the average citizen. Venoms from wasps, spiders, scorpions, and reptiles can be fatal to humans. Each year, in the Orient, tetrodotoxin from improperly prepared puffer fish makes this dish the last delicacy consumed by some unfortunate diners. The stories of Socrates' execution from being forced to drink an extract of the deadly poisonous spotted hemlock plant and Cleopatra's suicide at the fangs of a venomous asp are rooted in antiquity.

It is beyond the scope of this book to discuss toxic and otherwise hazardous natural products in detail. However, two classes of natural products—mycotoxins and alkaloids—are covered here briefly because in some respects their behavior as contaminants or atmospheric pollutants (such as aflatoxins in dust from moldy grain) is similar to that of hazardous waste chemicals.

Aflatoxins

Aflatoxins are produced by fungi on moldy food, particularly nuts, some cereal grains, and oil seeds. The most notorious of the aflatoxins is aflatoxin B_1, for which the structural formula is shown in Figure 10.1. Produced by *Aspergillus niger*, it is a potent liver toxin and liver carcinogen in some species. It is metabolized in the liver to an epoxide (see Section 4.8). The product is electrophilic with a strong tendency to bond covalently to protein, DNA, and RNA. Other common aflatoxins produced by molds are those designated by the letters B_2, G_1, G_2, and M_1.

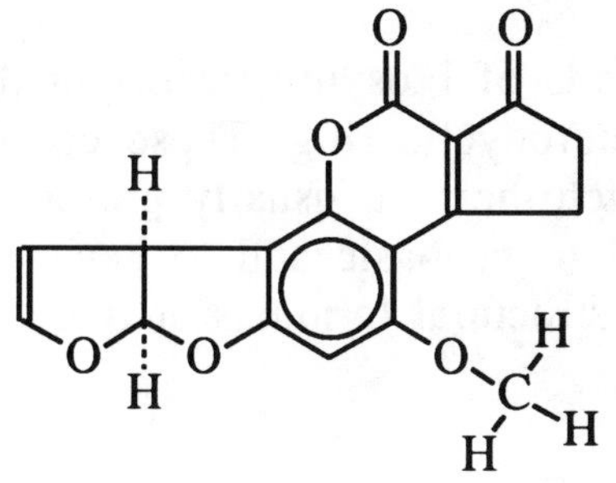

Figure 10.1. Structural formula of aflatoxin B_1, a mycotoxin.

Trichothecenes are composed of 40 or more structurally related compounds produced by a variety of molds, including *Cephalosporium*, *Fusarium*, *Myrothecium*, and *Trichoderma*,[3] which grow predominantly on grains. Much of the available information on human toxicity of trichothecenes was obtained from an outbreak of poisoning in Siberia in 1944. During the food shortages associated with World War II, the victims ate moldy barley, millet, and wheat. People who ate this grain suffered from skin inflammation; gastrointestinal tract disorders, including vomiting and diarrhea; and multiple hemorrhage. About 10 percent of those afflicted died.

Toxins from Mushrooms

Mushrooms are fungi that produce rather intricate spore-forming structures. Some mushrooms are edible, whereas others produce toxins that cause potentially fatal mushroom poisoning, a condition called **mycetism**. There are six major classes of toxic agents that occur in poisonous mushrooms.[6] These are ibotinic acid, muscarine, coprine, psilobycin, monomethyl hydralazine, and cyclopeptides. The last category is the most deadly and is produced by mushrooms with the colorful names of "Destroying Angel" and "Death Cap."

Despite the deadly nature of some mushrooms, relatively few people are killed by them. Ibotinic acid and psilobycin are psychoactive toxins that produce symptoms such as mood elevation and hallucinations. Mushrooms that contain these substances are sometimes ingested for "recreational" purposes.

Alkaloids

Alkaloids are compounds of biosynthetic origin that contain nitrogen, usually in a heterocyclic ring. These compounds are produced by plants in which they are usually present as salts of organic acids.[7] They tend to be basic and to have a variety of physiological effects. The structural formulas of five alkaloids are given in Figure 10.2.

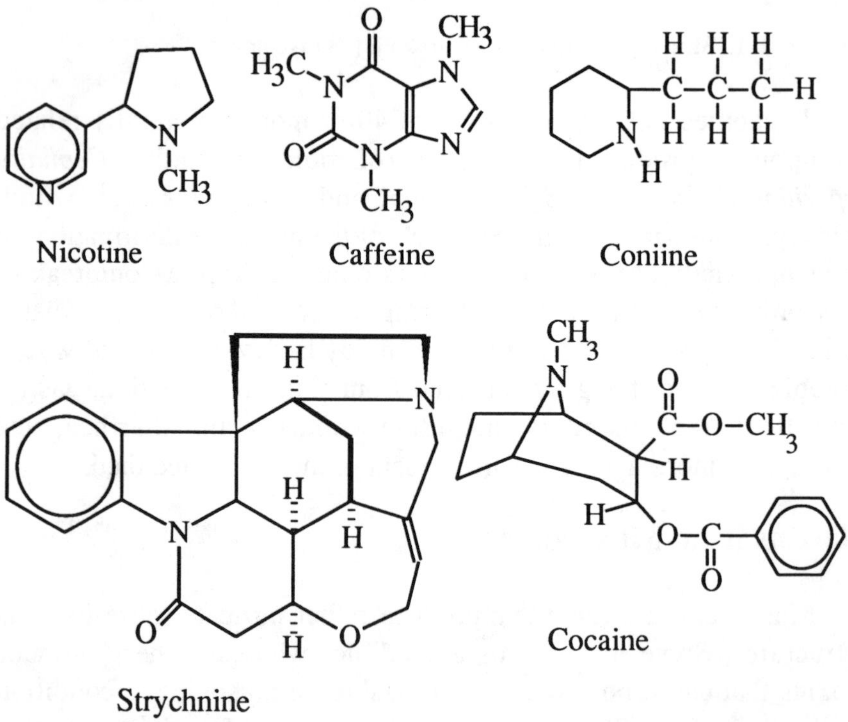

Figure 10.2. Structural formulas of typical alkaloids.

Among the alkaloids are some well-known (and dangerous) compounds. Nicotine is an agent in tobacco, toxic enough to be used as an insecticide, that has been described as "one of the most toxic of all poisons and (it) acts with great rapidity."[8] In 1988 the U.S. Surgeon General declared nicotine to be an addictive substance. Coniine is the major toxic agent in poison hemlock. Alkaloidal strychnine is a powerful, fast-acting convulsant. Cocaine in the concentrated form of "crack" is currently the illicit drug of greatest concern. Quinine and stereoisomeric quinidine are alkaloids that are effective antimalarial agents. Like some other alkaloids, caffeine contains oxygen. It is a stimulant that can be fatal to humans in a relatively high dose of about 10 grams.

10.5. BIOLOGICAL WARFARE AGENTS

Research has been conducted for decades on toxic and infectious agents for warfare. Such agents have a high potential to kill or disable people or animals and to damage crops. In recent years the development of recombinant DNA for "genetic engineering" has increased the potential to produce microorganisms with potential uses for biological warfare.

Historically, botulism and anthrax have received the most attention as potential biological weapons. *Clostridium botulinum* is essentially in a class by itself as a toxic substance. Just a "taste" of food contaminated with it can be fatal. Butolin "X," a chemical warfare agent, consists of botulism toxin dissolved in dimethyl-sulfoxide, DMSO. As mentioned in Section 9.6, DMSO acts as a carrier solvent and the toxin dissolved in it can be carried through skin with fatal results. Strains of *Clostridium botulinum* have been developed that are especially effective in producing botulism toxin.

Anthrax can infect both humans and animals through inhalation or absorption through breaks in the skin. The bacteria that cause anthrax, *Bacillus anthracis*, multiply very rapidly and produce hardy endospores that can last for many years. Without antibiotic therapy, anthrax has a 100 percent mortality rate. Accidental release of anthrax spores in a U.S. Army experiment in Utah killed several thousand sheep during the 1950s.

In addition to anthrax, several other infectious agents have been considered for biological warfare. These include tularemia, which causes rabbit fever; brucellosis, which causes undulant fever; and psittacosis, which causes parrot fever. In addition, glanders, a respiratory disease, was considered for biological warfare.

During the Vietnam war years concern over biological warfare was heightened by reports of "yellow rain" in Laos and Kampuchea. Some authorities believed that the material was a mycotoxin (fungal toxin) provided to Vietnam by the USSR. Persuasive arguments have since been presented that the mystery substance was bee feces discharged by the insects to dissipate heat!

LITERATURE CITED

1. Brunner, Calvin R., and Courtney H. Brown, "Hospital Waste Disposal by Incineration, *Journal of the Air Pollution Control Federation,* **38**, 1297-1309, (1988).
2. Stevens, William K., "Medical Waste is Piling Up, Generating New Concerns," *New York Times*, June 27, 1989, p. 17.
3. Cross, Frank L., Jr., and Rosemary Robinson, "Infectious Waste," Section 4.4 in *Standard Handbook of Hazardous Waste Treatment and Disposal*, Harry M. Freeman, Ed., McGraw Hill, New York, 1989, pp. 4.35-4.45.
4. U.S. Environmental Protection Agency, *EPA Guide for Infectious Waste Management*, U.S. EPA Office of Solid Waste, Washington, D.C., 1989.
5. Harris, John B., Ed, *Natural Toxins: Animal, Plant, Microbial*, Oxford University Press, New York, 1987.
6. Stallard, Donald J., Jr., "Mushroom Toxicity," *Missouri Medicine*, **85**, 674-677 (1988).
7. Cordell, Godfrey A., "Alkaloids," in *Kirk-Othmer Concise Encyclopedia of Chemical Technology*, Wiley-Interscience, New York, 1985, pp. 63-67.
8. Gosselin, Robert E., Roger P. Smith, and Harold C. Hodge, "Nicotine," in *Clinical Toxicology of Commercial Products,* 5th ed., Williams and Wilkins, Baltimore/London, 1984, pp. III-311–III-314.

11

Reduction, Recycling, and Resource Recovery

11.1. OVERVIEW OF WASTE TREATMENT FOR RESOURCE RECOVERY

As costs of hazardous waste disposal have mounted, the economics of putting wastes to some good use have improved. There are four very broad areas in which something of value may be obtained from wastes. They are the following:

- Direct recycle as raw material to the generator
- Transfer as a raw material to another process
- Utilization for pollution control or waste treatment
- Recovery of energy

Direct recycle is utilized frequently in the chemical industry when raw materials are not completely consumed in a synthesis process and are simply returned as feedstock. A substance that is a waste product from one process may serve as a raw material for another, sometimes in an entirely different industry. Some process wastes can be used to treat other wastes or for pollution control. For example, waste lime from the collection of dust in lime processing can be used to neutralize waste acid or to precipitate metals from wastewater. Whenever it is desirable to incinerate wastes, energy recovery should be considered, assuming that the quantities are large enough and that the production of waste is uniform enough to make energy recovery viable.

Recycling of materials, including hazardous substances, has a long history. This is especially true of economies stressed by poverty, wartime shortages, and, most recently, restrictions on the disposal of hazardous substances. Great Britain, a country that suffered severe material shortages due to destruction of shipping by enemy submarines during World War II, set up the National Industrial Recovery Association in 1942. In some respects this organization resembled modern day "waste exchanges."

Various organizations have been established to provide information about hazardous waste exchange and recycle. Several specific accomplishments of one such organization, the Minnesota Technical Assistance Program, have been described.[1] Among the examples cited were the following: (1) A user was located for waste redistilled dichloromethane produced by a process used to attach rubber gaskets to sewer pipes. (2) A process that produced large quantities of metal hydroxide sludge from the treatment of metals plating rinse water was replaced by an electrolytic and ion exchange metal recovery system that reduced sludge production by 90 percent. (3) Waste solvent from paint-thinning operations was reclaimed with a small still and reused for thinning paint and for cleaning equipment.

Numerous factors are involved in determining the feasibility of waste recycle and exchange. First of all, the material has to be good for something! It must not contain any undesirable impurities that cannot be dealt with safely and economically. For example, uranium mine tailings make good construction fill material except for the problem of radioactive radon gas emissions. Toxic lead may have to be removed from zinc oxide collected from air pollution control devices in zinc smelters if the zinc oxide is to be used to make zinc sulfate fertilizer. Transportation of the waste to an industry that can use it has to be feasible; use on site is most desirable.

Examples of Recycling

Recycling of scrap industrial impurities and products occurs on a large scale with a number of different materials. Most of these materials are not hazardous, but, as with most large-scale industrial operations, their recycle may involve the use or production of hazardous substances. Some of the more important examples are discussed briefly here.

Ferrous Metals

Ferrous metals are those that are composed primarily of iron. Ferrous metal scrap is recycled from steel mills, scrap steel (such as junked automobiles), and from municipal waste. An advantage of iron for recycling is that it can be separated magnetically from other wastes. Most ferrous scrap is recycled to steel making. In recent years electric-arc furnaces have become popular for high-quality steel manufacture, which has increased the demand for ferrous scrap because they use it exclusively for feedstock.

Nonferrous Metals

The commonly used nonferrous metals are produced in smaller quantities than iron, are more expensive, and are in shorter supply. A number of nonferrous metals are toxic in some form (lead and cadmium compounds, mercury metal vapor). All of these factors combined make nonferrous metal recycle highly desirable. Aluminum ranks next to iron in terms of quantities recycled. Major quantities of copper and copper alloys, zinc, and lead are recycled. Lesser amounts of cadmium, tin, and mercury are recycled. Silver is recycled from X-ray film and electronic applications using highly toxic cyanide solutions. Scrap copper is usually processed in a furnace, then refined electrolytically. Most scrap lead comes from spent automobile batteries. Recycled metal amounts to over 1/3 of the U.S. supply of aluminum and over half of domestic supplies of copper and lead.

Glass

Glass, which makes up about 10 percent of municipal refuse, is a popular material to recycle. A major hurdle with glass recycle is the sorting of glass by color after it has been separated from the municipal refuse. Few municipal glass reclamation systems accomplish more than 50 percent recycle of waste glass.

Paper

Paper, along with metals and glass is one of the items most commonly recycled from municipal refuse. The cellulose fibers in recycled paper tend to be shorter, more flattened, less strong and drier than fibers freshly produced from wood pulp. In addition, used paper is contaminated with constituents such as adhesives, inks, clay, and coatings, as well as dirt, grease, and other impurities. There have also been reports of paper contaminated with hazardous materials, such as PCBs. All of these factors tend to complicate paper recycle. Careful sorting of scrap paper at the source is important for successful paper recycling. New developments in paper recycling have made it more practical and applicable to a wider variety of raw material. Consideration of recycling in the composition of new paper products would also be helpful.

Plastics

Plastic is a term that describes a variety of polymeric materials that can be shaped or molded while in a liquid (plastic) state and formed as desired for containers and other objects. Thermosetting plastics are those that are synthesized and formed as part of the manufacturing process These plastics cannot be melted and repro- cessed, which limits the possibilites for recycling. Thermoplastic materials can be melted by heat and refabricated, which enables their recycle. Recycling of scrap thermoplastic produced in plastic fabrication is accomplished by grinding, adding more heat stabilizers, and mixing it with the feed of raw plastic.

Since World War II plastic has become a major constituent of municipal wastes. Most plastics do not biodegrade well or at all and they contribute a significant fraction of residual solid wastes remaining in landfills after biodegradation has occurred. Because of variable composition, impurities, pigment constituents and other factors, recycling of plastics in municipal solid wastes has proven difficult. As with paper, formulation of plastics with recycling as a major goal would prove helpful.

Rubber

Rubber is a recyclable material as was demonstrated by the rubber-short major powers in World War II. Now synthetic polymers provide an abundance of rubber, but the need to recycle rubber is greater than ever because of the literally mountains of used rubber tires that have accumulated in numerous locations. The magnitude of the problem was illustrated by a massive fire involving 14 million tires at a tire disposal site in Hagersville, Ontario, Canada in February, 1990.[2] Rubber is a hydrocarbon polymer that may contain other materials, such as carbon black filler (see Section 6.3); the Hagersville fire yielded about 160,000 gallons of oil byproduct that was collected and recycled in a nearby petroleum refinery. Rubber has a good fuel value of about 14,000 Btu/lb, which is comparable to the best grades of coal. Some kinds of rubber can be ground and recycled through the rubber fabrication process. A major complication in the burning and recycling of rubber in tires is the presence of strong steel wires in many radial tires. The steel often jams grinding equipment and handling mechanisms in both recycling and incineration operations.

11.2 WASTE MANAGEMENT FOR RESOURCE RECOVERY

Waste Reduction and Waste Minimization

Many hazardous waste problems can be avoided at early stages by **waste reduction**[3] and **waste minimization**. As these terms are most commonly used, waste reduction refers to source reduction— less waste-producing materials in, less waste out. Waste minimization can include treatment processes, such as incineration, which reduce the quantities of wastes requiring ultimate disposal. Reference is sometimes made to waste abatement in terms of the four Rs— reduction, reuse, reclamation, and recycling.[4]

Numerous factors, both economic and regulatory, favor waste reduction practices. Costs of treating and disposing of hazardous wastes have escalated along with problems in siting and getting permits for new hazardous waste storage and treatment units. Public opinion certainly favors reduced generation of wastes and there is a growing concern with liability associated with hazardous wastes. Furthermore, HSWA (1984) requires that "Wherever feasible, the generation of hazardous waste is to be reduced or eliminated . . .," the purpose of which is ". . . to minimize the present and future threat to human health and the environment."

Hierarchy of Waste Minimization

There exists a hierarchy of waste minimization, ranging from simple, readily-accomplished measures through those that involve relatively drastic measures.[5] These are the following:

- Increased diligence in housekeeping, such as using minimal amounts of water for washing equipment
- Substitution of less hazardous materials
- Recycling and reuse
- Process modification
- Disposal

There are several ways in which wastes can be minimized. These include source reduction and waste separation and concentration. Some wastes can be reused and recycled. Waste exchange between industries can be practiced. The most effective approaches to minimizing wastes center around careful control of manufacturing processes,[6] taking into consideration discharges and the potential for waste minimization at every step of manufacturing. When waste minimization is considered at a very early stage of process development or redesign, modifications may be made well upstream from discharge and treatment. Viewing the process as a whole (as outlined for a generalized chemical manufacturing process in Figure 11.1) often enables crucial identification of the source of a waste, such as a raw material impurity, which may be easier to eliminate from the feedstock than to treat as a waste. Similar approaches may be used with catalysts and process solvents.

Substantial waste reduction can be accomplished by careful inventory management.[7] This requires that excess amounts of raw materials not be ordered which require disposal as hazardous wastes when left over or out of date. Feedstocks and materials used in production processes can be examined for hazardous substance content and in some cases modified to reduce production of hazardous wastes. For this purpose the manufacturer's **Material Safety Data Sheets** can be very useful. Examples of hazardous waste reduction by substitution of materials used include replacement of cadmium in inks, pigments, and plating baths; substitution of chromium(III) for toxic chromate in plating baths; replacement of toxic cyanide in plating bath media; and substitution of water-based formulations for solvent-based formulations in paints, adhesives and degreasers.

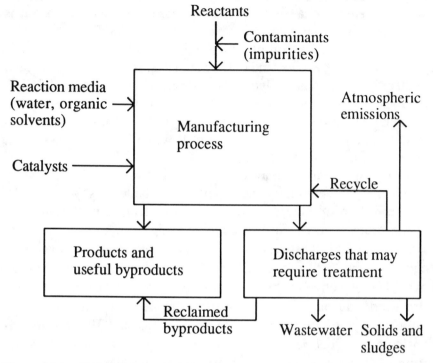

Figure 11.1. Chemical manufacturing process from the viewpoint of discharges and waste minimization.

Modification of the manufacturing process can yield substantial waste reduction. Some such modifications are of a chemical nature. Changes in chemical reaction conditions can minimize production of byproduct hazardous substances. In some cases potentially hazardous catalysts, such as those formulated from toxic substances, can be replaced by catalysts that are non-hazardous or that can be recycled rather than discarded.

Volume reduction can be important in minimizing wastes. An important means of volume reduction is dewatering and drying of sludge.

Wherever possible, recycling and reuse should be accomplished on-site; a process that produces recyclable materials is often the most likely to have use for them. Metals can be recovered from waste plating solutions and recycled in the plant, waste cleaning and rinsing solvents can be used in paint formulations, and solvents can be recovered by distillation or by condensation of solvent vapor.

Some very impressive reductions in amounts of wastes produced have been described in a model study of the electronics industry,[8] chosen for study because of its growth orientation and because it is among the top 20 waste solvent generating industries. The electronics industry uses metals for plating (copper, nickel, tin, and lead), cyanide, chelating agents, and other chemicals. In one facility studied, the quantity of metals sludge produced was reduced from 34,250 to 500 tons per year and in another from 700 to 22 tons per year. Another facility reduced its waste methyl chloroform solvent to 6 percent of its previous production and its waste Freon to 10 percent; the solvents saved were recycled.

A Monsanto plastics plant was able to cut loss of a resin to 1/4 by improved equipment maintenance and use of better gasket materials. Polymerization of excess resin converted it from a hazardous to a nonhazardous material, further reducing disposal costs.

Since 1975 Minnesota Mining and Manufacturing (3M) has been engaged in a "3P" program, standing for "Pollution Prevention Pays." The objective of this program has been to reduce the need for retrofitted add-on pollution control devices by eliminating or minimizing pollution at the source. In so doing, the 3P program has considered product development, engineering design, and manufacturing processes; it emphasizes product reformulation, changes in processes, resource recovery, and modifications of equipment design.[9] This program has succeeded in eliminating about 1.5 billion gallons of industrial wastewater containing about 10,000 tons of water pollutants each year. Annual production of air pollutants has been cut by an estimated 100,000 tons and the production of 150,000 tons of sludge is prevented each year. Examples of measures used include burning air contaminated with solvents from tape manufacture in an industrial oven to recover fuel from the solvents, replacement of a solvent-based coating for pharmaceutical tablets with a water-based coating that produces no solvent vapors, and replacing toxic chemical cleaning sprays with pumice scrubbers to clean copper sheeting.

Plans are underway to supplement the 3P program with a "3P +" plan designed to further curtail hydrocarbon emissions by 45,000 tons per year in the U.S. and 10,000 tons per year at 3M plants outside the U.S.

Dow Chemical Co. has its "WRAP" program standing for "Waste Reduction Always Pays." In addition to source reduction this emphasis is placed upon recycling and interchange of materials between manufacturing plants as well as reclamation of materials by measures such as air stripping or carbon adsorption. Where possible, combustible materials have been burned in industrial furnaces. Dow estimates that a 90 percent reduction of air and water emissions was achieved from 1974 to 1988, even though product output increased during that time.

In 1987 Chevron Corp. launched a waste minimization program called SMART, which stands for Save Money and Reduce Toxics.[10] The company saves about $300,000 annually by using triple rinse water from the cleaning of storage tanks as the aqueous medium in making emulsion asphalt. Disposal costs for the water had reached $1.75 per gallon prior to this application. Byproduct disulfide oil is recycled to another company that makes sulfuric acid. Spent caustic from a process used to remove hydrogen sulfide and organosulfur compounds from hydrocarbon gases is recycled to pulp and paper mills.

Waste Treatment

Waste treatment may occur at three major levels — **primary**, **secondary**, and **polishing**, somewhat analogous to the treatment of wastewater (see Section 3.7). Primary treatment is generally regarded as preparation for further treatment, although it can result in the removal of byproducts and reduction of the quantity and hazard of the waste. Secondary treatment detoxifies, destroys, and removes hazardous constituents. Polishing usually refers to treatment of water that is removed from wastes so that it may be safely discharged. However, the term can be broadened to apply to the treatment of other products as well so that they may be safely discharged or recycled. These three phases of waste treatment are illustrated in Figure 11.2.

Options for waste recovery may be divided among the following major categories:

- Physical separation
- Separation by phase transition of wastes
- Separation by transfer between phases
- Molecular separation
- Chemical separation

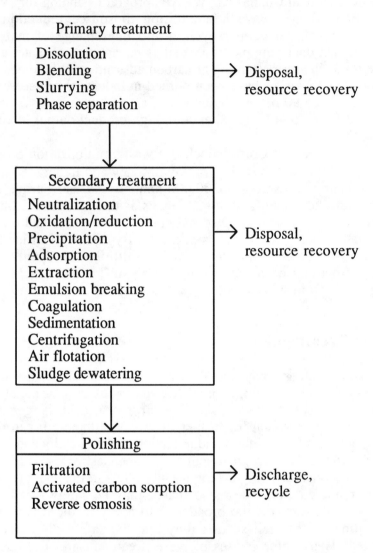

Figure 11.2. Major phases of waste treatment.

Many separation processes involve combinations of the above methods. Physical separation, which is discussed in detail in Chapter 12, includes settling and decanting, centrifugation, filtration, and flotation.[7] Separation by phase transition refers to non-chemical processes by which the material separated enters a different phase; it can be evaporated, sublimed, condensed from the vapor phase, distilled, or caused to precipitate by cooling a solution or evaporating solvent from it. A different approach is that of separation between phases, in which a substance in solution or bound to a solid is transferred to a similar state in another phase. An example would be a liquid-liquid extraction in which a heavy metal ion dissolved in aqueous solution is transferred to a separate organic solvent phase as a chelated metal. Other separations between phases are sorption onto solids (activated carbon or resins), ion exchange, and supercritical fluid extractions. Molecular separations occur across membranes that may be selective for particular types or sizes of species. These separations include reverse osmosis, electrodialysis, and ultrafiltration.

Chemical separations result from chemical reactions, such as the precipitation of cadmium ion in solution by hydrogen sulfide gas:

$$H_2S(g) \; + \; Cd^{2+} \; \longrightarrow \; CdS(s) \; + \; 2H^+ \qquad (11.2.1)$$

Other types of chemical reactions used for separations are reduction, oxidation (including electrolytic reduction and oxidation reactions), and cementation (replacement of a metal in the elemental state with a relatively less active metal originally present as an ion in solution). Chemical separations are discussed in detail in Chapter 13.

11.3. RECYCLING

Several classes of substances are commonly recycled. Prominent among these are metals and compounds of metals. Recycled inorganic substances include alkaline compounds (such as sodium hydroxide used to remove sulfur compounds from petroleum products), acids (steel pickling liquor where impurities permit reuse), and salts (for example, ammonium sulfate from coal coking used as fertilizer). The greatest quantities of recycled organic substances consist of solvents and oils, such as hydraulic and lubricating oils. In the chemical and petroleum industry catalysts are recycled. Commercially, chemicals that are surplus, are off-specification, or have expired shelf lives may be recycled. Some recycled substances have agricultural uses, such as waste lime or phosphate-containing sludges used to treat and fertilize acidic soils.

11.4. WASTE OIL UTILIZATION AND RECOVERY

Annual production of **waste oil** in the U.S. is of the order of 4 billion liters per year. Around half of this amount is burned as fuel and lesser quantities are recycled or disposed as waste. Much of the improper disposal of waste oil is from individuals who change oil in their own vehicles.

Properties of Waste Oil

Although waste oil is regarded as an organic material, it contains inorganic substances, such as metals. Waste oil is generated from lubricants for vehicle engines, lubricants in industrial processes, and spent hydraulic fluids. The collection, recycling, treatment, and disposal of waste oil are all complicated by the fact that it comes from diverse, widely dispersed sources.[11]

Waste oil contains several classes of potentially hazardous contaminants divided between organic constituents and inorganic constituents. Among the organic constituents are polycyclic aromatic hydrocarbons consisting of two rings (naphthalene) or more (including the procarcinogen, benzo(a)pyrene). Some of these come from the petroleum base stock used to make the oil and others are generated by pyrolysis during the exposure of the oil to high temperatures. Chlorinated hydrocarbons, such as trichloroethylene, get into waste oil as contaminants and possibly from chemical processes associated with the use of the oil.

Metals are the predominant contaminants of waste oil that are of concern. Aluminum, chromium, and iron may get into the oil from wear of metal parts. Barium and zinc are often present from oil additives. Lead contaminates motor oil from leaded gasoline, but is becoming a less common contaminant as leaded gasoline is phased out. Limits for arsenic and cadmium are specified in used, recycled oil.

Recycling Waste Oil

Several processes are used to convert waste oil to a feedstock hydrocarbon liquid for lubricant formulation. These processes may be divided into the three major steps shown in Figure 11.3.

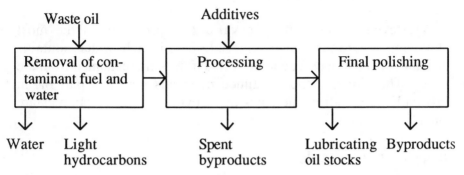

Figure 11.3. Major steps in reprocessing waste oil.

The first phase of waste oil treatment involves distillation to remove water and light ends that have come from condensation and contaminant fuel. The second phase, designated processing in Figure 11.3, varies considerably. It may be as simple as a vacuum distillation in which the three products are oil for further processing, a fuel oil cut, and a heavy residue. Another process uses a mixture of solvents including isopropyl and butyl alcohols and methylethyl ketone to dissolve the oil and leave contaminants as a sludge. Treatment with sulfuric acid followed by treatment with clay has been widely used. The sulfuric acid reacts with inorganic contaminants, which settle from the mixture as a sludge. Contact of the oil with clay removes acid and contaminants that cause odor and color.

The third step shown in Figure 11.3 employs vacuum distillation to separate lubricating oil stocks from a fuel fraction and heavy residue. This phase of treatment may also involve hydrofinishing, treatment with clay, and filtration.

Waste Oil Fuel

For economic reasons waste oil that is to be used for fuel is given minimal treatment of a physical nature, including settling, removal of water, and filtration. Metals in waste fuel oil become highly concentrated in its fly ash, which may be hazardous.

11.5. RECOVERY OF SOLUTES FROM WASTEWATER

Wastewater produced by a variety of processes is the most abundant hazardous waste material. Often such water contains contaminants that have economic value if they are reclaimed from the water. The value of the reclaimed materials may pay part or, in favorable cases, all of the costs of recovery. Several examples are cited in this section.

Recovery of Acids

Acidic solutions containing dissolved metals are common and troublesome wastes from a number of metal processing operations. Some examples of these solutions are listed below:

- HCl and Zn^{2+} from zinc stripping
- HNO_3 and Ni^{2+} from nickel stripping
- H_2SO_4 and Al^{3+} from aluminum anodizing
- H_2SO_4, HCl and Zn^{2+} from steel pickling
- HNO_3, HF, Fe^{2+}, and Cr^{3+} from stainless steel pickling
- HNO_3, H_2SO_4, Cu^{2+}, and Zn^{2+} from brass etching

These waste solutions must be treated to remove both metals and acids. Addition of lime works for this purpose as shown by the following reactions:

$$2HNO_3 + Ca(OH)_2 \longrightarrow Ca^{2+} + 2NO_3^- + 2H_2O \quad (11.5.1)$$

(Acid neutralization)

$$Cu^{2+} + Ca(OH)_2 \longrightarrow Ca^{2+} + Cu(OH)_2 \quad (11.5.2)$$

(Metal precipitation)

$$2HF + Ca(OH)_2 \longrightarrow CaF_2(s) + H_2O \quad (11.5.3)$$

(Precipitation of fluoride)

$$H_2SO_4 + Ca(OH)_2 \longrightarrow CaSO_4(s) + H_2O \quad (11.5.4)$$

(Acid neutralization, precipitation of sulfate)

However, lime produces large quantities of precipitate from which it is not possible to recover metals economically.

A system has been described[12] in which acid is absorbed by a resin and subsequently reclaimed by elution with water. The acid recovered can be recycled to the anodizing, pickling, plating, or etching operation from which it was recovered or it can be used for other purposes.

Acetic acid,

$$
\begin{array}{ccc}
& H & O \\
& | & \| \\
H- & C-C & -OH \quad \text{Acetic acid} \\
& | & \\
& H &
\end{array}
$$

is a common component of wastewater. Acetic acid can be extracted into organic solvents such as ethyl acetate or a mixture of ethyl acetate and benzene. After extraction, the volatile solvents may be separated from acetic acid by distillation.

Recovery of Phenol

Phenol is an important industrial chemical with many uses. Phenol and related compounds are produced in abundance in coal gasification and coal liquefaction. However, these industries are small on a global basis and the major source of phenol from coal is from coal coking where it occurs in a relatively concentrated form in byproduct water. Wastewaters from phenol resin manufacture and petroleum refining may also contain recoverable phenol.

Phenol is recovered from wastewater by solvent extraction, the principles of which are discussed in Section 12.8. Phenol is extracted from wastewater by a variety of water-immiscible organic solvents. Some of the solvents used for phenol extraction are hydrocarbons, including fractions distilled from crude oil, benzene, and toluene. Chlorinated hydrocarbons, alcohols (1-octanol) ethers (diisopropyl ether), esters (n-butyl acetate, tricresyl phosphate), and ketones (methylisobutyl ketone) will also extract phenol. In practice, mixtures of solvents may be most effective. One of the criteria in choosing a solvent is its selectivity for extraction of phenol in preference to other organic contaminants that are present in relatively high concentrations in water produced in petroleum refining and coal coking. Phenols extracted into the organic liquid are back-extracted to aqueous alkali, a process that occurs because of the ionization of phenol in contact with base:

$$\text{Phenol} \quad \bigcirc\!\!-OH(org) + OH^- \longrightarrow$$

$$\bigcirc\!\!-O^-(aq) + H_2O \qquad (11.5.5)$$

Phenolate anion

Acidification of the basic aqueous extract of phenolic compounds precipitates them in the solid form:

$$\bigcirc\!\!-O^-(aq) + H^+ \longrightarrow \bigcirc\!\!-OH(s) \qquad (11.5.6)$$

Phenolate anion   Phenol

Solvent that dissolves in water during extraction is recovered by steam distillation under vacuum.

Recovery of Metals

Metals from electroplating wastes and wastewater from other metal processing operations may be recovered for their economic value as part of the wastewater treatment process. In some cases metals may be recovered by electrochemical reduction (see Section 13.6) in which a direct current is applied between electrodes immersed in the wastewater. Metals plate out on the cathode:

$$Cu^{2+} + e^- \longrightarrow Cu \qquad (11.5.7)$$

An advanced electrolytic system has been described for the removal of zinc and cadmium from waste electroplating water containing cyanide.[13] The metals may be recycled to the electoplating process. This system was designed to remove metals from dilute solution with a high surface area cathode composed of carbon fibers. Cyanide was effectively oxidized at the anode and removed from the water.

Ion exchange (see Section 13.9) is widely used for the recovery of metals from wastewater. The metal-containing water is passed over a solid cation exchange resin where metal is exchanged for H^+ ion:

$$2H^+\{^-CatExRes\} + Zn^{2+} \longrightarrow$$
$$Zn\{^-CatExRes\}_2 + 2H^+ \quad (11.5.8)$$

The resin is regenerated and metal reclaimed in a concentrated form by treatment with a relatively concentrated solution of strong acid:

$$Zn\{^-CatExRes\}_2 + 2H^+ \longrightarrow$$
$$2H^+\{^-CatExRes\} + Zn^{2+} \quad (11.5.9)$$

Some wastewater solutions, such as those from some kinds of electroplating baths, contain cyanide. Cyanide forms negatively charged complex ions with various metals. For example, the cyanide complex of copper(I), $Cu(CN)_4^{3-}$, is very stable. Metal cyanide complexes may be removed from water by anion exchange resins. Regeneration of these resins may be difficult because of their strong affinity for the metal complex ions.

Chromium(VI) in water is in the form of a negatively charged species such as chromate ion, CrO_4^{2-}. It is recovered by passing the chromate-containing solution over an anion exchange resin in the base form. If other cationic metals are present, they must first be removed by cation exchange (see above) or other means to prevent precipitation of hydroxides in contact with the basic anion exchange resin. The reaction for the removal of chromate ion is the following:

$$2\{AnExRes^+\}OH^- + CrO_4 \longrightarrow$$
$$\{AnExRes^+\}_2CrO_4 + 2OH^- \quad (11.5.10)$$

The chromate retained by the anion exchange resin is eluted with strong base (NaOH) and reclaimed. This process regenerates the anion exchange column for further use.

11.6. RECOVERY OF WATER FROM WASTEWATER

It is often desirable to reclaim water from wastewater. This is especially true in regions where water is in short supply. Even where water is abundant, water recycle is desirable to minimize the amount of water that is discharged.

A little more than half of the water used in the U.S. is consumed by agriculture, primarily for irrigation. Steam generating power plants consume about one-fourth of the water, and other uses, including manufacturing and domestic uses, account for the remainder.

The three major manufacturing consumers of water are chemicals and allied products, paper and allied products, and primary metals.[14] These industries use water for cooling, processing, and boilers. Their potential for water reuse is high and their total consumption of water is projected to drop in future years as recycling becomes more common

The degree of treatment required for reuse of wastewater depends upon its application. Water used for industrial quenching and washing usually requires the least treatment, and wastewater from some other processes may be suitable for these purposes without additional treatment. At the other end of the scale, boiler makeup water, potable (drinking) water, water used to directly recharge aquifers, and water that people will contact directly (in boating, water skiing, and similar activities) must be of very high quality.

The treatment processes applied to wastewater for reuse and recycle depend upon both the characteristics of the wastewater and its intended uses. Solids can be removed by sedimentation and filtration. Biochemical oxygen demand (Table 3.2) is reduced by biological treatment, including trickling filters and activated sludge treatment (see Section 13.13 and Figure 13.4). For uses conducive to the growth of nuisance algae, nutrients may have to be removed. The easiest of these to handle is nutrient phosphate, which can be precipitated with lime. Nitrogen can be removed by denitrification processes.

Two of the major problems with industrial water recycle are heavy metals and dissolved toxic organic species. Heavy metals may be removed by ion exchange or precipitation by base or sulfide. The organic species are usually removed with activated carbon filtration. Some organic species are degraded biologically by bacteria in biological wastewater treatment.

The ultimate water quality is achieved by processes that remove solutes from water, leaving pure H_2O. A combination of activated carbon treatment to remove organics, cation exchange to remove dissolved cations and anion exchange for dissolved anions can provide very high quality water from wastewater. Reverse osmosis (see Section 12.4) can accomplish the same objective. However, these processes generate spent activated carbon, ion exchange resins that require regeneration, and concentrated brines (from reverse osmosis) that require disposal.

11.7. SOLVENT RECOVERY

Organic solvents of various kinds are used for many purposes, such as reaction media, extraction of fats, extraction of oils, degreasing, and dry cleaning. A large number of different compounds occur in hazardous waste solvents. Among the halogenated solvents listed in hazardous wastes are dichloromethane, carbon tetrachloride, trichlorofluoromethane, tetrachloroethylene, trichloroethylene, 1,1,1-trichloroethane, 1,1,2-trichloro-1,2,2-trifluoroethane, chlorobenzene, and 1,2-dichlorobenzene. Hydrocarbon solvents that occur in hazardous wastes include benzene, xylene, ethylbenzene and liquid alkanes. Other assorted compounds in waste solvents include acetone, methylisobutyl ketone, cyclohexanone, methanol, n-butyl alcohol, isobutanol, pyridine, carbon disulfide, 2-nitropropane, diethyl ether, and 2-ethoxyethanol.

For reasons of both economics and pollution control, many industrial processes that use solvents are equipped for solvent recycle.[15] The basic scheme for solvent reclamation and reuse is shown in Figure 11.4. Some loss of solvent occurs to the product and in purification, so fresh makeup solvent is added.

Solvents usually must be recovered from solids or parts that have been extracted or cleaned with solvent. This can be accomplished by heating the solids in vacuum or in a non-oxidizing atmosphere such as CO_2 or nitrogen. When solvent flammability is not a problem, liquids can be evaporated from solids with hot air. In either case solvents are recovered by condensation in a heat exchanger, adsorption with activated carbon, or scrubbing with another liquid, usually water. Solvents may also be separated from products by extraction with another solvent.

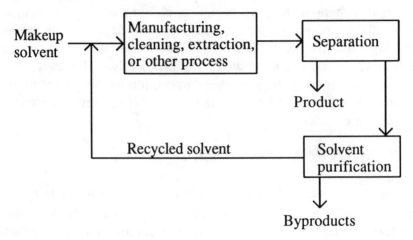

Figure 11.4. Overall process for recycling solvents.

A number of operations are used in solvent purification. Entrained solids are removed by settling, filtration, or centrifugation. Drying agents may be used to remove water from solvents and various adsorption techniques and chemical treatment may be required to free the solvent from specific impurities.

Fractional distillation, often requiring several distillation steps, is the most important operation in solvent purification and recycle. It is used to separate solvents from impurities, water, and other solvents. High efficiency columns, including packed, bubble-cap, and sieve plate columns may be required. The sophistication of the distillation apparatus required increases with the complexity of the mixture to be purified. A high degree of segregation (see Section 1.3 and Figure 1.2) of solvents is desirable and is a good reason for doing the solvent purification and recycle on site rather than shipping the solvent to another facility for purification, which may result in mixing with other solvents and impurites.

LITERATURE CITED

1. Thompson, Fay M., and Cindy A. McComas, "Technical Assistance for Hazardous Waste Reduction," *Environmental Science and Technology*, **21**, 1987, pp. 1154-1158.
2. Schneider, Keith, "Worst Tire Inferno Has Put Focus on Disposal Problem," *New York Times*, March 2, 1990, p. A8.

3. Bishop, Jim, "Waste Reduction," *Hazmat World*, October, 1988, pp. 56-61.

4. Brandt, Andrew S., "Canadian Perspectives," *Proceedings of the Thirty-First Ontario Waste Conference*, Ontario Ministry of the Environment, Toronto, Ontario, Canada, 1984, pp. 3-20.

5. Hanson, David J., "Hazardous Waste Management: Planning to Avoid Future Problems," *Chemical and Engineering News*, July 31, 1989, pp. 9-18.

6. "Waste Minimization – A New Term for a Tested Practice," *Impact*, **6**(1), IT Corporation, Monroeville, PA, 1987, pp. 1-3.

7. Hunt, Gary E., and Roger N. Schecter, "Minimization of Hazardous Waste Generation," Section 5.1 in *Standard Handbook of Hazardous Waste Treatment and Disposal*, Harry M. Freeman, Ed., McGraw Hill, New York, 1989, pp. 5.3-5.27.

8. Nunno, Thomas J., and Mark Arienti, "Waste Minimization Case Studies for Solvents and Metals Waste Streams," in *Land Disposal, Remedial Action, Incineration and Treatment of Hazardous Waste*, EPA/600/9-86/022, U. S. Environmental Protection Agency, Cincinnati, Ohio, 1986, pp. 278-284.

9. Balf, Tom, "Pollution Prevention Pays," *Hazmat World*, June, 1989, pp. 14-15.

10. Kimberly A. Roy, "SMART Management," *Hazmat World*, April, 1989, pp. 22-24.

11. McCabe, Mark M., "Waste Oil," Section 4.1 in *Standard Handbook of Hazardous Waste Treatment and Disposal*, Harry M. Freeman, Ed., McGraw Hill, New York, 1989, pp. 4.3-4.12.

12. Brown Craig J., "Ion Exchange," Section 6.5 in *Standard Handbook of Hazardous Waste Treatment and Disposal*, Harry M. Freeman, Ed., McGraw Hill, New York, 1989, pp. 6.59-6.75.

13. "Evaluation of the HSA Reactor for Metal Recovery and Cyanide Oxidation in Metal Plating Operations," EPA/600/S2-86/094, U.S. Environmental Protection Agency, Washington, D.C., 1986.

14. Montgomery, James M., *Water Treatment Principles and and Design*, John Wiley and Sons, New York, 1985, p. 306.

15. Cooper, C. M., "Solvent Recovery," in *Kirk-Othmer Concise Encyclopedia of Chemical Technology*, Wiley-Interscience, New York, 1985, pp. 1090-1091.

12

Physical Treatment

12.1. PHYSICAL TREATMENT PROCESSES

Physical Forms of Wastes

In discussing physical treatment of wastes it is useful to consider their physical nature. Hazardous wastes may occur as solids, liquids, or gases, as well as mixtures of these states of matter. Examples of mixtures are solids suspended in liquids, gases dissolved in liquids, and liquids held in solids, such as wet sludge material. In considering physical treatment methods, it is useful to classify wastes in the forms listed below:

- Volatile wastes (including gases, volatile solutes in water, gases or volatile liquids held by solids, such as catalysts)
- Liquid wastes
    Wastewater
    Organic solvents
- Dissolved or soluble wastes
    Water-soluble inorganic species
    Water-soluble organic species
    Compounds soluble in organic solvents
- Semisolids
    Sludges
    Greases
- Solids
    Dry solids (including granular solids with a significant water content, such as dewatered sludges)
    Solids suspended in liquids

Treatment Options

Although physical treatment processes are discussed separately in this chapter and chemical and biochemical treatment in Chapter 13, treatment options from these three areas are strongly interrelated. For example, a precipitate of a hazardous waste compound formed by a chemical reaction is separated from the reaction mixture by physical means, such as sedimentation and filtration. Activated sludge treatment of biodegradable wastes (see Figure 13.4) may be aided by the addition of nutrient chemicals and the biomass separated from the treatment effluent by physical means. Furthermore, in order to characterize wastes for physical treatment, it is necessary to know their chemical properties thoroughly. These include parameters such as the pH of wastewater or the chemical properties of solids that may be dangerously reactive when produced by the drying of sludges.

Biochemical properties also need to be known. Wastewater that contains readily biodegraded substances may result in the growth of bacteria on activated carbon. This not only interferes with the action of activated carbon in removing waste constituents, it suggests that prior biochemical treatment may be desirable.

In designing physical treatment and other treatment options, waste segregation (see Section 1.2) should have a high priority. For example, it is much less expensive and troublesome to treat and dispose of solids collected as such, rather than dissolving the solids in wastewater and treating the substances in a dilute solution.

Before considering specific physical treatment processes it is useful to have an overview of the various kinds of treatment to which hazardous substances may be subjected, including their advantages and disadvantages. Safety dictates that the first thing to do with hazardous waste is to consider any acute hazards that it may pose and take appropriate preventive action. Important examples are cyanides and sulfides, which can react with acid or hydrolyze to give toxic HCN and H_2S, respectively. Reactive wastes must be treated with the appropriate reagents to reduce their hazards. One of the first steps to consider in waste treatment is **separation**, which can save tremendous amounts of effort and expense by resegregating wastes in forms that can be treated most economically or which may not even be hazardous. Separation may be as simple as draining an aqueous layer from an organic one in a two-phase waste. It may involve mechanical operations such as filtration and centrifugation. Physical processes including distillation, evaporation, or precipitation by cooling may be used. Or chemical reactions, such as precipitation reactions may be applied. Examples of desirable products from separations include water pure enough to discharge, **waste-derived fuels** that can be used as a supplemental fuel in incinerators, and heavy metals that can be isolated in a small volume and reclaimed. Separation frequently results in **resource recovery** (Chapter 11) that yields material of economic value. For example, separation of hazardous substances sorbed to spent activated carbon can give a carbon product that can be reactivated and used again. **Biological treatment** (Chap. 3) is usually a good option for wastes amenable to it. These include dilute solutions and suspensions of biodegradable organics in water, as well as biodegradable solids and sludges. **Incineration** (Chap. 14) is often required, but should be minimized because of the costs and regulatory and institutional constraints imposed on it. Any wastes remaining after the above measures have been applied will require **stabilization** (Chapter 15) and land disposal.

As is the case with virtually any industrial operation, processes for the treatment of hazardous wastes may generate pollutants. The control of these pollutants may, itself, generate hazardous wastes, such as gaseous emissions, concentrates, brines, and sludges. Examples of pollutant control operations that may generate wastes include filters and precipitators for particulate matter, condensers for vapor, and activated carbon adsorption systems.

Methods of Physical Treatment

Physical treatment of wastes depends upon the physical properties of the material treated. These properties include state of matter, solubility in water and organic solvents, density, volatility, boiling point, and melting point. Knowledge of the physical behavior of wastes has been used to develop various unit operations for waste treatment that are based upon physical properties. These operations include the following:

- Phase separations
- Sedimentation
- Filtration

- Membrane separations
    Reverse osmosis
    Hyper- and ultrafiltration
- Sorption by activated carbon or resins
- Distillation and stripping
- Drying and evaporation
- Extraction

The most straightforward means of physical treatment involves separation of components of a mixture that are already in two different phases. **Sedimentation** and **decanting** are easily accomplished with simple equipment. In many cases the separation must be aided by mechanical means, particularly **filtration** or **centrifugation**. **Flotation** is used to bring suspended organic matter or finely divided particles to the surface of a suspension. In the process of **dissolved air flotation** (DAF), air is dissolved in the suspending medium under pressure and comes out of solution as minute air bubbles attached to suspended particles when the pressure is released. The buoyancy thus imparted to the particles causes them to float to the surface in a layer which is skimmed off to give a sludge waste or byproduct recovery material. One such material designated as a hazardous waste from specific sources is DAF float from the petroleum refining industry (K048). Emissions of volatile constituents to the air may cause problems in DAF operations.

Many wastes are composed of aqueous/organic mixtures in colloidal-sized **emulsions**. An important and often difficult waste treatment step consists of separating these components, a process called **emulsion breaking**. Acid (including waste acid), other chemical reagents, and heat may all be employed for this purpose. Once an emulsion is broken, centrifugation can be used to separate the phases.

A second major class of physical separation is that of **phase transfer** or **phase transition**. An important type of phase transfer process is **extraction**, including liquid-liquid and liquid solid extraction. One of the newer and most active areas of extraction is supercritical fluid extraction. Transfer of a substance from a solution to a solid phase is called **sorption**; an important example is sorption onto activated carbon. Sorption may be in part a chemical process, such as ion exchange in which ions are chemically exchanged at functional groups bound to solid cation exchangers as shown by the following example for the binding of lead from solution onto a cation exchanger:

$$2H^{+}\text{-(Cat. Exchr.)} + Pb^{2+} \longrightarrow Pb^{2+}\text{-(Cat. Exchr.)}_2$$
$$+ 2H^{+} \qquad (12.1.1)$$

In many phase separations the substance in question, itself, undergoes a phase transition which can occur in precipitation from a cooled solution, evaporation, condensation, and distillation. Phase transitions can be the result of a chemical reaction, for example, a solution reaction that forms a precipitate, or electrodeposition of a solid onto an electrode surface.

A third major class of physical separation is **molecular separation**, often based upon membrane processes in which dissolved contaminants or solvent are forced through a membrane. The most widely used membrane process is reverse osmosis in which water from wastewater is forced through a membrane that is selective for water, yielding purified water product and leaving behind a concentrated liquor containing impurities. Concentrated liquor from reverse osmosis may be treated by evaporation and solidification and the impurities recovered or disposed of in a solid form. Other membrane processes include ultrafiltration and electrodialysis in which ions in a solution subjected to electrolysis migrate selectively through alternate membranes that allow for the passage of cations and anions.

12.2. PHASE TRANSITION

Physical Precipitation

Physical precipitation is used here as a term to describe processes in which a solid forms from a solute in solution as a result of a physical change in the solution, as compared to chemical precipitation (see Section 13.3) in which a chemical reaction in solution produces an insoluble material. The major changes that can cause physical precipitation are cooling the solution, evaporation of solvent, or alteration of solvent composition. The most common type of physical precipitation by alteration of solvent composition occurs when a water-miscible organic solvent is added to an aqueous solution, so that the solubility of a salt dissolved in the solution is lowered below the point at which it precipitates. In rare cases solids precipitate from solution when the solution is heated. However, vapors and gases are evolved from heated solution, and heating can be used to help remove readily vaporized constituents from solution.

Freeze-Crystallization

Freeze-crystallization is a means of removing water from an aqueous mixture by production of ice, which is usually a relatively pure form of water. Therefore, rather than a solute coming out of solution as a precipitate, the water solvent is converted to a solid, leaving behind a more concentrated solution of contaminants. The ice is removed mechanically from the mixture in which it has formed. Although this technique is promising, it has not been widely used for hazardous waste treatment.

12.3. FILTRATION, SEDIMENTATION, AND PARTICLE REMOVAL

An important step in the treatment of hazardous wastewater is the removal of particles. A vast amount of experience has been gained in this area from municipal water and wastewater treatment, where most contaminants are either removed as particles or converted to particulate form prior to removal.[1] Often the particulate matter in wastewater is of colloidal size, of the order of a micrometer (μm) or less. The treatment of particulate matter is complicated by the fact that particle size may span two or more orders of magnitude. In many cases most of the <u>mass</u> of the particles is in the form of larger particles, typically of around 20 μm in diameter, whereas most of the <u>numbers</u> of particles are in smaller ones, typically less than 5 μm.

Sedimentation is about the simplest and least expensive treatment that can be applied to hazardous wastes. As the name implies, sedimentation involves leaving a liquid stand until solids settle from it. Sedimentation normally follows another treatment (see coagulation below) designed to facilitate settling of solids from the liquid.

Usually it is necessary to coagulate particles in water prior to their removal. **Coagulation** can be brought about by adding an organic polyelectrolyte or inorganic substances, such as an aluminum salt, accompanied by agitation of the suspension of particles. The chemical additive acts as a flocculating agent to cause the particles to stick together and settle out. Following coagulation, sedimentation, or settling, is allowed to occur. Finally the solids are removed by filtration or centrifugation.

Most solids removed from hazardous wastewater contain a high percentage of moisture. Removal of the water to give a smaller quantity of waste is called **dewatering**. The ease and degree of water removal are very important parameters in subsequent processing and disposal of wastes. Dewatering can often be improved with addition of a filter aid such as diatomaceous earth during the filtration step.

12.4. MEMBRANE PROCESSES

Membrane Separations

Membrane separations are based upon the selective passage of a solvent or solute through a thin membrane. The products are a relatively pure solvent phase (usually water) and a concentrate enriched in the solute impurities. The major types of membrane separations are outlined below.

Hyperfiltration/Ultrafiltration

The membranes in both **hyperfiltration** and **ultrafiltration** limit the sizes of solute species that can pass through a membrane in a manner analogous to a mechanical filter for solids. Hyperfiltration operates with both ions and organic solutes, allowing passage of species with molecular masses of about 100 to 500. Ultrafiltration is used for the separation of organic solutes with molecular masses of 500 to 1,000,000. With both of these techniques water and lower molecular mass solutes under pressure pass through the membrane as a stream of purified **permeate**, leaving a stream of **concentrate** containing impurities in solution or suspension.

Ultrafiltration and hyperfiltration are especially useful for concentrating suspended oil, grease, and fine solids in water. They also serve to concentrate solutions of large organic molecules and heavy metal ion complexes.[2]

Reverse Osmosis

Reverse osmosis is the most widely used of the membrane techniques. Although superficially similar to ultrafiltration and hyperfiltration, it operates on a different principle in that the membrane is selectively permeable to water and excludes ionic solutes. Reverse osmosis uses high pressures to force permeate through the membrane, producing a concentrate containing high levels of dissolved salts.

Cellulose acetate and some kinds of aromatic polyamides have the properties of high water, low ion, permeability required for use as reverse osmosis membranes. Composites that consist of a thin layer of salt-impermeable substance coated onto a porous support layer of polymer are also used for reverse osmosis membranes.

As with other membrane separation technologies, fouling is a major problem with reverse osmosis used for treating hazardous wastewater. As the levels of solutes rise in the concentrate solution during reverse osmosis, salts may precipitate from the concentrate and plug the membrane surface. The same may occur with suspended materials, products of iron corrosion, and deposits caused by the growth of bacteria. These problems can be reduced by measures such as pretreatment of the water to remove solids, activated carbon filtration of the feed, pH control to prevent precipitation reactions, and chlorination to prevent bacterial growth.

Reverse osmosis has been used for the treatment of hazardous wastes primarily in the metal plating industry. The major application has been purification of water used to rinse electroplated parts and contaminated by zinc, copper, nickel, cyanide, and sulfate ions. The concentrate solution (sometimes further concentrated by the evaporation of water) may be recirculated to the plating bath and the purified water returned to the rinse baths.

Electrodialysis

Electrodialysis is a membrane purification technique in which membranes permeable to cations alternate with membranes permeable to anions and the driving force for the separation is provided by electrolysis with a direct current between two electrodes. Alternate layers between the membranes contain concentrate (brine) and purified water. Among its other applications, electrodialysis can be used to concentrate plating wastes. Like reverse osmosis, electrodialysis is subject to fouling by suspended matter or salts that precipitate out of the concentrate.

12.5. SORPTION

Sorption by Activated Carbon

Activated carbon is a widely used sorbent applied to the removal of organic vapors from air, organic solutes from water, and, to a certain extent, inorganic solutes from water.[3] Activated carbon is used for several purposes in waste treatment. In some cases it is adequate for complete treatment. It can also be applied to pretreatment of waste streams going into processes such as reverse osmosis (see Section 12.4) to improve their efficiencies. Or it can be used to polish effluents from other treatment processes, such as biological treatment of degradable organic solutes in water.

Activated carbon is made by heating organic matter such as wood, coal, and lignite, to leave a carbon residue, which is then activated by partial oxidation with carbon dioxide, steam, or oxygen in air as illustrated by the reaction

$$C + H_2O \xrightarrow{\text{heat}} H_2 + CO \qquad (12.5.1)$$

The activation step introduces pores in the carbon, increases its surface area (to as much as 2,000 m^2/g), and introduces chemical groups onto the surface that have an affinity for sorbed materials, particularly nonpolar organic compounds. Activated carbon comes in the forms of **granular activated carbon** (GAC) and **powdered activated carbon** (PAC). Activated carbon sorption is reversible, which allows reclamation of sorbed materials and regeneration of the carbon for further use.

For the treatment of wastewater, activated carbon can be used in batch and continuous flow modes. It can be mixed with wastewater in reactors (usually with PAC) or held in columns (GAC in fixed beds) through which the liquid flows. When activated carbon is used in columns, a **breakpoint** is reached at which the carbon begins to become saturated with the material that it removes from water and the impurity concentration in the effluent becomes excessive, requiring replacement of the carbon in the column.

Activated carbon sorption is effective for removing a number of hazardous waste materials from water, especially those that are poorly water soluble and have high molecular masses. The major ones of these are summarized below:

- Hydrocarbons
  Aromatic hydrocarbons: Monocyclic aromatic hydrocarbons commonly used as solvents and present in fuels (benzene, toluene, xylene), naphthalene (U165), biphenyl, polycyclic aromatic hydrocarbons
  Aliphatic hydrocarbons: Constituents of gasoline, kerosene, jet fuel; lubricating oils; solvents such as cyclohexane (U056).
- Chorinated hydrocarbons: Trichloroethylene (U228), 1,1,1-trichloroethane, chlorobenzene (U037), polychlorinated biphenyls, chlorinated hydrocarbon insecticides, such as toxaphene
- Phenolic compounds
  Phenol (U188), cresols (U052), 1-naphthol, pentachlorophenol (F027)
- Organonitrogen compounds
  Amines: Aniline (U012), diaminotoluene (U221)
- Dyes
- Surfactants

Activated carbon has some disadvantages for waste treatment. It does not work well for organic compounds that are highly water-soluble or polar. It is expensive, which makes its use economically more advantageous following other treatment processes. Treatment operations that often precede activated carbon treatment include filtration to remove solids, skimming of water-insoluble oil and grease, biological treatment to eliminate biodegradable solutes, and chemical oxidation of organics (such as with ozone, O_3). Aeration can be used to remove volatile organic solutes prior to treatment with activated carbon. If iron(II) is present, or is added deliberately, it may oxidize during aeration,

$$4Fe^{2+} + 10H_2O + O_2 \longrightarrow 4Fe(OH)_3 + 8H^+ \qquad (12.5.2)$$

to produce gelatinous iron(III) hydroxide. This may prevent subsequent precipitation of iron(III) on the activated carbon. Furthermore, the $Fe(OH)_3$ acts as a coagulant to remove solids from the wastewater and to coprecipitate some inorganic species, thereby improving the efficiency of the activated carbon.

Pollution problems involved with the use of activated carbon to treat hazardous wastes are usually minimal. However, the disposal or regeneration of spent carbon can present major problems in some cases. Air emissions may be produced when the carbon is handled or regenerated.

Sorption by Other Solids

Solids other than activated carbon can be used for sorption of contaminants from liquid wastes. Some synthetic resins composed of organic polymers are effective for removal of organic solutes from wastewater and can be regenerated. Among the solutes removed from water by resins are aromatic hydrocarbons, chlorinated hydrocarbons, aldehydes, ketones, amines, and phthalate esters. Mineral substances can also be used for sorption. For example, clay is employed to remove impurities from waste lubricating oils in some oil recycling processes (see Section 11.4).

12.6. DISTILLATION AND STRIPPING

Distillation

Distillation consists of evaporating components of a mixture in a still and condensing the vapor, producing a **distillate** relatively richer in volatile components and leaving a bottoms fraction enriched in less volatile or nonvolatile components. **Flash distillation** is a single-step process in which the still is fed continuously as evaporation occurs; it is a simple, low-cost means of doing a crude separation. When greater separation is needed, multiple stage distillation is used in which a portion of the distillate is returned to a reflux column where it re-equilibrates with vapor, giving the equivalent of numerous simple distillations in series.

The overall process of distillation can be used with hazardous wastes to separate liquids of different volatilities or to remove volatile components from a relatively less volatile medium. Separation of liquids from solids by heating is classified as evaporation, which is discussed in Section 12.7. Distillation is most useful in separating mixtures of liquids; the presence of solids tends to cause fouling problems. The greatest single use of distillation in resource recovery is in recycling solvents (see Section 11.7). The application of distillation to waste oil reclamation is discussed in Section 11.4. Some other specific wastes amenable to treatment by distillation include the following:[4] (1) Aqueous phenolic wastes, (2) xylene contaminated with paraffin from histological laboratories, (3) mixtures of ethylbenzene and styrene, and (4) wastes from penicillin manufacture containing butyl acetate.

Wastes containing reactive constituents, such as peroxides, usually should not be distilled. Some organic compounds may polymerize during distillation, producing solids in the apparatus and sometimes generating excessive heat.

A relatively manageable potential pollution problem with distillation is that of air emissions, usually of volatile organic compounds from vents in condensers and storage tanks. Measures such as auxiliary condensers and activated carbon traps can be employed to minimize this kind of release. A greater problem is presented by **distillation bottoms** (still bottoms) consisting of unevaporated residues from distillation. Distillation bottoms are made up of solids, semisolid tars, and sludges and require disposal or destruction by incineration. Landfill disposal of distillation bottoms used to be widely practiced but is now becoming severely curtailed, as shown by the following specific examples of distillation bottoms wastes prohibited from landfill disposal as of June 8, 1989:[5]

- (K009) Distillation bottoms from the production of acetaldehyde from ethylene
- (K036) Still bottoms from toluene reclamation distillation in the production of disulfoton (see structure below)

Disulfoton, an organophosphate insecticide

- (K094) Distillation bottoms from the production of phthalic anhydride from orthoxylene
- (K095) Distillation bottoms from the production of 1,1,1-trichloroethane (nonwastewaters)

Stripping

Stripping is a means of separating volatile components from less volatile ones in a liquid mixture by the partitioning of the more volatile materials to a gas phase of air or steam. The gas phase is introduced into the mixture in a stripping tower that is packed or equipped with trays to provide maximum turbulence and contact between the liquid and gas phases. The two major products are condensed vapor and a stripped bottoms residue. Usually the waste stripped is an aqueous solution or suspension. Separation of the volatile materials from the residue commonly exceeds 99%.

Air stripping is usually confined to the removal of highly volatile substances present at low concentrations.[6] Examples of two volatile components that can be removed from water by air stripping are benzene and dichloromethane. Air stripping can also be used to remove ammonia from water that has been treated with a base to convert ammonium ion to volatile ammonia:

$$NH_4^+ + OH^- \longrightarrow NH_3 + H_2O \qquad (12.6.1)$$

Water from aquifers contaminated by volatile organic compounds has been purified by air stripping. In some cases of groundwater cleanup, the effluent has been discharged to the atmosphere. Odorous and toxic compounds should be removed from the effluent air by measures such as activated carbon sorption.

In steam stripping, both the gas phase and elevated temperature are provided by steam. The process is used for the same applications as air stripping. However, it is applicable to mixtures with relatively high concentrations of volatile constituents and also works with less volatile and more soluble constituents. It can also be used to strip organic solvents from mixtures containing nonvolatile constituents. Condensation of the distillate stream yields either an aqueous solution of the stripped constituents or, when the stripped components are water-immiscible, a two-phase mixture of organic species and water. Less volatile, more water soluble, constituents, such as acetone or pentachlorophenol, are more readily removed from water by steam stripping than by air stripping.

12.7. DRYING AND EVAPORATION

A process that works to purify some of the kinds of wastes that are not suitable for distillation is **evaporation** usually employed to remove water from an aqueous waste to concentrate it.[7] A special case of this technique is **thin-film evaporation** in which volatile constituents are removed by heating a thin layer of liquid or sludge waste spread on a heated surface. This process can be used to selectively remove relatively more volatile organic constituents from water or to remove water from high-boiling organic constituents. In the latter case it can be used to upgrade the heating value of a waste to enable its incineration. It has an advantage over steam stripping in that it does not introduce additional water into the product.

Evaporation can be combined with other waste treatment processes. Chemical Waste Management of Oak Brook, Illinois, is reported to market an add-on evaporation system for wastewater that also destroys organic contaminants.[8] A proprietary catalyst is used to convert organic solutes to carbon dioxide, water, and other inorganic species, such as hydrochloric acid from organohalides.

Drying is a very important industrial operation that is applicable to the treatment of various kinds of hazardous wastes.[9] Drying refers to the removal of solvent or water from a solid or semisolid (sludge) or the removal of solvent from a liquid or suspension, leaving a dried residue. A different kind of operation is **gas drying**, which is the removal of organic solvent vapor or water vapor from a gas stream using adsorbents, drying agents, or low temperatures. In **freeze drying**, the solvent, usually water, is sublimed from a frozen material.

Hazardous waste solids and sludges are dried to reduce the quantity of waste, to remove solvent or water that might interfere with subsequent treatment processes, and to remove hazardous volatile constituents. One of the more important and challenging steps in the treatment of hazardous wastes is the removal of water from aqueous sludges (dewatering).

There are some hazards to be considered in drying. Toxic or explosive organic vapors may be removed in drying, thereby creating a hazard. When water is removed from some sludges, reactive constituents in them may burn or even explode.

Drying can consume large amounts of energy for heating the solid, circulating gas over it, or freezing it for freeze-drying. Therefore, it is always desirable to remove as much liquid as possible by sedimentation, filtration, and centrifugation before drying is employed.

The heat required to dry a solid can be transferred to the material by several means. The major means of heat transfer are from heated trays or steam coils in contact with the material to be dried, heated air or gas circulating over the material, and radiant heat from heaters located over the solid. Microwave heating of substances that contain water may also be employed. When flammable or explosive vapors are removed by drying, a nonflammable gas, such as nitrogen, may be passed through the drying system.

Mechanically, batch and continuous dryers are available in a large variety of forms. The substance to be dried may be held in trays, rotating drums, fluid-beds, or almost any other conceivable device. Liquids or semisolids may be dried with a spray dryer. The substance is injected into a drying chamber in an atomized form and contacts hot gas, which carries away the liquid and leaves a granular dried product.

Drying Agents

Drying agents, or desiccants, are substances that absorb water from semisolids, liquids, or gases. In a broader sense, drying agents may also be regarded as substances that absorb organic solvents from semisolids, liquids, or gases. We refer to drying agents here as substances with an affinity for water.

There are two general ways in which drying agents act. One of these is physical adsorption in which water is taken up by a solid or hygroscopic (water-seeking) liquid. The other involves a chemical reaction, such as the conversion of anhydrous copper sulfate, $CuSO_4$, to the pentahydrate, $CuSO_4 \cdot 5H_2O$. In some cases both of these two mechanisms are involved with the use of a drying agent.

Numerous kinds of wastes may be treated with drying agents. These wastes include the following:

- Moist gases in which the presence of water vapor is not desired
- Organic solutions containing traces of water
- Aqueous-based sludges

12.8. EXTRACTION

Solvent Extraction and Leaching

Solvent extraction is the process in which a substance is transferred from solution in one solvent (usually water) to another (usually an organic solvent) without any chemical change taking place. When solvents are used to leach substances from solids or sludges, the process is called **leaching**. Because of the similarities between solvent extraction and leaching they are discussed together here. Solvent extraction and the major terms applicable to it are summarized in Figure 12.1. The same terms and general principles apply to leaching.

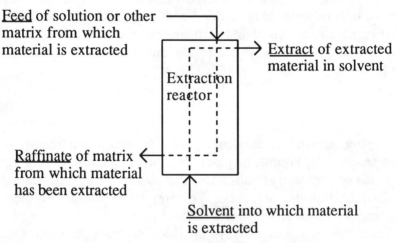

Figure 12.1. Outline of solvent extraction/leaching process with important terms underlined.

Although solvent extraction has excellent potential for the treatment of hazardous wastes, it has not been widely used for that purpose. The major application in waste treatment has been in the removal of phenol from byproduct water produced in coal coking, petroleum refining, and chemical syntheses that involve phenol. Phenol is extracted into an organic solvent such as methylisobutyl-ketone or light oil and subsequently back-extracted into aqueous base as the water-soluble phenolate salt as discussed in Section 11.5.

Polychlorinated biphenyls can be extracted from transformer oils with N,N-dimethyl formamide (DMF). Addition of water to the DMF extract results in separation of the PCBs. The DMF is subsequently recovered from its mixture with water by distillation.

Organic constituents of sludges contaminated with oils can be removed by solvent extraction. Examples of oily sludges include petroleum refinery waste sludge and mill scale. Alkyl amine solvents can be used to extract both oil and water from sludge. Subsequent heating and distillation enable separation of the amine, oil, and water components.

One of the more promising approaches to solvent extraction and leaching of hazardous wastes is the use of **supercritical fluids** as extraction solvents. A supercritical fluid is one that has characteristics of both liquid and gas and consists of a substance above its supercritical temperature and pressure. The most widely used substance for supercritical fluid extractions is carbon dioxide for which the supercritical temperature and pressure are 31.1°C and 73.8 atm pressure, respectively. After a substance has been extracted from a waste into a supercritical fluid at high pressure, the pressure can be released, resulting in separation of the substance extracted. The fluid can then be compressed again and recirculated through the extraction system.

The use of supercritical fluid extraction has been the subject of intense research efforts for the cleanup of hazardous wastes. Some possibilities for treatment of hazardous wastes by extraction with supercritical CO_2 are the following:

- Removal of organic contaminants from wastewater
- Extraction of organohalide pesticides from soil
- Extraction of oil from emulsions used in aluminum and steel processing
- Regeneration of spent activated carbon
- Purification of waste oils contaminated with PCBs, metals, and water using supercritical ethane

Soil Flushing and Washing

Extraction with water containing various additives can be used to cleanse soil contaminated with hazardous wastes. When the soil is left in place and the water pumped into and out of it, the process is called **flushing**; when soil is removed and contacted with liquid the process is referred to as **washing**. Here we use washing as a term applied to both processes.

The composition of the fluid used for soil washing depends upon the contaminants to be removed. The washing medium may consist of pure water or it may contain acids (to leach out metals or neutralize alkaline soil contaminants), bases (to neutralize contaminant acids), chelating agents (to solubilize heavy metals), surfactants (to enhance the removal of organic contaminants from soil and improve the ability of the water to emulsify insoluble organic species), or reducing agents (to reduce oxidized species).[10,11] Soil contaminants may dissolve, form emulsions, or react chemically. Often highly soluble soil contaminants are removed from soil by natural processes before soil washing is undertaken. Inorganic species commonly removed from soil by washing include heavy metals salts; lighter aromatic hydrocarbons, such as toluene and xylenes; lighter organohalides, such as trichloro- or tetrachloroethylene; and light-to-medium molecular mass aldehydes and ketones.

Extraction of hazardous species from soil frequently involves chemical reactions between these species and solutes in the water used. Such reactions are discussed in Section 13.8.

The **elutriate** water remaining from washing soil is treated and recycled to remove contaminants for subsequent disposal and to avoid discharge of polluted water. The treatment processes used include treatment with activated carbon to remove organic species, biological treatment to break down biodegradable organic solutes, treatment with sulfide to precipitate heavy metals, emulsion breaking, filtration, and reverse osmosis.

LITERATURE CITED

1. Lawler, Desmond F., "Removing Particles in Water and Wastewater," *Environmental Science and Technology*, **20**, 856-861 (1986).

2. MacNeil, Janet, and Drew E. McCoy, "Membrane Separation Technologies," Section 6.7 in *Standard Handbook of Hazardous Waste Treatment and Disposal*, Harry M. Freeman, Ed., McGraw Hill, New York, 1989, pp. 6.91-6.106.

3. Voice, Thomas C., "Activated Carbon Adsorption," Section 6.1 in *Standard Handbook of Hazardous Waste Treatment and Disposal*, Harry M. Freeman, Ed., McGraw Hill, New York, 1989, pp. 6.3-6.21.

4. Rogers, Tony N., and George Brant, "Distillation," Section 6.2 in *Standard Handbook of Hazardous Waste Treatment and Disposal*, Harry M. Freeman, Ed., McGraw Hill, New York, 1989, pp. 6..23-6.38.

5. "72 Wastes Are Added to the Land-Ban List," *Hazmat World*, February, 1989, pp. 16-17.

6. Boegel, Joan V., "Air Stripping and Steam Stripping," Section 6.8 in *Standard Handbook of Hazardous Waste Treatment and Disposal*, Harry M. Freeman, Ed., McGraw Hill, New York, 1989, pp. 6.107-6.118.

7. Delaney, B. Todd and Ronald J. Turner, "Evaporation," Section 7.7 in *Standard Handbook of Hazardous Waste Treatment and Disposal*, Harry M. Freeman, Ed., McGraw Hill, New York, 1989, pp. 7.77-7.84.

8. Hanson, David J., "Hazardous Waste Management: Planning to Avoid Future Problems," *Chemical and Engineering News*, July 31, 1989, pp. 9-18.

9. Ausikatis, Joseph P., "Drying," *Kirk-Othmer Concise Encyclopedia of Chemical Technology*, Wiley-Interscience, New York, 1985, pp. 372-374.

10. Raghavan, R., E. Coles, and D. Dietz, "Cleaning Excavated Soil Using Extraction Agents: A State-of-the-Art Review," EPA/600/S2-89/034, United states Environmental Protection Agency, Risk Reduction Engineering Laboratory, Cincinnati, Ohio 1990.

11. "Handbook of Remedial Action at Waste Disposal Sites," EPA/625/6-85/006, U.S. Environmental Protection Agency, Hazardous Waste Engineering Research Laboratory, Cincinnati, Ohio, 1985, pp. 9-45–9-46.

13

Chemical and Biochemical Treatment

13.1. INTRODUCTION

The applicability of chemical treatment to wastes depends upon the chemical properties of the waste constituents. These properties include acid-base, oxidation-reduction, precipitation, and complexation behavior; reactivity; flammability/combustibility; corrosivity; and compatibility with other wastes. The chemical behavior of wastes translates to various unit operations for waste treatment that are based upon chemical properties and reactions. These include the following:

- Acid/base neutralization
- Chemical precipitation
- Chemical flocculation
- Oxidation
- Reduction
- Chemical extraction and leaching
- Ion exchange

Chemical reactions of various types are used for the treatment and destruction of hazardous wastes as summarized in this chapter. An attractive feature of chemical treatment is the opportunity to treat wastes with other wastes. One of the largest such applications is the mutual neutralization of waste acids (such as acids from steel pickling liquor) and waste bases (for example, alkali remaining from the removal of sulfur from petroleum products). Waste ash containing high contents of calcium and magnesium oxides can be substituted for lime in waste treatment or stabilization. Combustible organic materials separated from waste oils, solvents, and sludges can be used as supplemental fuel in hazardous waste incinerators.

Methods for destroying some of the chemicals often used in laboratory procedures have been discussed in some detail.[1] Among the chemicals for which treatment procedures are given in this work are acid halides, acid anhydrides, amines, azo compounds, azoxy compounds, tetrazenes, boron trifluoride, calcium carbide, chloro-methylsilanes, chromates, hydrides, cyanides, dimethyl sulfate, ethidium bromide, hydrazines, mercaptans, inorganic sulfides, nitriles, peroxides, picric acid, and sodium amide.

13.2. ACID/BASE NEUTRALIZATION

Acids and bases are frequently encountered in hazardous wastes. Their presence in wastewater or waste leachate is manifested by extremes in pH, defined as,

$$pH = -\log[H^+]$$

where $[H^+]$ is the concentration of hydrogen ion in moles per liter (mol/L or M, see Section 2.5). A low pH indicates the presence of acid (acidity) and a high pH indicates the presence of base (alkalinity from OH^- ion). A pH of 7 is neutral and a range of pH from about 5 to 9 may be expected in "natural" waters. Such water may have a pH less than 7 from the presence of dissolved carbon dioxide because of the reaction,

$$CO_2 + H_2O \longrightarrow H^+ + HCO_3^- \tag{13.2.1}$$

which produces H^+ ion, or from the traces of strong acids from pollutant sources, such as acid rain. Groundwater may have a pH greater than 7 from the presence of alkalinity in the form of HCO_3^- ion. The bicarbonate ion is produced by the reaction of dissolved CO_2 with $CaCO_3$ in limestone:

$$CaCO_3 + CO_2 + H_2O \rightarrow Ca^{2+} + 2HCO_3^- \qquad (13.2.2)$$

It is important to realize that a high pH does not necessarily indicate an extremely high concentration of base and that a low pH does not necessarily indicate the presence of a high concentration of acid. For example, a solution of 1×10^{-3} mol/L NaOH has a high pH of 11, but a low concentration of base, which would require only 1×10^{-3} moles of strong acid for neutralization of 1 L of solution. A solution of 0.100 mol/L HCO_3^- has a pH of 8.3, much lower than that of the NaOH solution just described, but requiring 0.1 mol of acid per liter to neutralize it. Similarly, a relatively low concentration of HCl, a strong acid that exists in solution completely in the ionized forms of H^+ and Cl^-, will have a very low pH, whereas a much more concentrated solution of acetic acid, which is present in solution predominantly as unionized CH_3COOH, rather than H^+ and CH_3COO^- ions, would have a relatively high pH.

The process used to eliminate waste acids and bases is called **neutralization**, as shown by the following reaction:

$$H^+ + OH^- \rightarrow H_2O \qquad (13.2.3)$$

If too much base is present, acid is added to react with OH^- and, if too much acid is present, base is added to react with H^+. Although simple in principle, neutralization can present some problems in practice, such as evolution of volatile contaminants or mobilization of soluble substances. The heat generated from the above reaction when the wastes involved are relatively concentrated can result in dangerously hot solutions and even spattering. Strongly acidic or basic solutions are corrosive to pipes, containers, and mixing apparatus. There is a danger of adding too much acid or base and producing a product that is too acidic or basic.

The sources of acid or base used to treat alkaline or acidic wastes are determined by cost and safety. Ideally, wastes, such as waste acid metal pickling liquor or waste base from petroleum refining can be used. Lime, $Ca(OH)_2$, is a widely used base for treating acidic wastes, and it has the advantage of limited solubility, which prevents solutions of excess lime from reaching extremely high pH values. Sulfuric acid, H_2SO_4, is a relatively inexpensive acid for treating alkaline wastes. However, addition of too much sulfuric acid can produce highly acidic products; for some applications, acetic acid, CH_3COOH, is preferable. As noted above, acetic acid is a weak acid and an excess of it does little harm. It is also a natural product and biodegradable. These characteristics make its use desirable for soil flushing (see Section 12.8) where it may be difficult to control the quantity of acid needed to neutralize alkali in contaminated soil.

Neutralization, or pH adjustment, is often required prior to the application of other waste treatment processes.[3] These include activated carbon sorption, oxidation/reduction, wet air oxidation, stripping, and ion exchange. Microorganisms usually require a pH in the range of 6-9, so neutralization may be required prior to biochemical treatment.

13.3. CHEMICAL PRECIPITATION

Precipitation is used in hazardous waste treatment primarily for the removal of heavy metal ions from water as shown below for the chemical precipitation of cadmium:

$$Cd^{2+}(aq) + HS^-(aq) \rightarrow CdS(s) + H^+(aq) \qquad (13.3.1)$$

Physical precipitation in which solutes are removed by cooling the solution, evaporation, or alteration of solvent composition was mentioned in Sections 12.2 and 12.7. This section deals with **chemical precipitation** where a chemical reaction in solution is used to form an insoluble species.

Precipitation of Metals

Hydroxides and Carbonates

The most widely used means of precipitating metal ions is by the formation of hydroxides such as chromium(III) hydroxide:

$$Cr^{3+} + 3OH^- \rightarrow Cr(OH)_3 \qquad (13.3.2)$$

The source of hydroxide ion, OH^-, is a base (alkali), such as lime ($Ca(OH)_2$), sodium hydroxide (NaOH), or sodium carbonate (Na_2CO_3). The base, itself, may be a waste material, such as sodium hydroxide used to remove sulfur compounds from petroleum products. Most metal ions tend to produce basic salt precipitates, such as basic copper(II) sulfate, $CuSO_4 \cdot 3Cu(OH)_2$, formed as a solid when hydroxide is added to a solution containing Cu^{2+} and SO_4^{2-} ions.[4] The solubilities of many heavy metal hydroxides reach a minimum value, often at a pH in the range of 9-11, then increase with increasing pH values due to the formation of soluble hydroxo complexes, as illustrated by the following reaction:

$$Zn(OH)_2(s) \; + \; OH^-(aq) \; \rightarrow \; Zn(OH)_3^-(aq) \qquad (13.3.3)$$

The chemical precipitation method that is used most is precipitation of metals as hydroxides and basic salts with lime. Sodium carbonate can be used to precipitate hydroxides, carbonates, or basic carbonate salt precipitates. The carbonate anion produces hydroxide by virtue of its hydrolysis reaction with water:

$$CO_3^{2-} \; + \; H_2O \; \rightarrow \; HCO_3^- \; + \; OH^- \qquad (13.3.4)$$

Carbonate, alone, does not give as high a pH as do alkali metal hydroxides, which may have to be used to precipitate metals that form hydroxides only at relatively high pH values. A typical carbonate salt formed in the presence of carbonate ion is cadmium carbonate, $CdCO_3$, and a typical basic carbonate is one formed with lead ion, $2PbCO_3 \cdot Pb(OH)_2$. Some carbonate precipitates are more filterable than hydroxides.

Sulfides

The solubilities of some heavy metal sulfides are extremely low, so sulfide precipitation (see Reaction 13.3.1) can be a very effective means of treatment. Sources of sulfide ion include sodium sulfide (Na_2S), sodium hydrosulfide (NaHS), hydrogen sulfide (H_2S), and iron(II) sulfide (FeS). Hydrogen sulfide is a toxic gas that is, itself, considered to be a hazardous waste (U135). Iron(II) sulfide (ferrous sulfide) can be used as a safe source of sulfide ion to produce sulfide precipitates with other metals that are less soluble than FeS. Since iron(II) sulfide is only slightly soluble, itself, it presents a relatively low hazard due to the presence of sulfide ion in solution and the potential to form volatile hydrogen sulfide. However, H_2S can be a problem when metal sulfide wastes contact acid, which results in the production of H_2S gas by the following reaction:

$$MS + 2H^+ \rightarrow M^{2+} + H_2S \qquad (13.3.5)$$

Precipitation as Metals

Some metals can be precipitated from solution in the elemental metal form by the action of a reducing agent. A reducing agent used for this purpose is sodium borohydride. This reagent can be used, for example, to precipitate copper metal from plating solutions in which the metal is stabilized as copper(I) in the form of a complex ion:

$$8Cu^+ + NaBH_4O + 2H_2O \rightarrow 8Cu +$$
$$NaBO_2 + 8H^+ \qquad (13.3.6)$$

The volume of sludge from sodium borohydride precipitation is usually much less than that from precipitation of metals with lime.

Metal ions can be converted to the elemental form and removed from solution by reaction with more active metals by a process called **cementation**. An example is shown below for the reduction of toxic cadmium with relatively harmless zinc.

$$Cd^{2+} + Zn \rightarrow Cd + Zn^{2+} \qquad (13.3.7)$$

13.4. CHEMICAL FLOCCULATION

Many solids formed by chemical reactions do not produce precipitates that settle readily. Instead, many precipitates are colloidal suspensions which do not settle or which form a very bulky precipitate with a high water content. Therefore, it is often desirable to add a chemical **flocculating agent** that binds to the colloidal particles, as bridging groups between them that enable particles to join together and settle as a relatively dense, filterable precipitate.[5]

Flocculants are polymeric species, often **polyelectrolytes** that consist of large molecules with functional groups, such as SO_3^- or NH^+ attached. Examples of polymeric flocculating agents are illustrated in Figure 13.1.

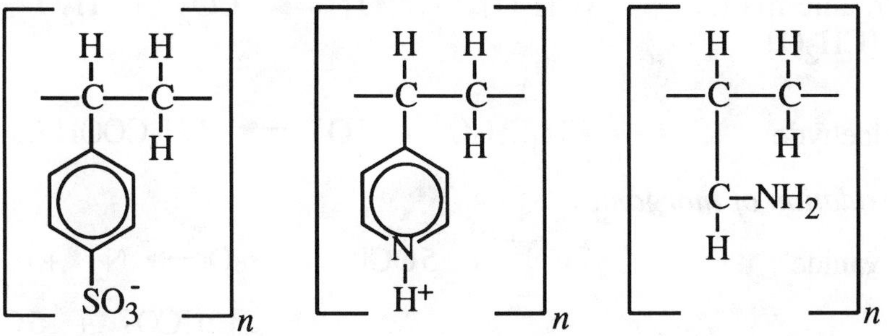

Polystyrene   Polyvinyl   Polyacrylamide
sulfonate   pyridinium   (neutral)
(cationic)   (anionic)

Figure 13.1. Polymeric species used as flocculants. Examples are shown of anionic and cationic polyelectrolytes and a nonionic polymer.

13.5. OXIDATION/REDUCTION

Oxidation and **reduction** (see Section 2.3) can be used for the treatment and removal of a variety of inorganic and organic wastes.[6] Some waste oxidants can be used to treat oxidizable wastes in water and cyanides. The net result of an oxidation or reduction reaction used to treat a waste constituent in aqueous solution is the conversion of the waste to a nonhazardous form or to a form that can be isolated physically. Important examples of oxidation/reduction processes for the treatment of hazardous wastes are given in Table 13.1.

Ozone as an Oxidant

Ozone, O_3, is a strong oxidant that can be generated on-site by an electrical discharge through dry air or oxygen as shown by the following reaction

$$3O_2 \xrightarrow[\text{discharge}]{\text{Electrical}} 2O_3 \tag{13.5.1}$$

Table 13.1. Oxidation/Reduction Reactions Used to Treat Wastes

Waste Substance	Reaction with Oxidant or Reductant

Oxidation of Organics

Organic matter, $\{CH_2O\}$ $\{CH_2O\} + \{O\} \rightarrow CO_2 + H_2O$

Aldehyde $CH_3CH_2O + \{O\} \rightarrow CH_3COOH$ (acid)

Oxidation of Inorganics

Cyanide $2CN^- + 5OCl^- + H_2O \rightarrow N_2 +$
$$2HCO_3^- + 5Cl^-$$

Iron(II) $4Fe^{2+} + O_2 + 10H_2O \rightarrow$
$$4Fe(OH)_3 + 8H^+$$

Sulfur dioxide $2SO_2 + 2O_2 + H_2O \rightarrow 2H_2SO_4$

Reduction of Inorganics

Chromate $2CrO_4^- + 3SO_2 + 4H^+ \rightarrow$
$$Cr_2(SO_4)_3 + 2H_2O$$

Permanganate $MnO_4^- + 3Fe^{2+} + 7H_2O \rightarrow$
$$MnO_2(s) + 3Fe(OH)_3(s) + 5H^+$$

Ozone is employed as an oxidant gas at levels of 1-2 wt% in air and 2-5 wt% in oxygen.[7] It has been used to treat a large variety of oxidizable contaminants, effluents, and wastes in the following categories:

- Municipal drinking water
    Disinfection; color, taste, and odor removal at more than 1,000 installations globally
- Wastewater
    Disinfects and removes oxidizable chemical contaminants from municipal and industrial wastewater. Oxidizes

organic compounds, including unsaturated alcohols, phenols, aldehydes; inorganic species, including H_2S, nitrite (to less toxic nitrate), cyanide, and Fe^{2+} (to insoluble Fe(III)).

- Sludges containing oxidizable constituents
- Gas streams containing toxic gases and odor-causing organic compounds

A safety advantage with ozone is its generation on site so that oxidant does not have to be stored and shipped. Its toxicity and status as an air pollutant are disadvantages, and scrupulous measures must be employed to prevent its release.

13.6. ELECTROLYSIS

Electrolysis consists of the electrochemical reduction and oxidation of chemical species in solution by means of electricity applied to electrodes from an external source. One species in solution (usually a metal ion) is reduced by electrons at the **cathode** and another gives up electrons to the **anode** and is oxidized there. Examples of cathodic and anodic half-reactions and the general concept of electrolysis are illustrated in Figure 13.2.

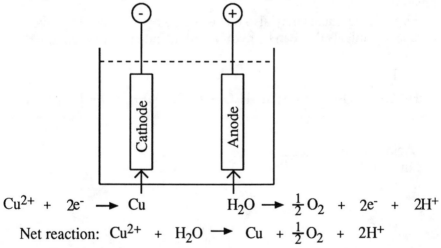

$$Cu^{2+} + 2e^- \longrightarrow Cu \qquad\qquad H_2O \longrightarrow \tfrac{1}{2}O_2 + 2e^- + 2H^+$$

Net reaction: $Cu^{2+} + H_2O \longrightarrow Cu + \tfrac{1}{2}O_2 + 2H^+$

Figure 13.2. Electrolysis of copper solution.

In hazardous waste applications electrolysis is most widely used in the recovery of metals. An obvious application is metal recovery from spent electroplating media. Other types of media to which electrolysis is applicable include wastewaters and rinsewaters from the electronics industries and from metal finishing operations. The metals that are most commonly recovered by electrolysis are cadmium, copper, gold, lead, silver, and zinc.[7] Metal recovery by electrolysis is made more difficult by the presence of cyanide ion which stabilizes metals in solution as the cyanide complexes, such as $Ni(CN)_4^{2-}$. Recovered metals are usually recycled to the process that produced the wastes.

13.7. REACTION WITH WATER

Many inorganic and organic chemicals are hazardous because of their reactions with water, commonly called **hydrolysis**. In some cases the reaction is so vigorous that a fire or even an explosion may result. In others, a hazardous gas, such as explosive H_2 or corrosive, toxic HCl is evolved. One of the ways to dispose of chemicals that are reactive with water is to subject them to hydrolysis under controlled conditions.[8]

Inorganic chemicals that can be treated by hydrolysis include metals that react with water; metal carbides, hydrides, amides, alkoxides, and halides; and nonmetal oxyhalides and sulfides. Examples of the treatment of these classes of inorganic species are given in Table 13.2.

Organic chemicals may also be treated by hydrolysis. The types of organic chemicals that can be hydrolyzed include the following:

$$H-\overset{\overset{\displaystyle H}{|}}{\underset{\underset{\displaystyle H}{|}}{C}}-\overset{\overset{\displaystyle O}{\|}}{C}-O-\overset{\overset{\displaystyle O}{\|}}{C}-\overset{\overset{\displaystyle H}{|}}{\underset{\underset{\displaystyle H}{|}}{C}}-H + H_2O \longrightarrow 2H-\overset{\overset{\displaystyle H}{|}}{\underset{\underset{\displaystyle H}{|}}{C}}-\overset{\overset{\displaystyle O}{\|}}{C}-O-H \quad (13.7.1)$$

Acid anhydride (acetic anhydride)

$$H-\overset{\overset{\displaystyle H}{|}}{\underset{\underset{\displaystyle H}{|}}{C}}-\overset{\overset{\displaystyle O}{\|}}{C}-Cl + H_2O \longrightarrow H-\overset{\overset{\displaystyle H}{|}}{\underset{\underset{\displaystyle H}{|}}{C}}-\overset{\overset{\displaystyle O}{\|}}{C}-OH + HCl \quad (13.7.2)$$

Acid halides (acetyl chloride)

$$H-\underset{\underset{H}{|}}{\overset{\overset{H}{|}}{C}}-N=C=O \;+\; H_2O \;\longrightarrow\; \text{Hydrolysis products} \qquad (13.7.3)$$

Isocyanates (methyl isocyanate)

Table 13.2. Inorganic Chemicals That May be Treated by Hydrolysis

Class of Chemical	Reaction with Water
Active metals (calcium)	$Ca + 2H_2O \longrightarrow H_2 + Ca(OH)_2$
Hydrides (sodium aluminum hydride)	$NaAlH_4 + 4H_2O \longrightarrow 4H_2 + NaOH + Al(OH)_3$
Carbides (calcium carbide)	$CaC_2 + 2H_2O \longrightarrow Ca(OH)_2 + C_2H_2$
Amides (sodium amide	$NaNH_2 + H_2O \longrightarrow NaOH + NH_3$
Halides (silicon tetrachloride)	$SiCl_4 + 2H_2O \longrightarrow SiO_2 + 4HCl$
Alkoxides (sodium ethoxide)	$NaOC_2H_5 + H_2O \longrightarrow NaOH + C_2H_5OH$

13.8. CHEMICAL EXTRACTION AND LEACHING

Chemical extraction or **leaching** in hazardous waste treatment is the removal of a hazardous constituent by reaction with an extractant in solution. The extraction of hazardous wastes and, specifically, their extraction from soil (soil washing or flushing) were discussed in Section 12.8. Some examples of chemical phenomena in extraction are cited here.

Acidic solutions dissolve poorly soluble heavy metal salts by reaction of the salt anions with H^+ as illustrated by the following:

$$PbCO_3 \ + \ H^+ \ \longrightarrow \ Pb^{2+} \ + \ HCO_3^- \qquad\qquad (13.8.1)$$

Acids also dissolve basic organic compounds such as amines and aniline. Extraction with acids should be avoided if cyanides or sulfides are present to prevent formation of toxic hydrogen cyanide or hydrogen sulfide. Nontoxic weak acids are usually the safest to use. These include acetic acid, CH_3COOH, and the acid salt, NaH_2PO_4.

Chelating agents, such as dissolved ethylenedinitrilotetraacetate (EDTA, HY^{2-}), dissolve insoluble metal salts by forming soluble species with metal ions:

$$FeS \ + \ H_2Y^{2-} \ \longrightarrow \ FeY^{2-} \ + \ H_2S \qquad\qquad (13.8.2)$$

Heavy metal ions in soil contaminated by hazardous wastes may be present in a coprecipitated form with insoluble iron(III) and manganese(IV) oxides, Fe_2O_3 and MnO_2, respectively. These oxides can be dissolved by reducing agents, such as sodium dithionate/citrate or hydroxylamine, which result in the production of soluble Fe^{2+} and Mn^{2+}. Heavy metal species, such as Cd^{2+} or Ni^{2+} are released and removed with the water.

13.9. ION EXCHANGE

Ion exchange is a means of removing cations or anions from solution onto a solid resin as illustrated by the two following reactions showing cadmium ion in solution being taken up by a cation exchanger and sulfate ion by an anion exchanger:

$$2H^{+-}(Cat.\ Exchr.) \ + \ Cd^{2+} \ \longrightarrow \ Cd^{2+-}(Cat.\ Exchr.)_2$$
$$+ \ 2H^+ \qquad (13.9.1)$$
$$2OH^{-+}(An.\ Exchr.) \ + \ SO_4^{2-} \ \longrightarrow \ SO_4^{2-+}(An.\ Exchr.)_2$$
$$+ \ 2OH^- \qquad (13.9.2)$$

Ions taken up by an ion exchange resin may be removed by treating the resin with concentrated solutions of acid, base, or salt (NaCl). The net result is to concentrate the ions originally removed from dilute solution in water into a much more concentrated solution.

The greatest use of ion exchange in hazardous waste treatment is for the removal of low levels of heavy metal ions from wastewater. Ion exchange is employed in the metal plating industry to purify rinsewater and spent plating bath solutions. Cation exchangers are used to remove cationic metal species, such as Cu^{2+}, from such solutions. Anion exchangers remove anionic cyanide metal complexes (for example, $Ni(CN)_4^{2-}$) and chromium(VI) species, such as CrO_4^{2-}.

Radionuclides may be removed from radioactive wastes and mixed waste (see Section 1.2) by ion exchange resins. High levels of radioactivity cause deterioration of ion exchange resins, although in some applications inorganic ion exchangers may be used.

Several characteristics of wastes must be considered prior to application of ion exchange. These include physical form of the wastes, presence of oxidants, and contents of metals, organic consituents, and suspended solids.[9]

13.10 CHEMICAL DESTRUCTION OF PCBs

Polychlorinated biphenyls (PCBs) and dielectric fluid made from PCBs (askarel) were discussed in Section 8.8. Because of their extreme chemical and biological stabilities, PCBs are difficult to destroy. These compounds can be burned in high-efficiency incinerators meeting stringent requirements for residence time, temperature, and excess oxygen.[10] Proprietary processes that use metallic sodium dissolved in various solvents can be used to selectively destroy PCBs dissolved at levels up to 10 parts per thousand in transformer fluids and other media. The reaction involves attack on C-Cl bonds as shown by the following generalized reaction:

$$(PCB)\text{–}Cl \;+\; Na \;\longrightarrow\; Biphenyl\ polymer \;+\; NaCl \quad (13.10.1)$$

The sodium chloride solid and a sludge consisting of polymer formed from the dechlorinated biphenyl constituents and other reaction byproducts is filtered from the liquid, which can be recycled as a dielectric fluid.

13.11. Photolytic Reactions

Photolytic reactions or the process of **photolysis** are those in which **photons** of electromagnetic radiation consisting of short-wavelength visible light or ultraviolet radiation are absorbed by a molecule, causing a chemical reaction to occur.[11] The energy, E, of the photon is equal to the product Planck's constant, h, and the frequency, ν, of the radiation and is represented in photochemical reactions as $h\nu$. An example of an important photochemical reaction is the photodecomposition of nitrogen dioxide, which occurs in the earth's atmosphere and produces reactive oxygen atoms that start the processes by which photochemical smog is formed:

$$NO_2 + h\nu \rightarrow NO + O \qquad (13.11.1)$$

Photolysis can be used to destroy a number of kinds of hazardous wastes.[12] In such applications it is most useful in breaking chemical bonds in refractory organic compounds. The irradiation of TCDD (see Section 9.5) by ultraviolet light in the presence of hydrogen atom donors {H} results in reactions such as the following:

As photolysis proceeds, more H-C bonds are broken, the C-O bonds are broken, and the final product is a harmless organic polymer.

An initial photolyis reaction can lead to the generation of reactive intermediates that participate in **chain reactions** that lead to the destruction of a compound. One of the most important reactive intermediates is the hydroxyl radical, HO·, where the dot represents an unpaired electron. Molecular fragments with unpaired electrons are highly reactive species called **free radicals**. In some cases substances are added to the reaction mixture to absorb radiation and generate reactive species that destroy wastes. Such a substance is called a **sensitizer**.

In addition to TCDD, photolysis has been used to destroy several other kinds of hazardous waste substances. Included are herbicides (atrazine), 2,4,6-trinitrotoluene (TNT), and polychlorinated biphenyls (PCBs). The addition of a chemical oxidant, such as potassium peroxidisulfide, $K_2S_2O_8$, enhances destruction by oxidizing active photolytic products.

13.12. BIODEGRADATION OF WASTES

The latter part of this chapter covers biological treatment of hazardous wastes. It does not consider directly the widely practiced biological treatment of municipal wastewaters, food processing byproducts, and other wastes for which biological treatment has long been established. Although it has some shortcomings in the degradation of complex chemical mixtures,[13] biological treatment offers a number of significant advantages and has considerable potential for the degradation of hazardous wastes, even *in situ*.[14]

Biodegradation of wastes is their conversion by biological processes to simple inorganic molecules and, to a certain extent, to biological materials. The complete bioconversion of a substance to inorganic species such as CO_2, NH_3, and phosphate is called **mineralization. Detoxification** refers to the biological conversion of a toxic substance to a less toxic species, which may still be a relatively complex, or even more complex material. An example of detoxification is illustrated below for the enzymatic conversion of paraoxon (a highly toxic organophosphate insecticide) to *p*-nitrophenol, which has only about 1/200 the toxicity of the parent compound:

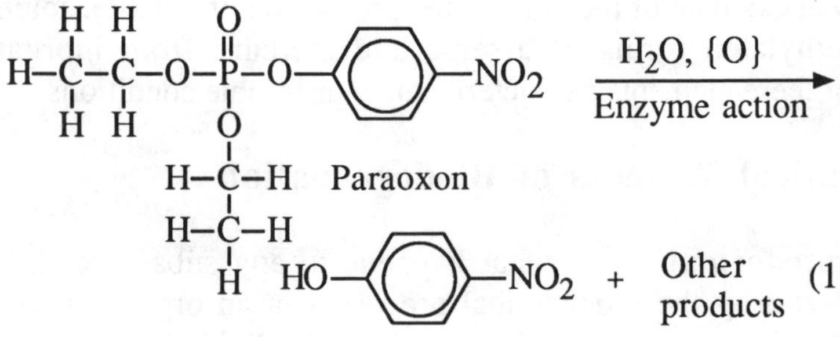

$$\frac{H_2O, \{O\}}{\text{Enzyme action}}$$

Paraoxon

$$HO-\!\!\!\bigcirc\!\!\!-NO_2 \;+\; \text{Other products} \qquad (13.12.1)$$

Usually the products of biodegradation are molecular forms that tend to occur in nature and that are in greater thermodynamic equilibrium with their surroundings. The definition of bio-degradation is illustrated by an example in Figure 13.3. Bio-degradation is usually carried out by the action of microorganisms, particularly bacteria and fungi.

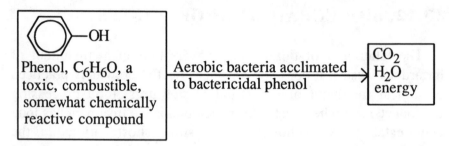

Figure 13.3. Illustration of biological treatment of a hazardous waste constituent.

Biodegradation of municipal wastewater and solid wastes in landfills occurs by design. Biodegradation of any kind of waste that can be metabolized takes place whenever the wastes are subjected to conditions conducive to biological processes. The most common type of biodegradation is that of organic compounds in the presence of air, that is, **aerobic processes**. However, in the absence of air, **anaerobic biodegradation** may also take place. Furthermore, inorganic species are subject to both aerobic and anaerobic biological processes.

Although biological treatment of wastes is normally regarded as degradation to simple inorganic species such as carbon dioxide, water, sulfates, and phosphates, the possibility must always be considered of forming more complex or more hazardous chemical species. An example of the latter is the production of volatile, soluble, toxic methylated forms of arsenic and mercury from inorganic species of these elements by bacteria under anaerobic conditions.

Biochemical Aspects of Biodegradation

Biotransformation is what happens to any substance that is **metabolized** by the biochemical processes in an organism and is altered by these processes. **Metabolism** is divided into the two general categories of **catabolism**, which is the breaking down of more complex molecules, and **anabolism**, which is the building up of life molecules from simpler materials. The substances subjected to biotransformation may be naturally occurring or *anthropogenic* (made by human activities). They may consist of *xenobiotic* molecules (see Chapter 4) that are foreign to living systems.

An important biochemical process that occurs in the biodegradation of many synthetic and hazardous waste materials is **cometabolism**. Cometabolism does not serve a useful purpose to an organism in terms of providing energy or raw material to build biomass, but occurs concurrently with normal metabolic processes. An example of cometabolism of hazardous wastes is provided by the white rot fungus, *Phanerochaete chrysosporium*, which degrades a number of kinds of organochlorine compounds, including DDT, PCBs, and chlorodioxins, under the appropriate conditions. The enzyme system responsible for this degradation is one that the fungus uses to break down lignin in plant material under normal conditions.

Enzymes in Waste Degradation

Enzyme systems (see Section 4.4) hold the key to biodegradation of hazardous wastes. For most biological treatment processes currently in use, enzymes are present in living organisms in contact with the wastes. However, in some cases it is possible to extract enzymes from organisms and use cell-free extracts of enzymes to treat hazardous wastes. For this application the enzymes may be present in solution or, more commonly, immobilized in biochemical reactors.[15]

13.13. FACTORS INVOLVED IN BIOLOGICAL TREATMENT

Biological treatment of hazardous wastes usually is carried out by microorganisms—bacteria and sometimes fungi. In considering waste biodegradation, it is useful to keep in mind that microorganisms require sources of energy, carbon, macronutrients and micronurients. Except for algae and photosynthetic bacteria, which utilize sunlight for energy, microorganisms extract energy from the reactions that they mediate or "catalyze." Carbon can come from inorganic sources (CO_2, HCO_3^-) or from organic sources, including the substances undergoing biodegradation. Macronutrients required include sources of elements in addition to carbon used in building biomass—hydrogen, oxygen, nitrogen as well as smaller quantities of sulfur, and phosphorus. Trace amounts of micronutrients are needed to support biological processes and as constituents of enzymes. Important micronutrients are calcium, magnesium, potassium, sodium, chlorine, cobalt, iron, vanadium, and zinc. Sometimes sulfur, phosphorus, and micronutrients must be added to media in which microorganisms are used to degrade hazardous wastes in order for optimum growth to occur.

Organisms

A crucial factor in the biodegradation of hazardous waste substances is the selection of organisms that will degrade a particular type of material. Genetic engineering holds considerable promise for the development of microorganisms that can readily degrade toxic and recalcitrant wastes. A more practical approach at the present time is selection of indigenous microorganisms at a hazardous waste site that have the ability to degrade particular kinds of molecules. Often the populations of such microorganisms can be increased and conditions (such as those of nutrition) enhanced for their growth.

Biodegradability

An obvious requirement for the biological treatment of wastes is **biodegradability**. The biodegradability of a compound is influenced by its physical characteristics, such as solubility in water and vapor pressure, and by its chemical properties, including molecular mass, molecular structure, and presence of various kinds of functional groups, some of which provide a "biochemical handle" for the initiation of biodegradation. Many hazardous waste materials are biocidal and are often therefore not even considered candidates for biodegradation. However, with the appropriate organisms and under the right conditions, even substances that are biocidal to most microorganisms can undergo biodegradation. For example, as shown in Figure 13.3, normally bactericidal phenol is readily metabolized by the appropriate bacteria acclimated to its use as a carbon and energy source.

Recalcitrant or biorefractory substances are those that resist biodegradation and tend to persist and accumulate in the environment. Such materials are not necessarily toxic to organisms, but simply resist their metabolic attack. Even some compounds regarded as biorefractory may be degraded by microorganisms adapted to their biodegradation. Examples of such compounds and the types of microorganisms that can degrade them include endrin (*Arthrobacter*), DDT (*Hydrogenomonas*), phenylmercuric acetate (*Pseudomonas*), and raw rubber (*Actinomycetes*). Chemical pretreatment, especially by partial oxidation, can make some kinds of recalcitrant wastes much more biodegradable.

Properties of hazardous wastes can be changed to increase biodegradability. This is especially true of wastes that consist of several constituents, one or more of which inhibit biological processes. Sometimes a waste substance that is toxic to microorganisms at a relatively high concentration is degraded well in more dilute media. Biodegradation that is inhibited by extremes of pH in the waste may occur when excess acid or base is neutralized. Toxic organic and inorganic substances, such as heavy metal ions, can be removed in some cases prior to biological treatment. Substances that are precursors to formation of solid precipitates that form during biodegradation and inhibit it can be removed. For example, soluble iron(II) need to be removed because it forms bacteria-inhibiting deposits of gelatinous iron(III) hydroxide during aerobic treatment of wastes:

$$2\{CH_2O\} \;\rightarrow\; CO_2 \;+\; CH_4 \qquad\qquad (13.13.1)$$

Conditions

The conditions under which biodegradation is carried out have a strong effect on the efficiency of biological treatment. Temperature affects the rate of biodegradation which, to a point, usually increases in rate with increasing temperature. The pH of the medium needs to be controlled, usually to within a range of 6-9 (see Section 13.2). The need to remove toxic impurities was mentioned above. Stirring and mixing can be factors in biodegradation. The load of material to the process is important; sudden fluctuations can be detrimental. Oxygen is important; an adequate supply is needed for aerobic treatment, whereas oxygen is excluded for anaerobic processes.

Bioreactors

A **bioreactor** is a container in which biological treatment processes take place. Bioreactors may have a continuous flow of material or may be operated in a batch mode. Efficient bio-degradation occurs when a large mass of microorganisms is maintained relative to the amount of waste present at any particular time. This can be done with an attached fixed film of organisms. One such system is the **trickling filter** bioreactor in which layers of microorganisms are established on media, such as rocks, over which aqueous waste is sprayed. The microorganisms may also be suspended as a sludge in the bioreactor and recirculated as is done in the **activated sludge** process (see Publicly Owned Treatment Works, POTW, discussed in Section 3.7) illustrated in Figure 13.4. Widely used in sewage treatment, the activated sludge process enables buildup of a large mass of microorganisms and is very efficient.

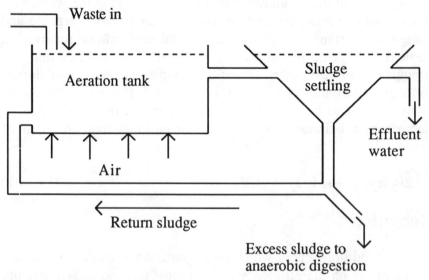

Figure 13.4. Activated sludge process.

13.14. ANAEROBIC TREATMENT OF ORGANIC WASTES

Aerobic waste treatment processes utilize aerobic bacteria and fungi that require molecular oxygen, O_2. These processes are often favored by microorganisms, in part because of the high energy yield obtained when molecular oxygen reacts with organic matter. Aerobic waste treatment is well adapted to the use of an activated sludge process (see Figure 13.4). It can be applied to hazardous wastes such as chemical process wastes and landfill leachates. Some systems use powdered activated carbon as an additive to absorb nonbiodegradable organic wastes.

Contaminated soils can be mixed with water and treated in a bioreactor to eliminate biodegradable contaminants in the soil. It is possible in principle to treat contaminated soils biologically in place by pumping oxygenated, nutrient-enriched water through the soil in a recirculating system, a biochemical variation of soil flushing (see Section 12.8). Aerobic treatment of contaminated soil is relatively expensive, in part because of the high amount of energy required.

The off-gases from aerobic treatment of hazardous wastes often contain volatile organic compounds. Because of the potential toxicities of these compounds and their odors, the off-gases may require activated carbon filtration or other treatment.

13.15. ANEROBIC TREATMENT OF ORGANIC WASTES

Anaerobic waste treatment in which microorganisms degrade wastes in the absence of oxygen can be practiced on a variety of organic hazardous wastes. Compared to the aerated activated sludge process, anaerobic digestion requires less energy; yields less sludge byproduct; generates sulfide (H_2S), which precipitates toxic heavy metal ions; and produces methane gas, CH_4, which can be used as an energy source.[16]

The overall process for the anaerobic digestion of biomass is a fermentation process in which organic matter is both oxidized and reduced. Where biomass is represented by the formula, $\{CH_2O\}$, the simplified reaction for anaerobic fermentation producing carbon monoxide and methane is the following:

$$2\{CH_2O\} \rightarrow CO_2 + CH_4 \qquad (13.15.1)$$

In practice the microbial processes involved are quite complex. Most of the wastes for which anaerobic digestion is suitable consist of oxygenated compounds, including those given in Figure 13.5.

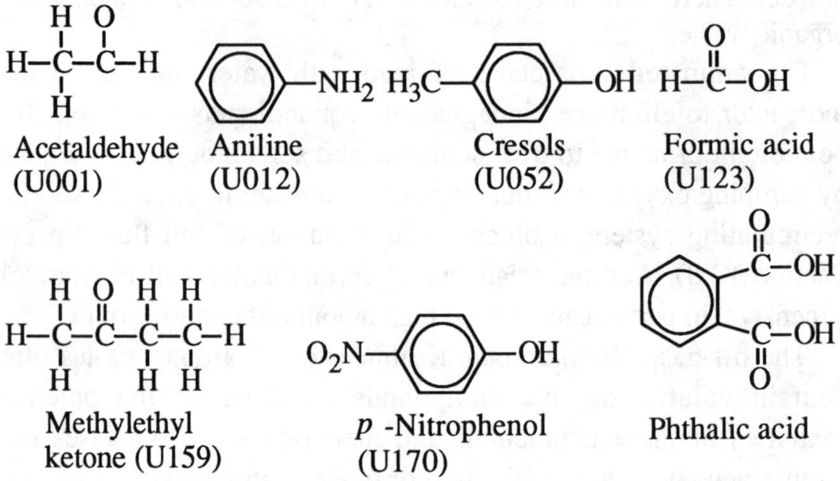

Figure 13.5. Examples of waste compounds that can be degraded by anaerobic digestion.

13.16. COMPOSTING AND LAND TREATMENT

Composting and land treatment are two means of treating solid hazardous wastes utilizing the biological activity of naturally occurring organisms. These two techniques have many similarities. In land treatment the waste to be treated is mixed with soil, whereas composting does not involve soil as a major part of the medium. Both of these techniques are useful for solid or solidified wastes.

Land Treatment

A variety of enzyme activities are exhibited by microorganisms in soil. Even sterile soil may show enzyme activity due to extracellular enzymes secreted by microorganisms in soil. Some of these enzymes are hydrolase enzymes (see Section 4.4), such as those that catalyze the hydrolysis of organophosphate compounds as shown by the reaction,

$$R-O-\overset{\displaystyle\overset{X}{\|}}{\underset{\displaystyle\underset{R}{|}}{\underset{|}{O}}}-O-Ar \xrightarrow[\text{Phosphatase enzyme}]{H_2O}$$

$$R-O-\overset{\displaystyle\overset{X}{\|}}{\underset{\displaystyle\underset{R}{|}}{\underset{|}{O}}}-OH \; + \; HOAr \qquad (13.16.1)$$

where R is an alkyl group, Ar is a substituent group that is frequently aromatic, and X is either S or O. Another example of a reaction catalyzed by soil enzymes is the oxidation of phenolic compounds by diphenol oxidase:

$$\text{(structure)} \; + \; \{O\} \xrightarrow[\text{oxidase}]{\text{Diphenol}} \text{(structure)} \; + \; H_2O \qquad (13.16.2)$$

Soil may be viewed as a natural filter for wastes. Soil has physical, chemical, and biological characteristics that can enable waste detoxification, biodegradation, chemical decomposition, and physical and chemical fixation. A number of soil characteristics are important in determining its use for land treatment of wastes. These characteristics include physical form, ability to retain water, aeration, organic content, acid-base characteristics, and oxidation-reduction behavior.[17] Soil is a natural medium for a number of living organisms that may have an effect upon biodegradation of hazardous wastes. Of these, the most important are bacteria, including those from the genera *Agrobacterium*, *Arthrobacteri*, *Bacillus*, *Flavobacterium*, and *Pseudomonas*. Actinomycetes and fungi are important in decay of vegetable matter and may be involved in biodegradation of wastes. Other unicellular organisms that may be present in or on soil are protozoa and algae. Soil animals, such as earthworms, affect soil parameters such as soil texture. The growth of plants in soil may have an influence on its waste treatment potential in areas such as uptake of soluble wastes and erosion control.

Wastes that are amenable to land treatment are biodegradable organic substances. However, soil bacterial cultures may develop that are effective in degrading normally recalcitrant compounds through acclimation over a long period of time. This occurs particularly at contaminated sites, such as those where soil has been exposed to crude oil for many years. Land treatment is most used for petroleum refining wastes and is applicable to the treatment of fuels and wastes from leaking underground storage tanks. It can also be applied to biodegradable organic chemical wastes, including some organohalide compounds. Land treatment is not suitable for the treatment of wastes containing acids, bases, toxic inorganic compounds, salts, heavy metals, and organic compounds that are excessively soluble, volatile, or flammable.[18]

Composting

Composting of hazardous wastes is the biodegradation of solid or solidified materials in a medium other than soil. Bulking material, such as plant residue, paper, municipal refuse, or sawdust may be added to retain water and enable air to penetrate to the waste material. Composting can be carried out in a container, often with air pumped through it to aid biodegradation. Alternatively, the waste material may be placed in open piles or rows (windrows) which are reformed periodically to facilitate penetration of air.

Successful composting of hazardous waste depends upon a number of factors,[19] including those discussed in Section 13.13. The first of these is the selection of the appropriate microorganism or **inoculum**. Once a sucessful composting operation is underway, a good inoculum is maintained by recirculating spent compost to each new batch. Other parameters that must be controlled include oxygen supply, moisture content (which should be maintained at a minimum of about 40%), pH (usually around neutral), and temperature. The composting process generates heat so, if the mass of the compost pile is sufficiently high, it can be self-heating under most conditions. Some wastes are deficient in nutrients, which must be supplied from commercial sources, such as commercial fertilizer. One of the nutrients most likely to be deficient is nitrogen, and the carbon/nitrogen level is a critical parameter for many wastes. This ratio is very high in some plant wastes such as straw or sawdust, so that nitrogen should be added.

Other wastes, such as manure and, particularly, urine, have a low C/N ratio reflecting a high nitrogen content, and these wastes may be used in composting to increase the nitrogen content to required levels.

As with other treatment technologies, composting requires consideration of potential environmental pollution factors. Volatile toxic or noxious organic compounds may be emitted to the air from composting operations. Leachate from composting may contain undegraded organic substances, heavy metals, and other inorganic solutes, so it must be collected, analyzed and treated.

13.17. BACTERIAL DESTRUCTION OF PCBs

Potentially, bacteria can be used for bioremediation of sites contaminated with organic wastes such as polychlorinated biphenyls (PCBs). Usually the first place to search for bacteria capable of degrading organic wastes is in sites where the wastes have been in contact with soil for some time so that the bacteria have had an opportunity to develop strains capable of biodegrading the wastes. Because of their prevalence in soil, these bacteria are often of the *Pseudomonas* or *Alciligenes* genuses. For wastes such as PCBs that consist of many different molecular variations of the same class of compounds, bioremediation is complicated by the high selectivity of various strains of bacteria for molecular type. For example, aerobic bacteria generally are required to degrade PCBs with 3 or 4 Cl atoms per molecule, whereas anaerobic bacteria are not effective for these compounds.[20] Some anaerobic bacteria metabolize PCBs with 5 or more Cls per molecule. It is not possible to devise a system that is hospitable for both types of bacteria at the same time. Different isomers of PCB molecules that have the same number of Cl atoms per molecule may show big differences in biodegradability by a particular strain of bacteria.

LITERATURE CITED

1. Lunn, George, and Sansone, Eric B. *Destruction of Hazardous Chemicals in the Laboratory*, John Wiley and Sons, Somerset, New Jersey, 1990.
2. Daley, Peter S., "Cleaning up Sites with On-Site Process Plants," *Environmental Science and Technology*, **23**, 912-916 (1989).

3. "Handbook of Remedial Action at Waste Disposal Sites," EPA/625/6-85/006, U.S. Environmental Protection Agency, Hazardous Waste Engineering Research Laboratory, Cincinnati, Ohio, 1985, pp. 10-45–9-47.
4. Manahan, Stanley E., *Quantitative Chemical Analysis*, Brooks/Cole Publishing Co., Pacific Grove, CA, 1986, p. 192.
5. Stowe, Elizabeth, "Flocculants: Flocculants Provide a Potpouri of Water and Industrial Wastewater Treatment Solutions," *Hazmat World*, March, 1990, 26-31.
6. Roy, Kimberly A., "Interox America," *Hazmat World*, March, 1990, 26-31.
7. Ficek, Kenneth J, and Wolfram Hartwig, "New Ozone System Designed specifically for Small Systems,"*Waterworld News*,Waterworld News, **6**, American Water Works Association, Denver, CO, March/April 1990, 18-19.
8. Wilds, Alan, "Hydrolysis," Section 6.4 in *Standard Handbook of Hazardous Waste Treatment and Disposal*, Harry M. Freeman, Ed., McGraw Hill, New York, 1989, pp. 6.47-6.58.
9. Rastegar, Hamid, James Lu, and Chris Conroy, "Assessment of Chemical Treatment Technologies and their Restrictive Waste Characteristics," in *Water and Residue Treatment*, Volume II, Hazardous Materials Control Research Institute, Silver Spring, Maryland, 1987, pp. 61-63.
10. McCoy, Drew E., "PCB Wastes," Section 4.2 in *Standard Handbook of Hazardous Waste Treatment and Disposal*, Harry M. Freeman, Ed., McGraw Hill, New York, 1989, pp. 4.13-4.23.
11. Wayne, Richard P., *Principles and Applications of Photochemistry*, Oxford University Press, New York, 1988.
12. Kearney, Hilip C., and Paul H. Mazzocchi, "Photolysis," Section 7.3 in *Standard Handbook of Hazardous Waste Treatment and Disposal*, Harry M. Freeman, Ed., McGraw Hill, New York, 1989, pp. 7.33-4.40.
13. Mueller, James, G., Peter J. Chapman, and P. Hap. Pritchard, "Creosote-Contaminated Sites: Their Potential for Bioremediation," *Environmental Science and Technology*, **23**, 1197-1201 (1989).

14. J. M. Thomas and C. H. Ward. *"In Situ* Biorestoration of Organic Contaminants in the Subsurface," *Environmental Science and Technology*, **23**, 760-766 (1989).

15. Glaser, John A., "Enzyme Systems and Related Reactive Species," Section 9.2 in *Standard Handbook of Hazardous Waste Treatment and Disposal*, Freeman, Harry M., Ed., McGraw-Hill Book Company, New York, 1989, pp. 9.61-9.73.

16. Torpy, Michael F., "Anaerobic Digestion," Section 9.2 in *Standard Handbook of Hazardous Waste Treatment and Disposal*, Freeman, Harry M., Ed., McGraw-Hill Book Company, New York, 1989, pp. 9.19-9.28.

17. Sposito, Garrison, *The Chemistry of Soils*, Oxford University Press, New York, 1989.

18. Phung, Tan, "Land Treatment of Hazardous Wastes," Section 9.4 in *Standard Handbook of Hazardous Waste Treatment and Disposal*, Freeman, Harry M., Ed., McGraw-Hill Book Company, New York, 1989, pp. 9.41-9.51.

19. Doyle, Richard C., Jenefir D. Isbister, George A. Anspach, David Renard, and Judith F. Kitchens, "Composting Explosives Contaminated Soil," in *Water and Residue Treatment*, Volume II, Hazardous Materials Control Research Institute, Silver Spring, Maryland, 1987, pp. 90-95.

20. Brown, Malcolm W., "Toxic Feast for Microbes Isn't so Easy to Arrange," *New York Times*, May 23, 1989, p. 25.

Thermal Treatment and Incineration

14.1. THERMAL TREATMENT METHODS

Thermal treatment of hazardous wastes can be used to accomplish most of the common objectives of waste treatment. For example, a reactive, combustible material can be burned to produce a much lower volume of ash with recovery of heat from the waste. Volatile, mobile organic materials can be eliminated. Toxic compounds and pathogens can be destroyed. In some cases, however, thermal processes produce even more dangerous byproducts, and proper control measures must be employed.

The most widely applied means of thermal treatment of hazardous wastes is **incineration**. Incineration utilizes high temperatures, an oxidizing atmosphere, and often turbulent combustion conditions to destroy wastes. Methods other than incineration make use of high temperatures to destroy or neutralize hazardous wastes. These include **pyrolysis** in which the material is heated strongly in the absence of oxygen, **gasification** in which the waste is treated in a reducing atmosphere that produces combustible gas products, **wet oxidation** involving reaction with oxygen of substances dissolved in superheated water, and a number of less conventional technologies, such as those that use molten salt or glass or high temperature plasmas of ionized gas.

Incineration

Discussion of incineration in this chapter is largely restricted to hazardous waste rather than incineration of municipal refuse. However, both have much in common. Furthermore, the production of hazardous byproducts must be considered in municipal waste incineration.

Hazardous waste incineration will be defined here as a process that involves exposure of the waste materials to oxidizing conditions at a high temperature, usually in excess of 900°C. Normally the heat required for incineration comes from the oxidation of organically-bound carbon and hydrogen contained in the waste material or in supplemental fuel:

$$C(organic) \; + \; O_2 \; \longrightarrow \; CO_2 \; + \; heat \qquad (14.1.1)$$

$$2H(organic) \; + \; \tfrac{1}{2}O_2 \; \longrightarrow \; H_2O \; + \; heat \qquad (14.1.2)$$

These reactions destroy organic matter and generate heat required for endothermic reactions, such as the breaking of C-Cl bonds in organochlorine compounds.

As of 1989, there were about 100 hazardous waste incinerators operating in the U. S. and another 100 were going through the permitting process.[1] In addition, many liquid hazardous wastes were burned as recycled material in cement kilns. Classified as **blended fuel supplements**, these materials could be burned in cement kilns that did not have to meet the same requirements for emissions of hydrocarbons, acid gases (HCl), and particulate matter as were required for hazardous waste incinerators, although this exemption was slated to expire some time after 1989.

In Europe, where landfill space has long been in short supply, hazardous waste incineration followed by landfilling of ash and slag is practiced more widely than in the U. S. Switzerland has plans to build four regional incinerators for industrial wastes and some hazardous constituents of municipal refuse. Swiss laws pertaining to incinerator emissions are more stringent than corresponding laws in the U. S. and require that particulate matter, oxygen, NO_x, HF, and HCl all be monitored. At Biebesheim in the West German state of Hessen two solid waste incinerators (the first built in 1917) dispose of 50 million tons of waste per year. West German law prohibits landfill disposal of volatile organic substances, which must be incinerated. Denmark has a single 100-million-ton-per-year solid waste incinerator in Nyborg. The growth of the Green Party in Europe has been accompanied by questions about the safety of hazardous waste incineration, particularly in respect to emissions of heavy metals and TCDD.

Several combustion systems and a variety of associated equipment are used in hazardous waste incineration. These are discussed in Section 14.3

14.2. INCINERABLE WASTES

Ideally, incinerable wastes are predominantly organic materials that will burn. Wastes with a heating value over 8,000 Btu/lb usually burn well, whereas those with a heating value less than 5,000 Btu/lb generally do not support combustion. In some cases, however, it is desirable to incinerate wastes that will not burn alone and which require auxiliary fuel. Examples of such wastes are nonflammable organochloride wastes, some aqueous wastes, or soil in which the elimination of a particularly troublesome contaminant is worth the expense and trouble of incinerating it. The availability of an incinerator at a waste disposal site may make it the method of choice for disposing of wastes that would not otherwise be treated by incineration.

Large quantities of combustible manufacturing **byproduct liquids** suitable for incineration are produced each year. Many of these are cleaning and degreasing solvents, including both nonchlorinated and chlorinated compounds (Figure 14.1), as well as mixtures of organic liquids with water. As of the early 1980s the ten organic compounds most commonly incinerated in the U.S. were methanol, acetonitrile, toluene, ethanol, amyl acetate, acetone, xylene, methylethyl ketone, adipic acid, and ethyl acetate.[2]

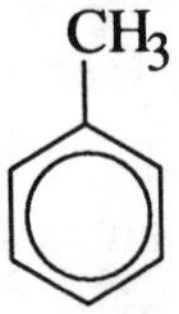

Toluene,
a nonhalogenated
hydrocarbon solvent

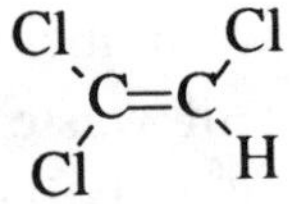

Trichloroethylene,
a halogenated solvent

Figure 14.1. Examples of nonhalogenated and halogenated solvents that may be incinerated.

Inorganic matter, water, and organic hetero element contents of liquid wastes are important in determining their incinerability. Numerous manufacturing processes produce incinerable pumpable liquids or sludges. These materials can be categorized as **process residuals** and often have relatively high water and ash contents.

The distinction between liquid and **solid hazardous wastes** is that the latter cannot be pumped or atomized. Some sludges, including some with relatively high water contents, are classified as solids for incineration purposes. Many solid hazardous wastes can be ground very fine for incineration. However, grinding consumes large quantities of energy and results in greater emissions of particulate matter that must be removed as fly ash.

A 1983 study of the types of hazardous wastes that are incinerated[3] showed that 44% of the wastes consisted of spent, nonhalogenated solvents and organic-contaminated corrosive and reactive wastes (EPA waste codes F003, D002, and D003). Hydrocyanic acid (highly toxic HCN) was also a major contributor to the waste stream, as were acrylonitrile still bottoms and contaminated water. Commonly incinerated industrial wastes include pesticides, plasticizers and oils. Off-specification chemicals are often incinerated.

Many chemical compounds are contained in wastes that are burned in hazardous waste incinerators. These include acetone, methylene chloride, carbon tetrachloride, trichloroethylene, perchloroethylene, chloroform, chloromethane, chlorobenzene, dichlorobenzene, ethyl chloride, hexachlorocyclopentadiene, benzyl chloride, benzene, toluene, *ortho*-xylene, cumene, phenol, aniline, nitrobenzene, methylethyl ketone, phthalic anhydride, naphthalene, toluene diisocyanate methyl acetate, diphenylamine, phosgene, and Freon 113 (see chlorofluorocarbons in Section 8.5).

Hazardous Waste Fuel

Many industrial wastes, including hazardous wastes, are burned as fuel for energy recovery in industrial furnaces and boilers and in incinerators for nonhazardous wastes, such as sewage sludge incinerators. This process is called **coincineration**, and more combustible wastes are utilized by it than are burned solely for the purpose of waste destruction. In addition to heat recovery from combustible wastes, it is a major advantage to use an existing on-site facility for waste disposal rather than a separate hazardous waste incinerator. Wastes used in this manner are called hazardous waste fuel. In 1983 there were approximately 1,300 facilities in the U.S. burning hazardous waste fuel at a rate of about 230 million gallons per year.[4]

Liquid hazardous waste fuel should meet several important criteria as follows:[5] It should contain no more than 1.0% separable water and have a heating value of at least 10,000 Btu/gal. The viscosity should be low enough to enable pumping and injection into the burners. Solids should not exceed 20% by volume and the content of metals should not contribute excessively to fly ash volume or toxicity.

Wastes with ignitability characteristics that are not classified as hazardous (EPA "D" wastes) can be used as hazardous waste fuel, as can some of the generic non-source-specific "F" wastes. Several materials classified as source-specific wastes (EPA "K" wastes) are potentially suitable as hazardous waste fuel. Listed in order of EPA hazardous waste number these are K015, still bottoms from distillation of benzyl chloride; K018, heavy ends from fractionation in ethyl chloride production; K022, distillation bottom tars from production of phenol/acetone from cumene; K023, distillation light ends from the production of phthalic anhydride from naphthalene; K024, distillation bottoms from production of phthalic anhydride from naphthalene; K025, distillation bottoms from the production of nitrobenzene by nitration of benzene; K027, centrifuge and distillation residues from toluene diisocyanate production; K030, heavy ends from combined production of trichloroethylene and perchloroethylene; K049, slop-oil-emulsion solids from the petroleum refining industry; K093, distillation light ends from the production of phthalic anhydride from orthoxylene; and K094, distillation bottoms from the production of phthalic anhydride from *ortho*-xylene.

Supplemental Fuels

Several types of fuels are used as necessary to add to the heating value of incinerated hazardous wastes. The most important of these is natural gas consisting mostly of methane, CH_4. Methane is the cleanest burning hydrocarbon fuel and adds few contaminants to those from the waste. Liquid fuel oils used as supplemental fuels consist of a variety of petroleum products ranging from heavy residual fuel to lighter distillates. In some cases, the sulfur content of fuel oil can result in excessive sulfur dioxide emissions. Residual sodium chloride and vanadium can cause some difficulties. Although a very effective supplement to wastes in incineration, coal produces substantial quantities of ash and usually contains sufficient sulfur to produce excessive amounts of sulfur dioxide. However, in incinerators equipped to control these emissions at the levels produced by coal, it may be the supplemental fuel of choice. This is especially true of fluidized-bed incinerators.

Characterization of Incinerable Wastes

The efficiency of incineration, incinerator capacity, control of emissions, reliability, control of emissions, and protection of incinerator equipment and personnel require careful characterization of wastes prior to incineration. The properties to be characterized include both physical and chemical characteristics. Some of the more important properties of hazardous wastes that must be characterized prior to incineration are given in Table 14.1.[6,7]

From an operational and pollution control perspective it is often convenient to classify organic wastes for incineration on the basis of the major elements present in the organic compounds, most of which contain a preponderance of carbon and hydrogen. The simplest wastes to handle are those consisting only of C and H or of C, H, and O. A second class may contain these elements plus N, but little or no chlorine. A third class of incinerable organic wastes consists of C, H, and Cl, perhaps some O, but little or no nitrogen. A fourth class of wastes may contain any of the elements just mentioned, as well as F, Br, S, P, Si, Na, and other elements.

Table 14.1. Important Properties of Incinerable Wastes

Properties	Importance
Morphology, density, rheology	Feeding characteristics
Heating value	Energy balance in combustion chamber, need for auxiliary fuel
Carbon, hydrogen, oxygen, water	Heating value, combustibility
Organohalides	Require high temperatures for destruction, halogen product emissions (particularly HCl)
Organic sulfur, nitrogen, and phosphorus	Sulfur, nitrogen, and phosphorus oxide emissions
Ash	Amount and hazard of unburned residue
Ash particle size, fusion characteristics	Slagging, plugging of equipment
Hazardous trace elements including heavy metals	Hazard from release of heavy metals and other toxic elements during incineration and from ash residue
Salts	Ash characteristics, damage to incinerator refractory linings, especially by sodium

When burned in pure oxygen, the unsaturated organohalide compound trichloroethylene reacts as follows to produce Cl_2:

$$C_2HCl_3 + 2O_2 \rightarrow 2CO_2 + HCl + Cl_2 \qquad (14.2.1)$$

Incinerated wastes having a low H/Cl ratio tend to produce reactive, toxic, oxidant Cl_2 along with more benign HCl gas, so the ratio may have to be adjusted upward in some cases by the addition of hydrogen-rich waste or fuel.

Significant levels of toxic trace metals, such as lead, cadmium, or arsenic, in the waste may prevent it from being incinerated. The incineration of such metals may require stringent, specialized air pollution control measures and result in classification of the ash as a hazardous waste, which can make its disposal prohibitively expensive.

Wastes that pose particular chemical or biological hazards may be poor candidates for incineration. Included in this category are wastes that are particularly toxic or carcinogenic, or likely to become so during handling or incineration. Highly reactive wastes may pose fire or explosion hazards. Although a uniform waste feed is generally desirable, mixing of wastes can cause problems when they are chemically incompatible.

The analysis of hazardous wastes for incineration poses special challenges. It is difficult to obtain representative samples of many kinds of mixed wastes. Variations in waste feedstock may require frequent analysis and characterization.

Species Interaction during Incineration

A significant number of test burns of hazardous waste incineration have been conducted on a large scale. The results of these tests have been used to a large extent to establish operating criteria for hazardous waste incinerators. Comparatively little information is available on the fundamental processes that take place during incineration.[8] Although criteria for hazardous waste destruction are usually set on the basis of thermodynamics, the combustion and destruction processes are often kinetically controlled.

An important consideration in hazardous waste incineration is the possibility of inhibitory interactions. An example of this is the inhibition of the breakdown of trichloroethylene (C_2HCl_3) under oxidizing conditions by the presence of methane. In the absence of inhibiting species an initial step appears to be the formation of free Cl atoms which take part in chain reactions resulting in the destruction of C_2HCl_3. Methane, however, scavenges the Cl atom,

$$CH_4 + Cl \rightarrow CH_3 + HCl \qquad (14.2.2)$$

and inhibits the oxidation of trichloroethylene to the extent that temperatures of about 950°C are required for its destruction. It is possible that supplemental fuels added to hazardous waste incinerators may have inhibitory effects, such as the one just described.

14.3. INCINERATION SYSTEMS

As shown in Figure 14.2 hazardous waste incineration systems may be divided into four major components. Each of these will be discussed briefly here.

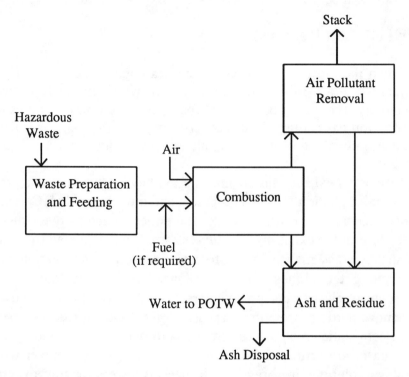

Figure 14.2. Major components of a hazardous waste incinerator system.

Waste Preparation and Feeding

Waste preparation obviously depends upon the nature of the waste being processed. Liquid wastes may require filtration, settling to remove solid material and water, blending to obtain the optimum incinerable mixture, or heating to decrease viscosity. Solids may require shredding and screening.

Several means are employed to feed wastes into the incinerator. Atomization is commonly used for liquid wastes. Several mechanical devices, such as rams and augers, are used to introduce solids into the incinerator.

Combustion Chambers

There are four major types of combustion chambers plus several new types in various stages of development. The most common kinds of combustion chambers are liquid injection, fixed hearth, rotary kiln, and fluidized bed. These are discussed in more detail in Section 14.4.

Air Pollutant Removal

Often the most complex part of a hazardous waste incineration system is the air pollution control system, which involves several operations. These may be divided broadly into the four categories of combustion gas cooling, heat recovery, and conditioning; particulate matter removal; acid gas removal; and byproduct treatment and handling.

Incinerator exhaust gas streams are too hot for the materials used in most air pollution control equipment and must be cooled. Cooling may be accomplished with a waste heat boiler, which recovers much of the heat of the exhaust gas to raise steam. Heat recovered from the combustion gas may be used to heat incoming combustion air. Quenching techniques can also be used to cool exhaust gas.

Venturis, electrostatic precipitators, and fabric filters may be used to remove particles from the combustion gas. Ionizing wet scrubbers and electrostatic precipitators remove both particles and acid gases. Acid gases are scrubbed from the combustion gas by contact with scrubber solutions in spray towers, packed towers, or tray towers. Demisters are used to remove liquid droplets before the combustion gas is released to the stack. Solid, sludge, and liquid residues from air pollution control may require chemical treatment, such as neutralization.

Ash and Residue Treatment

Hot ash is often quenched in water. Prior to disposal it may require dewatering and chemical stabilization. A major consideration with hazardous waste incinerators and the types of wastes that are incinerated is the disposal problem posed by the ash, especially in respect to potential leaching of heavy metals.

14.4. TYPES OF INCINERATORS

Hazardous waste incinerators of "conventional design" are divided into four major types as described below. In the United States, rotary kiln and liquid injection incinerators each account for approximately 40 percent of capacity, hearth incinerators 11 percent, and other types, including fluidized bed, for the remaining capacity.[9]

Conventional Design

Hazardous waste incinerators are classified according to the configuration of the combustion chamber.[10] It is beyond the scope of this work to go into the engineering design of these types in detail. However, a brief description of each is necessary for later discussion of the chemical and environmental aspects of hazardous waste incineration.

Liquid injection incinerators (Figure 14.3) burn pumpable liquid wastes. Maximum fuel surface area and conversion to combustible gas are attained by dispersing the liquid waste mechanically into small droplets of around 40 μm diameter as it is injected into the incinerator, and it is burned in an atomized form. Fuel and waste may be injected separately. Combustion efficiency is reduced by droplets that are too large or nonvolatile to vaporize completely and by unburned vapor.

Fixed-hearth incinerators have single or multiple hearths upon which combustion occurs. Such incinerators may be operated in a two-stage mode where the first stage has a stoichiometric deficiency of oxygen and the second stage operates with excess air at a higher temperature. They may be used for liquid or solid wastes.

Multiple hearth incinerators in a vertical configuration have a series of hearths over which the wastes are stirred mechanically. The wastes are introduced into the top of the incinerator and fall through drop holes in the hearths. Combustion air is introduced into the bottom of the incinerator. In the top part of the incinerator, wastes are dried, in the middle portion, which is the hottest, the wastes burn, and in the lower portion the ash is cooled while heating incoming combustion air. Multiple hearth incinerators have been used to a large extent to incinerate sewage sludge.

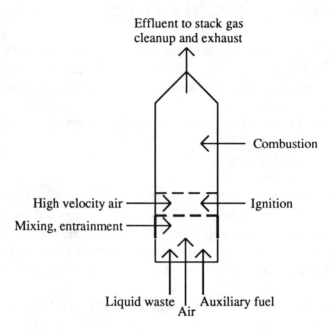

Figure 14.3. Liquid injection incinerator.

Rotary kiln incinerators (Figure 14.4) are those in which the primary combustion chamber is a rotating cylinder lined with refractory materials. An afterburner is employed downstream from the kiln to complete destruction of the wastes. Rotary kiln incinerators operate with a wide variety of liquid, semisolid (sludge), and solid wastes, which may be burned simultaneously. Good mixing is achieved by rotation of the kiln. A disadvantage of rotary kiln incinerators that may be especially troublesome with hazardous wastes is the need to maintain gastight seals at both ends of the kiln. Leakage of these seals results in fugitive emissions that can cause air pollution problems and hazards to personnel working on the unit. Leakage can also occur from sudden excursions in pressure resulting from rapid burning of portions of the waste. To minimize these problems, rotary kilns are always operated under negative pressure.

Rotary kiln incinerators have found wide use in the destruction of hazardous wastes. According to a Canadian study the rotary kiln is considered to be the most versatile kind of hazardous waste incinerator.[11]

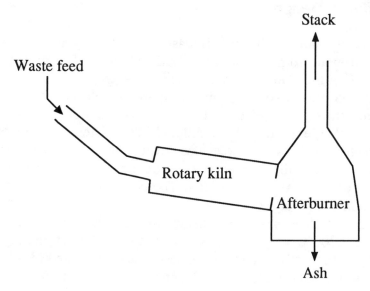

Figure 14.4. Rotary kiln incinerator.

Fluidized-bed incinerators have a bed of granular solid (such as sand) maintained in a suspended state by injection of air.[12] These devices can be used for liquid wastes or solid wastes of relatively uniform particle size. Advantages of fluidized-bed incinerators are excellent mixing, minimal requirement for excess air, and potential to retain waste gases in the bed material. Preheating the air injected into the bed reduces its cooling effect and enables incineration of wastes with lower heating values.

Circulating fluidized-bed combustion (CBC) is an advanced form of fluidized-bed incinerator that has been adapted to use with mobile units used on-site.[13] This system circulates wastes and sorbent solids (limestone) through a combustion chamber that is in a loop configuration. The second stage of the loop is a cyclone in which solids settle to the bottom for recirculation to the first stage and from the top of which flue gases are exhausted. Combustion occurs at a relatively low 900°C (which minimizes NO_x formation and ash slagging) with very complete mixing due to high turbulence. Efficient combustion is achieved by injection of secondary combustion air at three different places in the combustion chamber. In addition to the combustion chamber, the major components of this system are waste, fuel, and sorbent feeding systems; ash removal system; flue gas cooler; baghouse filter; and stack.

Advanced Design

Several promising hazardous waste incinerator types are substantially different from the four types outlined above. The most prominent ones of these are mentioned here.

Plasma incinerator systems make use of a plasma of ionized air injected through an electrical arc into which liquid wastes are injected. The extremely high 5,000-15,000°C temperature of the plasma breaks the waste molecules down to their constituent atoms, which remain in the elemental form or recombine to form simple molecules, such as CO and HCl. The combustible gas product is burned in a later stage[14].

Electric reactors use resistance-heated incinerator walls at around 2,200°C to heat wastes by radiative heat transfer. This pyrolyzes the waste leading to its destruction. A related idea is contained in **infrared systems** which generate intense infrared radiation by passing electricity through silicon carbide resistance heating elements. The infrared radiation heats the waste and drives off volatile matter that is burned in a secondary combustion chamber. These systems have shown promise for the incineration of contaminated soil.

Molten salt combustion uses a bed of molten sodium carbonate at about 900°C to destroy the wastes. Gaseous HCl, NO_x, and SO_2 wastes are retained in the molten salt, thereby reducing the need for downstream air pollution control. Sulfur from sulfur compounds incinerated in the molten salt process are retained as Na_2SO_4 or Na_2SO_3, phosphorus as Na_3PO_4, and chlorine as NaCl. Silicon is converted to sodium silicate, Na_2SiO_3.

Molten glass processes use a pool of molten glass to transfer heat to the waste and to retain products in a poorly leachable glass form. These devices show promise for the incineration of wastes containing heavy metals, which can cause severe leaching problems in ash.

Numerous kinds of small-scale and specialty incinerators are under development. One example is an incinerator designed to accommodate individual drums and incinerate their contents.[15] Called a Drummed Waste Disposal System, the apparatus is designed to pyrolyze wastes in a drum and burn the vapors in a secondary combustion chamber.

Although not necessarily classified as a hazardous waste incinerator, another type of incinerator should be mentioned here. This type is the **fume incinerator** used to destroy vapor and particulate matter emissions. Fume incinerators are used as a second stage of pyrolytic thermal treatment of hazardous wastes.[16]

Combustion Conditions

The key to effective incineration of hazardous wastes lies in the combustion conditions. These require (1) sufficient free oxygen in the combustion zone; (2) turbulence for thorough mixing of waste, oxidant, and (where used) supplemental fuel; (3) high combustion temperatures above about 900°C to ensure that thermally resistant compounds do react; and (4) sufficient residence time (at least 2 seconds) to allow reactions to occur.

Effectiveness of Incineration

EPA standards for hazardous waste incineration are based upon the effectiveness of destruction of the **principal organic hazardous constituents** (POHC). Measurement of these compounds before and after incineration gives the **destruction removal efficiency** (DRE) according to the formula,[17]

$$DRE = \frac{W_{in} \quad W_{out}}{W_{in}} \times 100 \qquad (14.4.1)$$

where W_{in} and W_{out} are the mass flow rates of the principal organic hazardous constituent (POHC) input and output (at the stack downstream from emission controls), respectively. Incinerator standards are based on the DRE with a current requirement in the U.S. of 99.99%. In addition at least 99% of hydrogen chloride product must be removed from stack gas if stack gas emission levels exceed 4 lb/h, and standards are specified for particulate matter stack emissions.

14.5. SOIL INCINERATION

Incineration of contaminated soil is a practice that is growing in importance. Among the most common contaminants that may require incineration of soil are petroleum hydrocarbons, dioxins, and PCBs.

One operational system for soil incineration makes use of a "turbulent combustion loop" in which soil particles are repeatedly subjected to oxidizing conditions at 790-900°C in a combustion chamber that is 30 feet high and 3 feet in diameter with a 1-foot thick ceramic liner.[18] Using this system, soil is incinerated at a rate of 2-4 tons per hour. Dry soil with a heating value from combustible contaminants of at least 3,000 Btu/lb may be incinerated without auxiliary fuel. Because of recirculation in the combustion chamber an afterburner is not required. Limestone added to the soil retains acid gases (HCl and SO_2), eliminating the need for wet scrubbing.

14.6. EMISSIONS AND RESIDUES FROM INCINERATION

Four major potential contaminant problems from the incineration of hazardous wastes are undestroyed organic constituents, organic byproducts of thermochemical processes, heavy metal emissions, and acid gas emissions.[2] Complete destruction of organic constituents requires a sufficiently high temperature (900°C), adequate residence time (2 s), and good mixing (turbulence). The more volatile metals that are likely to be emitted include mercury, arsenic, beryllium, cadmium, chromium, nickel, and lead. The most common acid gas emission is hydrogen chloride, HCl, which reacts with water vapor at lower temperatures and in the atmosphere to produce highly corrosive hydrochloric acid:

$$HCl(g) + H_2O \longrightarrow H^+(aq) + Cl^-(aq) \qquad (14.6.1)$$

When substantial amounts of sulfur are present in the fuel, formation of sulfur dioxide must be considered.

Exhaust Gas Cooling and Quenching

Air pollution control equipment usually must operate at 300°C or less which requires cooling the hot exhaust gases from an incinerator. In larger, more efficient installations, a waste heat boiler is used to cool exhaust gases and recover heat. Exhaust gas heat can also be used to increase the temperature of incoming combustion air.

Quenching consists of spraying water adiabatically into the exhaust gas so that the water evaporates and the heat of vaporization cools the gas to its **adiabatic saturation temperature.**

Air pollutants

Air pollutants from hazardous waste incineration may be produced as (1) **pollutant gases** from combustion of elements other than C and H, (2) **fly ash** from inorganic matter and unburned carbon, and (3) **products of incomplete combustion** (PIC). Emissions from these sources depend upon the nature of the waste, the type of incinerator, how the incinerator is operated, and the type and operation of the air pollution control equipment. Each type of air pollutant is discussed briefly here.

Pollutant gases

Most pollutant gases emitted by hazardous waste incineration are acidic gases. Two types of priority pollutant gases that may be produced in hazardous waste incineration are oxides of sulfur and oxides of nitrogen. The major sulfur oxide emission is sulfur dioxide, SO_2, produced from sulfur in organic or inorganic matter. Small quantities of sulfur trioxide, SO_3, may also be present. Nitrogen oxide emissions are designated collectively as NO_x and consist predominantly of NO and NO_2. They are produced from combustion of nitrogen in the fuel and by reaction of O_2 and N_2 in the combustion chamber by reactions such as the following:

$$2N(\text{waste/fuel}) + O_2 \longrightarrow 2NO \qquad\qquad (14.6.2)$$

$$N_2 + O_2 \longrightarrow 2NO \qquad\qquad (14.6.3)$$

The production of NO_x is minimized by low excess air and low combustion temperatures. Although nitrous oxide, N_2O, may be produced as a transient species during combustion of nitrogen-containing wastes, it normally is not emitted because of its very rapid reaction with hydrogen atoms under combustion conditions:[19]

$$N_2O + H\cdot \longrightarrow N_2 + HO\cdot \tag{14.6.4}$$

The combustion of organochloride compounds in the waste produces hydrogen chloride and under some conditions small quantities of elemental chlorine (Cl_2). High combustion temperatures and hydrogen-rich fuel favor HCl production. A final type of acidic air pollutant that may be emitted is phosphorus pentoxide (P_2O_5 or, more correctly, P_4O_{10}) and its hydrated form, phosphoric acid (H_3PO_4).

Fly ash

Fly ash (also called particulate matter or "particulates") consists of solids that are entrained with the stack gas and is distinguished from bottom ash, which is removed as a solid residue from the incinerator. Fly ash contains a wide variety of materials, including unburned carbon, metal oxides, silicates, and salts. The factor that most influences the composition of fly ash is the composition of the waste that is incinerated, but incineration conditions also play a role. Oxidation of metal compounds yields metal oxides. Salts are produced by chemical processes, as well as from the evaporation of solvents from salt solutions. Many of the constituents of fly ash are the result of condensation of substances from the vapor phase, which is especially conducive to the production of very small particles that are difficult to remove from exhaust gas.

Products of Incomplete Combustion

Products of incomplete combustion (PIC) occur as a result of inadequate combustion conditions and from species that are particularly resistant to combustion. A major class of these substances consists of carbon-rich substances. Solid carbon residue is such a material, which contributes to fly ash. Carbon monoxide, CO, results from partial oxidation of organic compounds. Polycyclic aromatic hydrocarbons (PAH) are formed by the preferential oxidation of hydrogen in an oxygen deficient environment. Consider the example of pyrene shown in Figure 14.5. The ratio of hydrogen to carbon is much lower in pyrene than in the corresponding 16-carbon alkane.

$C_{16}H_{34}$

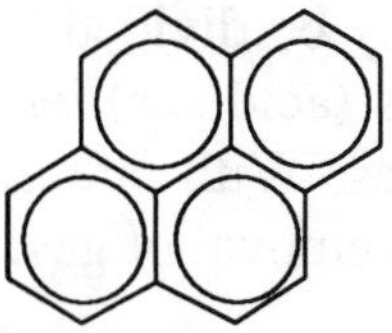

16–Carbon alkane   Pyrene, formula $C_{16}H_{10}$

Figure 14.5. Alkane and polycyclic aromatic hydrocarbon with the same number of carbon atoms. The PAH has a much lower H:C ratio.

Partially oxidized hydrocarbons are also products of incomplete combustion. Such compounds include alcohols, aldehydes, ketones, and organic acids. There are also some miscellaneous products of incomplete oxidation, such as amines.

The organic products that are perhaps of greatest concern in both municipal and hazardous waste incineration are the polychlorinated dibenzofurans (PCDFs) and polychlorinated-p-dioxins (PCDDs).[20] The basic structures of these two classes of compounds are shown in Figure 14.6. Each has 8 carbon atoms on its 2 outer rings to which chlorine atoms may bond so that it is possible to form many different compounds (congeners), some of which are extremely toxic. Both PCDFs and PCDDs are thought to form as reaction products of the combustion of organochloride compounds, such as chlorinated phenols and polychlorinated biphenyls (PCBs).

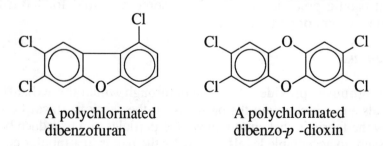

A polychlorinated   A polychlorinated
dibenzofuran   dibenzo-p -dioxin

Figure 14.6. Structures of polychlorinated dibenzofurans and polychlorinated-p-dioxins. Chlorine atoms may be bonded to any of the available carbon atoms on the benzene rings to give a wide variety of different compounds.

14.7. EMISSIONS REMOVAL

The removal of emissions from hazardous waste incinerator exhaust gas may be divided between the two categories of particle removal and gas (acid gas) removal. Several devices are available to remove particles. With present technology scrubber systems are favored for the removal of gases.

Particle Removal

To have a RCRA permit, a hazardous waste incinerator facility must not emit particulate matter at a rate exceeding 180 mg per standard cubic meter corrected to 7 percent oxygen in the stack gas. The correction factor used is

$$\text{Correction factor} = \frac{14}{21 - Y} \tag{14.7.1}$$

where Y is the percentage of oxygen on a dry basis in the stack gas. It is multiplied times the concentration of particulate matter measured in the stack to get a corrected value of particulate matter concentration.

The selection of a particle removal system for hazardous waste incinerator exhaust gases depends upon the particle loading, nature of particles (especially size distribution), and type of gas scrubbing system used. Most hazardous waste incinerators use baghouses, venturi scrubbers, or ionizing wet scrubbers for particle removal. Dry electrostatic precipitators also have some potential for hazardous waste incinerator particle control.

Baghouses

Baghouses provide fabric filters through which the exhaust gas travels and which retain the particles on the fabric surface. Periodically the fabric is shaken to remove the particles and to reduce backpressure to acceptable levels. Typically the bag is in a tubular configuration as shown in Figure 14.7. Numerous other configurations are possible. Collected particulate matter is removed from bags by mechanical agitation, blowing air on the fabric, or rapid expansion and contraction of the bags.

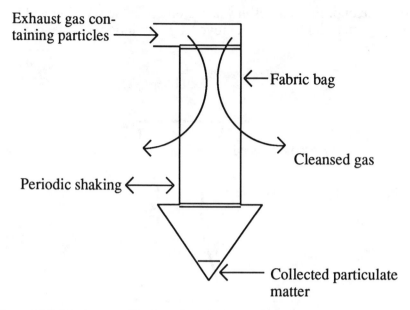

Figure 14.7. Baghouse collection of particulate emissions.

Although simple, baghouses are generally effective in removing particles from exhaust gas. Particles as small as 0.01 μm in diameter are removed, and removal efficiency is relatively high for particles down to 0.5 μm in diameter.

Venturi Scrubbers

A venturi scrubber passes gas through a converging section, throat, and diverging section as shown in Figure 14.8. Injection of the scrubbing liquid at right angles to incoming gas breaks the liquid into very small droplets, which are ideal for scavenging particles from the gas stream. In the reduced-pressure (expanding) region of the venturi, some condensation can occur, adding to the scrubbing efficiency. In addition to removing particles, venturis may serve as quenchers to cool exhaust gas and as scrubbers for pollutant gases.

Ionizing Wet Scrubbers

Ionizing wet scrubbers place an electrical charge on particles upstream from a wet scrubber. Larger particles and some gaseous contaminants are removed by scrubbing action. Smaller particles tend to induce opposite charges in water droplets in the scrubber and in its packing material and are removed by attraction of the opposite charges.

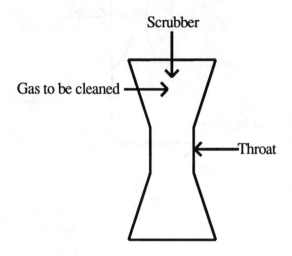

Figure 14.8. Venturi scrubber.

Gas Removal

For most hazardous waste incinerators the major pollutant gas produced is hydrogen chloride. In order to receive a RCRA permit, any incinerator that emits more than 1.8 kg/hour of HCl in the exhaust gas must be equipped to remove at least 99 percent of the HCl.

Pollutant gases are usually removed from hazardous waste incinerator exhaust by absorption in a liquid. For most intimate contact, the exhaust gas and liquid usually flow in opposite directions through a column. Typical of such absorbers is a countercurrent packed column through which gas is introduced into the bottom and moves upward to contact liquid introduced through the top. The most contaminated liquid contacts the most contaminated gas and *vice versa*. The most common application of this sort uses water as a scrubber liquid to remove HCl gas. This requires neutralization of the water and construction of the scrubber from acid-resistant materials. For removal of other acid gases such as SO_2, alkali scrub solutions must be used. Lime water containing up to 30% $Ca(OH)_2$ by weight can be used in some types of scrubbers, although scaling and plugging often result.

Treatment of scrubber water to remove scavenged particles, precipitated salts, and other impurities may be expensive and leave large quantities of sludges and contaminated water for disposal. For that reason various dry scrubbing processes are attractive. In these processes scrubber material may be introduced as a liquid solution that is dried by the heat of the exhaust gas to produce particles that can be collected in a baghouse. For sulfur dioxide removal, for example, a lime slurry is injected into the gas stream through a spray dryer and sulfur dioxide is absorbed by the reaction

$$SO_2 + Ca(OH)_2 \rightarrow CaSO_3 + H_2O \qquad (14.7.2)$$

and part of the calcium sulfite is oxidized to solid calcium sulfate as follows:

$$2CaSO_3 + O_2 \rightarrow \ + 2CaSO_4 \qquad (14.7.3)$$

Residue and Ash Handling and Disposal

A hazardous waste incineration facility may produce bottom ash, as well as solids (fly ash), sludges, and aqueous byproducts from stack gas pollutant control. Ash is allowed to cool or quenched with water. Prior to its disposal in a secure land disposal facility ash may require chemical treatment for fixation/stabilization. The content and leachability of heavy metals is a matter of increasing concern in respect to ash disposal

Aqueous waste streams are produced by quenching hot exhaust gas and by processes that remove particulate matter and acid gases from the combustion gas. These streams contain suspended particulate matter that may have to be removed, acids (HCl) that must be neutralized, and some organic compounds, such as PICs. These streams are allowed to settle, neutralized with base, and recirculated until the total dissolved solids level becomes too high.

14.8. WET AIR OXIDATION

Organic compounds and oxidizable inorganic species can be oxidized by oxygen in aqueous solution.[21] The source of oxygen usually is air. Rather extreme conditions of temperature and pressure are required, with a temperature range of 175-327°C and a pressure range of 300-3,000 psig (2070-20,700 kPa). The high pressures allow a high concentration of oxygen to be dissolved in the water and the high temperatures enable the reaction to occur. As a first step in the process, the wastewater and air are brought up to pressure. An initial heating starts the reaction, after which exothermic oxidation reactions in the reaction mixture help to generate heat to keep the reaction going. The length of time that wastes are exposed to oxidation is controlled by reactor residence time. After discharge from the reactor, the treated wastewater is cooled to 35-60°C with incoming water/air mixture (which is preheated in the process) or with cooling water. Gases are removed in a separator vessel. Hydrocarbon contaminant may be present in these gases. It is removed in part by wet scrubbing used to cool the gases, as well as by adsorption columns and afterburning. The oxidized liquor product may require additional treatment, such as addition of sulfide to precipitate metals.

Wet air oxidation has been applied to the destruction of cyanides in electroplating wastewaters.[22] The oxidation reaction for sodium cyanide is the following:

$$2Na^+ + 2CN^- + O_2 + 4H_2O \longrightarrow$$
$$2Na^+ + 2HCO_3^- + 2NH_3 \qquad (14.8.1)$$

Above about 390°C temperature and 218 atm pressure water becomes supercritical and the hydrogen bonds between water molecules in the liquid state are no longer present. Under these conditions, water reverses its solvent behavior so that organic compounds dissolve in it more readily and salts precipitate out. Addition of oxygen to supercritical water containing organic compounds can result in their destruction or conversion to lower molecular mass products.

UV-Enhanced Wet Oxidation

Another approach to the oxidation of organics and oxidizable inorganic species in water involves the use of hydrogen peroxide (H_2O_2) as an oxidant assisted by ultraviolet radiation (hv).[23] For the oxidation of organic species represented in general as {CH_2O}, the overall reaction is

$$2H_2O_2 + \{CH_2O\} + hv \longrightarrow CO_2 + 3H_2O \qquad (14.8.2)$$

The ultraviolet radiation breaks chemical bonds and serves to form reactive oxidant species, such as the hydroxyl radical, $HO\cdot$. As a result, most organic compounds can be oxidized to simple inorganic species, including carbon dioxide and water. Major disadvantages are the costs of hydrogen peroxide reagent and electrical energy required to produce a high flux of ultraviolet radiation.

LITERATURE CITED

1. Hanson, David J., "Hazardous Waste Management: Planning to Avoid Future Problems," *Chemical and Engineering News*, July 31, 1989, pp. 9-18.
2. Travis, R. C., G. A. Holton, E. L. Etnier, S. C. Cook, F. R. O'Donnell, D. M. Hetrick, and E. Dixon "Incineration of Hazardous Waste: A Critical Review," *Journal of Hazardous Materials*, **14**, 309-20 (1987).
3. Vogel, G., "Composition of Hazardous Waste Streams Currently Incinerated," Mitre Corporation Report, U. S. Environmental Protection Agency, Washington, D.C., 1983.
4. Oppelt, E. Timothy, "Incineration of Hazardous Waste: A Critical Review," *Journal of The Air Pollution Control Association*, **37**, 558-586 (1987).
5. Henz, Donald, "Know *All* That's Involved Before Burning Hazardous Wastes, *Power*, March, 1987, pp. 45-48.
6. Louis, Theodore, and Joseph Reynolds, *Introduction to Hazardous Waste Incineration*, John Wiley and Sons, New York, 1987.
7. Stumbar, James P., Robert H. Sawyer, Gopal D. Gupta, Joyce M. Perdek, and Frank J. Freestone, "Effect of Feed Characteristics on the Performance of EPA's Mobile Incineration System," presented at the 15th Annual Research Symposium on Remedial Action, Treatment, and Disposal of HazardousWaste, Cincinnati, Ohio, April 10-12, 1989.
8. Senkan, Selim M., "Thermal Destruction of Hazardous Waste," *Environmental Science and Technology*, **22**, 368–370 (1988).

9. Vogel, Gregory A., "Hazardous-Waste Management Capacity," Section 2.2 in *Standard Handbook of Hazardous Waste Treatment and Disposal*, Freeman, Harry M., Ed., McGraw-Hill Book Company, New York, 1989, pp. 2.21-2.31.

10. Oppelt, E. Timothy, "Hazardous Waste Destruction," *Environmental Science and Technology*, **22**, 403–404 (1988).

11. "Facilities Development Process Phase 4A Report," Ontario Waste Management Corporation, Toronto, Ontario, 1985, p. 45.

12. Brunner, Calvin R., "Incineration: Today's Hot Option for Waste Disposal," *Chemical Engineering*, October 12, 1987, pp. 96-106.

13. Shu, Abraham C., "Thermal Treatment," *Hazmat World*, August, 1989, pp. 50-55.

14. Krieger, James, "Plasma Technology to Tackle Toxic Wastes," *Chemical and Engineering News*, December 22, 1986, pp. 20-21.

15. "McGill Develops 5-Star Disposal Technology,"*Impact* **6**, April, 1989, pp. 3-4, International Technology Corporation, Monroeville, PA.

16. Shah, J. K., T. J. Schultz, and V. R. Daiga, "Pyrolysis Processes," Section 8.7 in *Standard Handbook of Hazardous Waste Treatment and Disposal*, Freeman, Harry M., Ed., McGraw-Hill Book Company, New York, 1989, pp. 8.91-8.104.

17. Mournighan, Robert E., Marta K. Richards, and Howard Wall,, "Incinerability Ranking of Hazardous Organic Compounds," presented at the 15th Annual Research Symposium on Remedial Action, Treatment, and Disposal of Hazardous Waste, Cincinnati, Ohio, April 10-12, 1989.

18. Piellisch, A. Richard, "Mobile Incineration," *Hazmat World*, June, 1989, pp. 28-29.

19. Lyon, Richard K., John C. Kramlich, and Gerald A. Cole, "Nitrous Oxide: Sources, Sampling, and Science Policy," *Environmental Science and Technology*, **23**, 392–393 (1989)

20. Hutzinger, Otto, Ghulam Ghaus Choudhry, Brock G. Chittim, and Les E. Jonston, "Formation of Polychlorinated Dibenzofurans and Dioxins during Combustion, Electrical Equipment Fires, and PCB Incineration," *Environmental Health Perspectives*, **60**, 3-9 (1985).

21. "On Site Engineering Report of Treatment Technology Performance and Operation for Wet Air Oxidation of F007 at Zimpro/Passavant, Incorporated in Rothschild, Wisconsin," Office of Solid Waste, U. S. Environmental Protection Agency, Washington, D. C., 1988.
22. Warner, H. Paul, "Destruction of Cyanides in Electroplating Wastewaters Using Wet Air Oxidation," presented at the 15th Annual Research Symposium on Remedial Action, Treatment, and Disposal of Hazardous Waste, Cincinnati, Ohio, April 10 12, 1989.
23. Peroxidation Systems, Inc., "Organic Oxidation On-Site," *Hazmat World*, May, 1989, p. 26.

15

Immobilization, Fixation, and Disposal

15.1. IMMOBILIZATION

Immobilization and **stabilization** are used here as general terms to describe techniques whereby hazardous wastes are placed in a form suitable for long term disposal. **Fixation** is sometimes used to mean much the same thing.[1] **Solidification** describes a process in which a liquid or semisolid sludge waste is converted to a monlithic solid form or granular solid material.

Immobilization serves several purposes. It usually improves the handling and physical characteristics of wastes. It isolates the wastes from their environment, especially groundwater, so that they have the least possible tendency to migrate. This is accomplished by physically isolating the waste, reducing its solubility, and decreasing its surface area. Immobilization includes physical and chemical processes that reduce surface areas of wastes to minimize leaching. Immobilization may involve both chemical and physical processes, as well as combinations of two or more processes. The major immobilization processes are the following:

- Chemical fixation of waste
- Physical fixation of waste
    Solidification with cement (Portland cement)
    Solidification with silicate materials
    Sorption to a solid matrix material

Imbedding in thermoplastics
Imbedding in organic polymers
Surface encapsulation
Vitrification (in glassy material)

Stabilization

Stabilization means the conversion of a waste to a physically and chemically more stable material. Stabilization may include chemical reactions that produce products that are less volatile, soluble, and reactive. Solidification by chemical means (formation of a precipitate by a chemical reaction) or physical means (evaporation of water from aqueous wastes or sludges) is a means stabilization. Stabilization is required for land disposal of wastes.

Solidification

Solidification may involve chemical reaction of the waste with the solidification agent, mechanical isolation in a protective binding matrix, or a combination of chemical and physical processes. It can be accomplished by sorption onto solid material, reaction with cement, reaction with silicates, encapsulation, and imbedding in polymers or thermoplastic materials.

The simplest solidification process is one in which water is evaporated from an aqueous or sludge waste to leave a solid residue. As an example consider a sludge of iron(III) hydroxide containing heavy metal contaminants and some free water. Evaporation of the water results in formation of a solid product which, with further heating and drying is converted to solid iron(III) oxide, Fe_2O_3, a very insoluble material that holds heavy metal ions.

Solidification of aqueous sludges is often accomplished by the addition of a substance that absorbs water and produces a solid. Portland cement is effective for this purpose for many wastes. **Pozzolanic material** consisting of hygroscopic silicates, such as those produced in fly ash, is an effective drying agent for solidification processes.

Encapsulation

As the name implies, **encapsulation** is used to coat wastes with an impervious material so that they do not contact their surroundings.

For example, a water-soluble waste salt encapsulated in asphalt would not dissolve, so long as the asphalt layer remains intact. Coating of individual small particles with an impervious material is called **microencapsulation**, whereas imbedding larger aggregates of material in such a matrix is **macroencapsulation**. A common means of encapsulation is with heated, molten materials that solidify when cooled. Thermoplastics, asphalt, and waxes can be used for this purpose. A more sophisticated approach to encapsulation is to form polymeric resins from monomeric substances in the presence of the waste. The substances that form the polymers can thoroughly penetrate the waste material, then bind it strongly in a polymer matrix. Problems can occur when the wastes adversely affect the polymerization process or the quality of the product formed.

15.2. CHEMICAL FIXATION

Chemical fixation is a process that binds a hazardous waste substance in a less mobile, less toxic form by a chemical reaction that alters the waste chemically. As an example consider the reaction of toxic, soluble chromium(VI) on a carbon surface when the carbon is partially gasified by reaction with O_2 and steam:

$$4C + 4Na_2CrO_4 + O_2 \xrightarrow[\text{Carbon matrix}]{\text{Heat}} 2Cr_2O_3 + 4Na_2CO_3 \quad (15.2.1)$$

The chromium(III) product, Cr_2O_3, is insoluble, relatively nontoxic and attached firmly to the carbon matrix.

Although chemical fixation and physical fixation are discussed separately here, they often occur together. Polymeric inorganic silicates containing some calcium and often some aluminum are the inorganic materials most widely used as a fixation matrix. Many kinds of heavy metals are chemically bound in such a matrix, as well as being held physically by it. Similarly, some organic wastes are bound by reactions with matrix constituents. For example, humic acid wastes react with calcium in a solidification matrix to produce insoluble calcium humates.[2]

15.3. PHYSICAL FIXATION

Solidification with Cement

Portland cement is widely used for solidification of hazardous wastes. In this application, Portland cement provides a solid matrix for isolation of the wastes, chemically binds water from sludge wastes, and may react chemically with wastes (for example, the calcium and base in Portland cement react chemically with inorganic arsenic sulfide wastes to reduce their solubilities). However, most wastes are held physically in the rigid Portland cement matrix and are subject to leaching. There are several different kinds of Portland cement; the one most widely used for waste solidification is Type I. Wastes containing sulfate (SO_4^{2-} ion) or sulfite (SO_3^{2-} ion) are more successfully solidified with Types II and V Portland cement.

When Portland cement is used for solidification, it is usually mixed directly with wastes, which may provide part, or all of the water required to set the cement. Ideally, the product is a monolithic solid with a low exposed surface area, although often it is a granular material with a relatively high surface/volume ratio. Portland cement is often used in relatively small quantities as a setting agent for other immobilization reagents such as silicates.

As a solidification matrix, Portland cement is most applicable to inorganic sludges containing heavy metal ions that form insoluble hydroxides and carbonates in the basic carbonate medium provided by the cement. Addition of Portland cement to an arsenic sulfide sludge produces a concrete material in which the waste is held in a matrix of calcium silicates and aluminum hydrates. The success of solidification with Portland cement strongly depends upon whether or not the waste adversely affects the strength and stability of the concrete product. A number of substances are incompatible with Portland cement because they interfere with its set and cure and cause deterioration of the cement matrix with time.[3] These substances include organic matter such as petroleum or coal; some silts and clays; sodium salts of arsenate, borate, phosphate, iodate, and sulfide; and salts of copper, lead, magnesium, tin, and zinc.

However, a reasonably good disposal form can be obtained by absorbing organic wastes with a solid material, which in turn is set in Portland cement. This approach has been used with hydrocarbon wastes sorbed by an activated coal char matrix.[4] The product was reported to be a reasonably strong monolithic solid material. Because of its carbon content, this material can be burned in incinerators, keeping that option open if required for later, more permanent disposal.

Advantages of Portland cement for waste immobilization include its nonhazardous nature, affinity for water in wastes, ability to form solids, universal availability at low cost, and well known procedures for working with it. However, it does add to the mass and volume of waste and often provides only minimal protection against leaching.

Solidification with Silicate Materials

The term, **silicate**, is used to denote a number of substances containing oxyanionic silicon such as SiO_3^{2-}. Water-insoluble silicates (pozzolanic substances) used for waste solidification include fly ash, flue dust, clay (an alumino silicate mineral), calcium silicates, and ground up slag from blast furnaces. Soluble silicates, such as sodium silicate, may also be used. Silicate solidification usually requires a setting agent, which may be Portland cement (see above), gypsum (hydrated $CaSO_4$), lime, or compounds of aluminum, magnesium, or iron. The product may vary from a granular material to a concrete-like solid. In some cases the product is improved by additives, such as emulsifiers, surfactants, activators, calcium chloride, clays, carbon, zeolites, and various proprietary materials.

Success has been reported for the solidification of both inorganic wastes and organic wastes (including oily sludges) with silicates. The advantages and disadvantages of silicate solidification are similar to those of Portland cement discussed above. One consideration that is especially applicable to fly ash is the presence in some silicate materials of leachable hazardous substances, which may include arsenic and selenium.

An Example of a Solidification Process

An example of a typical commercial solidification process is discussed here to illustrate the numerous parameters that must be considered. The SEALOSAFE® process uses insoluble silicates and cement to produce a solid that contains hazardous wastes. It is typical of many similar processes that use silicates and/or cement.[5]

The first step is to determine whether a waste can be solidified and effectively immobilized by solidification. This is done by chemical analysis, assessment of the materials that went into the waste (when known), and laboratory-scale solidification of samples of the waste. The major parameters analyzed include selected heavy metals, organic carbon, acidity, alkalinity, pH, organic carbon, cyanide, sulfide, and specific gravity.

A number of major kinds of wastes may not be amenable to treatment by cement and silicate. These include high levels of leachable soluble salts; wastes that produce noxious gas in reactions with water or base (carbides, phosphides, and ammonium salts); water-soluble anionic heavy metal species that are not well bound by the solid matrix (chromates, borates, sugar); substances, such as sulfate and borate salts, that prevent setting of cement; and carcinogenic, explosive, flammable, or volatile organic compounds. Wastes that are amenable to solidification by the SEALOSAFE or related processes include liquid electroplating wastes and electroplating sludges containing acids, bases, cyanide, chromium, cadmium, copper, nickel, and zinc; liquid waste from electronics manufacture containing chromium, copper, lead, selenium, and tin; solid spent catalysts containing cobalt, copper, chromium, manganese, molybdenum, nickel, and vanadium; solid residues from oil incineration containing vanadium; asbestos; chloralkali sludge containing mercury; sludge from paint manufacture containing antimony, cadmium, cobalt, and lead; and sludge from the manufacture of fungicides, insecticides, and preservatives containing arsenic, chromium, mercury, and sulfides.

Waste pretreatment is carried out to remove especially harmful constituents and to make the waste compatible with the solid matrix. Wastes exhibiting extremes of pH are neutralized to a pH in the range of approximately 6-9. Cyanides should be oxidized and toxic, soluble chromate reduced to nontoxic, insoluble chromium(III). Interference by some organic species may require addition of activated carbon to the waste mixture (see sorption below).

Following pretreatment, a slurry of the waste is prepared. It is important for the slurry to have a water content within a specified range; excess water may be removed by settling, centrifugation, or addition of dry wastes (fly ash, solids collected in air pollution control).

The final step in solidification is addition of the reagents that produce the solid, in this case cement and silicates. These materials react with water in the waste to form solids, such as hydrated silicates, which bind chemically to some of the waste constituents. After the solid product has been made and cured, it is desirable to test its properties to see if any special precautions are required in the final disposal of the wastes. Tests performed may include leaching, compressive strength, and permeability by water.

Thermal Processes in Solidification

In the solidification processes described above, water is an important ingredient of the hydrated solid matrix. Therefore, the solid should not be heated excessively or exposed to extremely dry conditions, which could result in diminished structural integrity from loss of water. In some cases, however, heating a solidified waste is an essential part of the overall solidification procedure. For example, an iron hydroxide matrix can be converted to highly insoluble, refractory iron oxide by heating. Organic constituents of solidified wastes may be converted to inert carbon by heating. Heating is an integral part of the process of vitrification (see below). Because of the cost of energy, heating of solidified wastes is not done unless it is required for special kinds of wastes.

Sorption to a Solid Matrix Material

Hazardous waste liquids, emulsions, sludges, and free liquids in contact with sludges may be sorbed by solid **sorbents**, including activated carbon (for organics), fly ash, kiln dust, clays, vermiculite, and various propietary materials. Sorption may be done to convert liquids and semisolids to dry solids, improve waste handling, and reduce solubility of waste constituents. Sorption can also be used to improve waste compatibility with substances such as Portland cement used for solidification and setting, as cited previously in this section for char used for sorption of organic wastes and set in Portland cement. Specific sorbents may also be used to stabilize pH and pE (a measure of the tendency of a medium to be oxidizing or reducing[6]).

The action of sorbents can include simple mechanical retention of wastes, physical sorption, and chemical reactions. It is important to match the sorbent to the waste. A substance with a strong affinity for water should be employed for wastes containing excess water and one with a strong affinity for organic materials should be used for wastes with excess organic solvents. The amount of sorbent required depends upon the levels of consituents to be removed.

Thermoplastics and Organic Polymers

Thermoplastics are solids or semisolids that become liquified at elevated temperatures. Hazardous waste materials may be mixed with hot thermoplastic liquids and immobilized in the cooled thermoplastic matrix, which is rigid but deformable. The thermoplastic material most used for this purpose is asphalt bitumen. Other thermoplastics, such as paraffin and polyethylene have also been used to immobilize hazardous wastes.

Among the wastes that can be immobilized with thermoplastics are those containing heavy metals, such as electroplating wastes. Organic thermoplastics repel water and reduce the tendency toward leaching in contact with groundwater. Compared to cement, thermoplastics add relatively less material to the waste.

An immobilization technique similar to that just described uses **organic polymers** produced in contact with solid wastes to imbed the wastes in a polymer matrix. Three kinds of polymers that have been used for this purpose include polybutadiene, urea-formaldehyde, and vinyl ester-styrene polymers. The reagents used to make the polymers (monomers) are mixed with the waste and a polymerization catalyst to enable polymerization to occur. The product is a polymeric solid in which the wastes are imbedded. This procedure is more complicated than is the use of thermoplastics but, in favorable cases, yields a product in which the waste is held more strongly.

Vitrification

Vitrification or **glassification** consists of imbedding wastes in a glass material. In this application, glass may be regarded as a high-melting-temperature inorganic thermoplastic. Molten glass can be used, or glass can be synthesized in contact with the waste by mixing and heating with glass constituents—silicon dioxide (SiO_2), sodium carbonate (Na_2CO_3), and calcium oxide (CaO). Other constituents may include boric oxide, B_2O_3, which yields a borosilicate glass that is especially resistant to changes in temperature and chemical attack. In some cases glass is used in conjunction with thermal waste destruction processes, serving to immobilize hazardous waste ash consituents. Some wastes are detrimental to the quality of the glass. Aluminum oxide, for example, may prevent glass from fusing.

Vitrification is relatively complicated and expensive, the latter because of the energy consumed in fusing glass. Despite these disadvantages, it is the best immobilization technique for some special wastes and has been promoted for solidification of radionuclear wastes because glass is chemically inert and resistant to leaching. However, high levels of radioactivity can cause deterioration of glass and lower its resistance to leaching.

15.4. ULTIMATE DISPOSAL

Regardless of the destruction, treatment, and immobilization techniques used, there will always remain from hazardous wastes some material that has to be put somewhere. A very favorable case is that of hydrocarbon wastes that are incinerated; the products are carbon dioxide and water that is discharged to the atmosphere. At the other extreme is contaminated soil; it can be detoxified, but little or nothing can be done to reduce its bulk. In some cases even the incineration product is still considered to be a hazardous waste (specifically, residues resulting from the incineration or thermal treatment of soil contaminated with some chlorinated benzene and phenol compounds (F028)). This section briefly addresses the ultimate disposal of ash, salts, liquids, solidified liquids, and other residues that must be placed where their potential to do harm is minimized.

Disposal Aboveground

In some important respects disposal aboveground, essentially in a pile designed to prevent erosion and water infiltration, is the best way to store solid wastes. Perhaps its most important advantage is that it avoids infiltration by groundwater that can result in leaching and groundwater contamination common to storage in pits and landfills.[7] In a properly designed aboveground disposal facility any leachate that is produced drains quickly by gravity to the leachate collection system, where it can be detected and treated.

Aboveground disposal can be accomplished with a storage mound, which must have several features to operate successfully. Such a mound begins with a layer of compacted soil or clay lain somewhat above the original soil surface and shaped to allow leachate flow and collection. Two flexible membrane liners with a layer of low-permeability clay are placed on top of the compacted soil and an appropriate leachate collection system is installed. Solid wastes are then placed atop the liners and covered with a flexible membrane, clay cap, and topsoil layer planted with vegetation. The slopes around the edges of the storage mound are sufficiently great to allow good drainage of precipitation, but gentle enough to deter erosion.

Landfill

Landfill historically has been the most common way of disposing of solid hazardous wastes and some liquids, although it is being severely limited in the U.S. by new regulations. Landfill involves disposal that is at least partially underground in excavated cells, quarries, or natural depressions. Usually fill is continued above ground to most efficiently utilize space and provide a grade for drainage of precipitation.

The greatest environmental concern with landfill of hazardous wastes is the generation of leachate from infiltrating surface water and groundwater with resultant contamination of groundwater supplies. Hazardous waste landfills constructed under current regulations provide elaborate systems to contain, collect, and control such leachate.[8]

There are several components to a modern landfill. A landfill should be placed on a compacted low-permeability medium, preferably clay, which is covered by a flexible-membrane liner consisting of water-tight impermeable material. This liner is covered with granular material in which is installed a secondary drainage system. Next is another flexible-membrane liner above which is installed a primary drainage system for the removal of leachate. This drainage system is covered with a layer of granular filter medium, upon which the wastes are placed. In the landfill, wastes of different kinds are separated by berms consisting of clay or soil covered with liner material. (In some installations only one liner is required.) Batches of wastes placed in the landfill at different times are covered with soil.

When the fill is complete, the waste is capped to prevent surface water infiltration and covered with compacted soil. In addition to leachate collection, provision may be made for a system to treat evolved gases, particularly when methane-generating biodegradable materials are disposed in the landfill.

The flexible-membrane liner is a key component of state-of-the-art landfills. It controls seepage out of, and infiltration into the landfill. Obviously, liners have to meet stringent standards to serve their intended purpose. In addition to being impermeable, the liner material must be strongly resistant to biodegradation, chemical attack, and tearing. A means must be available for joining liner segments together by impermeable seams. Liner materials used for hazardous waste landfills consist of rubber (including chlorosulfonated polyethylene) or plastic (including chlorinated polyethylene, high-density polyethylene, and polyvinylchloride).

Capping is done to cover the wastes, prevent infiltration of excessive amounts of surface water, and prevent release of wastes to overlying soil and the atmosphere. In cases where the wastes may generate gases, such as methane from anaerobic biodegradation, provision should be made for gas collection. Caps come in a variety of forms and are often multilayered. Some of the problems that may occur with caps are settling, erosion, ponding, damage by rodents and penetration by plant roots.

The description above applies to an idealized landfill constructed under current regulations. Most older landfills do not meet such stringent standards, which has caused many of the current problems with hazardous waste disposal.

Surface Impoundments of Liquids

Many liquid hazardous wastes, slurries, and sludges are placed in **surface impoundments**, which usually serve for treatment and often are designed to be filled in eventually as a landfill disposal site. Most liquid hazardous wastes and a significant fraction of solids are placed in surface impoundments in some stage of treatment, storage, or disposal.

A surface impoundment may consist of an excavated "pit," a structure formed with dikes, or a combination thereof. The construction is similar to that discussed above for landfills in that the bottom and walls should be impermeable to liquids and provision must be made for leachate collection. The chemical and mechanical challenges to liner materials in surface impoundments are severe so that proper geological siting and construction with floors and walls composed of low-permeability soil and clay are important in preventing pollution from these installations.

Deep-well Disposal of Liquids

Deep-well disposal of liquids consists of their injection under pressure to underground strata isolated by impermeable rock strata from aquifers. Early experience with this method was gained in the petroleum industry where disposal is required of large quantities of saline wastewater coproduced with crude oil. The method was later extended to the chemical industry for the disposal of brines, acids, heavy metal solutions, organic liquids, and other liquids.

A number of factors must be considered in deep-well disposal. Wastes are injected into a region of elevated temperature and pressure, which may cause chemical reactions to occur involving the waste constituents and the mineral strata. Oils, solids, and gases in the liquid wastes can cause problems such as clogging. Corrosion may be severe. Microorganisms may have some effects. Most problems from these causes can be mitigated by proper waste pretreatment.

The most serious consideration involving deep-well disposal is the potential contamination of groundwater. Although injection is made into permeable saltwater aquifers presumably isolated from aquifers that contain potable water, contamination may occur. Major routes of contamination include fractures, faults, and other wells. The disposal well, itself, can act as a route for contamination if it is not properly constructed and cased or if it is damaged.

15.5. IN-SITU TREATMENT

In-situ treatment refers to waste treatment processes that can be applied to wastes disposed aboveground, in a landfill, or in a surface impoundment without removing the waste. In those limited cases where it can be practiced, in-situ treatment is highly desirable as a waste site remediation option.[9] For in-situ treatment to work, any reagents must be transferred efficiently to wastes. Wastes held in porous sandy soil or coarse silt best meet this requirement. Several aspects of in-situ treatment are discussed here.

Immobilization

In-situ immobilization is used to convert wastes to insoluble forms that will not leach from the disposal site. Heavy metal contaminants can be immobilized by chemical precipitation as the sulfides. Sulfides of metals such as lead, cadmium, zinc, and mercury are very insoluble. There are several disadvantages to this technique. Although highly mobile hydrogen sulfide gas effectively precipitates heavy metals, it can be a dangerous reagent to use because of its toxicity. A solution of sodium sulfide, Na_2S, maintained at a high pH by excess sodium hydroxide can be used to immobilize heavy metals as the sulfides. It is important to have sufficient alkali to maintain a basic solution to prevent the production of hydrogen sulfide from waste acid that might be present by the reaction

$$2H^+ \quad + \quad S^{2-} \longrightarrow H_2S \qquad (15.5.1)$$

Gaseous hydrogen sulfide

Sulfide from precipitant

Hydrogen ion from waste acid present

Another disadvantage is that excess sulfide may become a groundwater pollutant. Although precipitated metal sulfides should remain as solids in the anaerobic conditions of a landfill, unintentional exposure to air can result in oxidation of the sulfide and remobilization of the metals as soluble sulfate salts.

Oxidation and reduction reactions can be used to immobilize heavy metals in-situ. Oxidation of soluble Fe^{2+} and Mn^{2+} to their insoluble hydrous oxides, $Fe_2O_3 \cdot xH_2O$ and $MnO_2 \cdot xH_2O$, respectively, can precipitate these metal ions and coprecipitate other heavy metal ions. However, subsurface reducing conditions could result in reformation of soluble reduced species. Reduction can be used in-situ to convert soluble, toxic chromate to insoluble chromium(III).

Chelation may convert metal ions to less mobile forms, although with most agents chelation has the opposite effect. A chelating agent called Tetran is supposed to form metal chelates that are strongly bound to clay minerals.

Solidification In-Situ

In situ solidification can be used as a remedial measure at hazardous waste sites. One approach is to inject soluble silicates followed by reagents that cause them to solidify. For example, injection of soluble sodium silicate followed by calcium chloride or lime would form solid calcium silicate.

Detoxification In-Situ

When only one, or a limited number of harmful constituents is present in a waste disposal site, it may be practical to consider detoxification in-situ. This approach is most practical for organic contaminants including pesticides (organophosphate esters and carbamates), amides, and esters. Among the chemical and biochemical processes that can detoxify such materials are chemical and enzymatic oxidation, reduction, and hydrolysis. Chemical oxidants that have been proposed for this purpose include hydrogen peroxide (for some wastes, oxidation with peroxide requires an Fe^{2+} catalyst), ozone, and hypochlorite.

Enzyme extracts collected from microbial cultures and purified have been considered for in-situ detoxification. One cell-free enzyme that has been used for detoxification of organophosphate insecticides is parathion hydrolase. The hostile environment of a chemical waste landfill, including the presence of enzyme-inhibiting heavy metal ions, is detrimental to many biochemical approaches to in-situ treatment. Furthermore, most sites contain a mixture of hazardous constituents, which might require several different enzymes for their detoxification.

Permeable Bed Treatment

Some groundwater plumes contaminated by dissolved wastes can be treated by a permeable bed of material placed in a trench through which the groundwater must flow. Acidic wastes and heavy metals are particularly amenable to such treatment. Limestone contained in a permeable bed neutralizes acid and precipitates some kinds of heavy metals as hydroxides or carbonates. Synthetic ion exchange resins can be used in a permeable bed to retain heavy metals and even some anionic species, although competition with ionic species present naturally in the groundwater can cause some problems with the use of ion exchangers. Activated carbon in a permeable bed will remove some organics, especially less soluble, higher molecular mass organic compounds.

Permeable bed treatment requires relatively large quantities of reagent, which argues against the use of activated carbon and ion exchange resins. In such an application it is unlikely that either of these materials could be reclaimed and regenerated as is done when they are used in columns to treat wastewater. Furthermore, ions taken up by ion exchangers and organic species retained by activated carbon may be released at a later time, causing subsequent problems. Finally, a permeable bed that has been truly effective in collecting waste materials may, itself, be considered a hazardous waste requiring special treatment and disposal.

In-Situ Thermal Processes

Heating of wastes in-situ can be used to remove or destroy some kinds of hazardous substances. Both steam injection and radio frequency heating have been proposed for this purpose. Volatile wastes brought to the surface by heating can be collected and held as condensed liquids or by activated carbon.

One approach to immobilizing wastes in-situ is high temperature vitrification using electrical heating. This process involves placing conducting graphite between two electrodes placed on the surface and passing an electrical current between the electrodes. In principle, the graphite becomes very hot and "melts" into the soil leaving a glassy slag in its path. Volatile species evolved are collected and, if the operation is successful, a nonleachable slag is left in place. It is easy to imagine problems that might occur, including difficulties in getting a uniform melt, problems from groundwater infiltration, and very high consumption of electricity.

15.6. LEACHATE AND GAS EMISSIONS

Leachate

The production of contaminated leachate is a possibility with most disposal sites. Therefore, new hazardous waste landfills require leachate collection/treatment systems and many older sites are required to have such systems retrofitted to them. In the U.S. under the Minimum Technological Requirements specified in Section 202 of the 1984 RCRA amendments, leachate collections are required for new hazardous waste landfills and expansions of existing ones. A "pump-and-treat" system is one of the more common remedial actions required. Even when more permanent remedies are contemplated, leachate treatment may be required as an intermediate measure to prevent imminent danger to groundwater supplies.

The design and construction of leachate collection systems are primarily engineering concerns and will not be addressed in detail here. Modern hazardous waste landfills typically have dual leachate collection systems, one located between the two impermeable liners required for the bottom and sides of the landfill and another just above the top liner of the double-liner system. The upper leachate collection system is called the primary leachate collection system, and the bottom is called the secondary leachate collection system. Leachate is collected in perforated pipes that are imbedded in granular drain material. In a properly designed system, leachate does not collect to a depth greater than 1 foot over the drainage pipes, so that hydraulic pressures on the landfill linings are low as is the potential for leakage. The landfill is designed so that the leachate flows by gravity to sumps, where it can be pumped out for treatment. It should be noted though that a leachate collection system is relatively easy to design for a new installation, but may be very difficult for an existing one from which leachate may flow in numerous and unpredictable paths.

Chemical and biochemical processes have the potential to cause some problems for leachate collection systems. One such problem is clogging by insoluble manganese(IV) and iron(III) hydrated oxides upon exposure to air. This phenomenon has occurred in drainage systems installed to remove water from waterlogged soils. Exposed to air in drainage pipes, dissolved soluble soluble Mn^{2+} and Fe^{2+} are oxidized to insoluble deposits of $Fe_2O_3 \cdot xH_2O$ and $MnO_2 \cdot xH_2O$, respectively, and clogging of the drainage pipes results.

Leachate consists of water that has become contaminated by wastes as it passes through a waste disposal site. It contains waste constituents that are soluble, not retained by soil, and not degraded chemically or biochemically. Some potentially harmful leachate constituents are products of chemical or biochemical transformations of wastes.

The best approach to leachate management is to prevent its production by limiting infiltration of water into the site. Rates of leachate production may be very low when sites are selected, designed, and constructed with minimal production of leachate as a major objective. A well-maintained, low-permeability cap over the landfill is very important for leachate minimization.

Hazardous Waste Leachate Treatment

The first step in treating leachate is to characterize it fully, particularly with a thorough chemical analysis of possible waste constituents and their chemical and metabolic products. The biodegradability of leachate constituents should also be determined.

The options available for the treatment of hazardous waste leachate are generally those that can be used for industrial wastewaters.[10] They are summarized briefly below:

One of two major ways of removing organic wastes is biological treatment by an activated sludge, or related process (see Section 13.13 and Figure 13.4). It may be necessary to acclimate microorganisms to the degradation of hazardous wastes constituents that are not normally biodegradable (see biodegradation of phenols, Section 13.12). Consideration needs to be given to possible hazards of biotreatment sludges, such as those containing excessive levels of heavy metal ions. The other major process for the removal of organics from hazardous waste leachate is sorption by activated carbon (see Section 12.5), usually in columns of granular activated carbon. Activated carbon and biological treatment can be combined with the use of powdered activated carbon in the activated sludge process. The powdered activated carbon sorbs some constituents that may be toxic to microorganisms and is collected with the sludge. A major consideration with the use of activated carbon on hazardous waste leachate is the hazard that spent activated carbon may present from the wastes it retains. These hazards may include those of toxicity or reactivity, such as those posed by explosives manufacture wastes sorbed to activated carbon. Regeneration of the carbon is expensive and can be hazardous in some cases.

Hazardous waste leachate can be treated by a variety of chemical processes, including acid/base neutralization, precipitation, and oxidation/reduction. In some cases these treatment steps must precede biological treatment; for example, leachate exhibiting extremes of pH must be neutralized in order for microorganisms to thrive in it. Cyanide in the leachate may be oxidized with chlorine and organics with ozone, hydrogen peroxide promoted with ultraviolet radiation, or dissolved oxygen at high temperatures and pressures. Heavy metals may be precipitated with base, carbonate, or sulfide.

Leachate can be treated by a variety of physical processes. In some cases, simple density separation and sedimentation can be used to remove water-immiscible liquids and solids. Filtration is frequently required and flotation (see Section 12.1) may be useful. Leachate solutes can be concentrated by evaporation, distillation, and membrane processes, including reverse osmosis, hyperfiltration, and ultrafiltration. Organic constituents can be removed by solvent extraction, air stripping, or steam stripping.

Synthetic resins are useful for removing hazardous solutes from hazardous waste leachate. Organophilic resins have proven useful for the removal of alcohols; aldehydes; ketones; hydrocarbons; chlorinated alkanes, alkenes, and aryl compounds; esters, including phthalate esters; and pesticides. Cation exchange resins are effective for the removal of heavy metals.

Gas Emissions

In the presence of biodegradable wastes, designated here as $\{CH_2O\}$, methane and carbon dioxide gases are produced in landfills by anaerobic degradation as represented by the following reaction:

$$2\{CH_2O\} \xrightarrow[\substack{\text{microbial} \\ \text{processes}}]{\text{Anaerobic}} CH_4 + CO_2 \qquad (15.6.1)$$

Gases may also be produced by chemical processes with improperly pretreated wastes, as would occur in the hydrolysis of calcium carbide to produce acetylene:

$$CaC_2 + 2H_2O \longrightarrow C_2H_2 + Ca(OH)_2 \qquad (15.6.2)$$

Odorous and toxic hydrogen sulfide, H_2S, may be generated by the chemical reaction of sulfides with acids or by the biochemical reduction of sulfate by anaerobic bacteria (*Desulfovibrio*) in the presence of biodegradable organic matter:

$$SO_4^{2-} + 2\{CH_2O\} + 2H^+ \xrightarrow[\text{bacteria}]{\text{Anaerobic}}$$

$$H_2S + 2CO_2 + 2H_2O \quad (15.6.3)$$

Gases such as these pose hazards of toxicity and flammability, as well as the potential to form explosive mixtures with air. Furthermore, gases permeating through landfilled hazardous waste may carry along waste vapors, such as those of volatile aryl compounds and low-molecular-mass chlorinated hydrocarbons. Of these, the ones of most concern are benzene, carbon tetrachloride, chloroform, 1,2-dibromoethane, 1,2-dichloroethane, dichloromethane, tetrachloroethane, 1,1,1-trichloroethane, trichloroethylene, and vinyl chloride.[11] Because of the hazards from these and other volatile species, it is important to minimize production of gases and, when gas production is likely, to provide for venting, and even treatment, of the gas. Landfill gas can be collected by systems of perforated or slotted pipes placed horizontally below the landfill cap and connected to vertical riser pipes through which the gas is released. An extraction blower can be installed to pull the gas from the landfill. If the gas is too hazardous to vent to the atmosphere, it may be flared or run through a gas incinerator. Activated carbon can be used to remove hazardous organic vapor constituents from the gas.

LITERATURE CITED

1. Wiles, Carlton C., "Solidification and Stabilization Technology," Section 7.8 in *Standard Handbook of Hazardous Waste Treatment and Disposal*, Freeman, Harry M., Ed., McGraw Hill Book Company, New York, 1989, pp. 7.85-7.101.

2. Manahan, Stanley E., "Humic Substances and the Fates of Hazardous Waste Chemicals," Chapter 6 in *Influence of Aquatic Humic Substances on Fate and Treatment of Pollutants*, American Chemical Society, Washington D.C., 1988.

3. "Direct Waste Treatment," Section 10 in *Remedial Action at Waste Disposal Sites*, EPA/625/6-85/006, U.S. Environmental Protection Agency Hazardous Waste Engineering Research Laboratory, Cincinnati, Ohio, 1985, pp.10-1–10-151.

4. "Destruction and Immobilization of Metal-Contaminated PCB Sludges by Gasification on an Activated Char from Subbituminous Coal," Stanley E. Manahan, Shubhender Kapila, Chris Cady, and David Larsen, preprint extended abstract of papers presented before the Division of Environmental Chemistry, American Chemical Society, Dallas, Texas, April, 1989.

5. Clements, J. A., and C. M. Griffiths, "Solidification Processes," Chapter 5 in *Hazardous Waste Management Handbook*, Butterworths, London, 1985, pp. 146-166.

6. "Redox Equilibria in Natural Waters," Chapter 3 in *Environmental Chemistry*, 4th ed., Stanley E. Manahan, Brooks/Cole Publishing Co., Pacific Grove, California, 1984, pp. 36-56.

7. Brown, K. W., and David C. Anderson, "Aboveground Disposal," Section 10.7 in *Standard Handbook of Hazardous Waste Treatment and Disposal*, Freeman, Harry M., Ed., McGraw-Hill Book Company, New York, 1989, pp. 10.85-10.91.

8. Wright, Thomas D., David E. Ross, and Lori Tagawa, "Hazardous-Waste Landfill Construction: The State of the Art," Section 10.1 in *Standard Handbook of Hazardous Waste Treatment and Disposal*, Freeman, Harry M., Ed., McGraw Hill Book Company, New York, 1989, pp. 10.3-10.23.

9. "In-Situ Treatment," Chapter 9 in *Remedial Action Technology for Waste Treatment Sites*, 2nd ed., Kathleen Wagner, Kevin Boyer, Roger Claff, Mark Evans, Susan Henry, Virginia Hodge, Shahid Mahmud, Douglas Sarno, Ellen Scopino, and Philip Spooner, Noyes Data Corporation, Park Ridge, New Jersey, 1986, pp. 367-437.

10. Shuckrow, Alan J., Andrew P. Pajak, and C. J. Touhill, *Hazardous Waste Leachate Management Manual*, Noyes Data Corporation, Park Ridge, New Jersey, 1982.

11. Mills, John, "Landfill Gas Analysis," *American Laboratory*, June, 1989, pp. 74-80.

INDEX

3